Series in BioEngineering

The Series in Bioengineering serves as an information source for a professional audience in science and technology as well as for advanced students. It covers all applications of the physical sciences and technology to medicine and the life sciences. Its scope ranges from bioengineering, biomedical and clinical engineering to biophysics, biomechanics, biomaterials, and bioinformatics.

Indexed by WTI Frankfurt eG, zbMATH.

Bamidele O. Awojoyogbe · Michael O. Dada

Digital Molecular Magnetic Resonance Imaging

Springer

Bamidele O. Awojoyogbe
Department of Physics
Federal University of Technology
Minna, Nigeria

Michael O. Dada
Department of Physics
Federal University of Technology
Minna, Nigeria

ISSN 2196-8861 ISSN 2196-887X (electronic)
Series in BioEngineering
ISBN 978-981-97-6369-6 ISBN 978-981-97-6370-2 (eBook)
https://doi.org/10.1007/978-981-97-6370-2

This Springer imprint is published by the registered company Springer Nature Singapore Pte Ltd.
The registered company address is: 152 Beach Road, #21-01/04 Gateway East, Singapore 189721, Singapore

Preface

In recent years, it has been discovered that heterogeneous diseases such as cardiovascular diseases, diabetes, hepatitis and cancer may be attacked effectively by developing specific treatment plans that are generally referred to as personalized treatments. This new approach in medicine involves using novel methodology as a bridge between diagnostic and therapeutic applications to form a single agent, allowing for diagnosis, drug delivery and treatment response monitoring at the molecular level. This method is time and cost effective because it is a one in all procedure which combines diagnosis and therapy. This novel methodology involving atomic nuclei could be studied with NMR/MRI since their unique resonant frequencies lead to the production of distinct spectra for diagnostic imaging. This book has been able to develop molecular imaging methods and computer programs for a computational analysis of non-communicable disease conditions based on digitally generated magnetic resonance signal using the fundamental Bloch NMR model equation, physics-informed neural networks (PNN) and artificial intelligence to design and deploy magnetic resonance systems with the capacity to work in a digital manner with features of low cost, high performance and accuracy. The computational techniques presented in this book point to the possibility of showing contrast between different stages of tissue diseases at the molecular level. The book is outlined as follows:

Chapter 1 introduces the possibility of advanced digital signal process technique and modular designed structure to generate arbitrary pulse sequence in order to meet the requirement of digital magnetic resonance imaging methods for medical use.

Chapter 2 briefly describes the principle of artificial intelligence (AI) in medicine, neural network in scientific computing, physics-informed neural networks and the use of partial differential equations (PDEs), neural network architecture in physics-informed neural network (PINNs) and evolution of physics-informed neural networks.

The partial differential equations employed in Chap. 3 are based on the Bloch NMR flow equations and NMR diffusion equation. Each physical problem solved in this chapter is described by a Bloch-related equation based on a close observation of how the system usually behaves in the presence of NMR magnetic fields. Diffusion in physiological flows is dependent on the magnitude of the fluid velocity that is consistent with the tissue under investigation. One of the goals of this chapter is to develop a computational model from the wave equation and optical Bloch equation and Maxwell's equation for cellular imaging. The formulation to achieve this objective is presented in detail. The formulation is for a single spin and is very useful for cellular imaging. The optical NMR signals have been obtained in terms of transverse polarization function. Using Maxwell's equations, the polarization functions have been related to the optical properties of the cellular component under investigation. This is very important in developing new techniques for imaging the causes of human diseases. The chapter also provides a means of measuring microscopic polarization function in laboratory settings.

Chapter 4 develops a novel approach for enhancing the precision of molecular imaging in the diagnosis and treatment of adult brain tumors. This leveraged the power of artificial intelligence, specifically physics-informed neural networks (PINNs), to address spatial and temporal resolution problems in magnetic resonance molecular imaging.

Chapter 5 develops machine learning models to improve the accuracy of predictions and decision-making in different models by sorting across big datasets to identify hidden trends, patterns and insights and explore the data in the developed models for diagnosis of human medical conditions.

The purpose of Chap. 6 is to construct a high-accuracy deep learning algorithm with the best performing model for the diagnosis and classification of medical pictures related to Alzheimer's disease. The CNN model was used to train MRI brain images. In order to construct and deploy the model, we used specialized application software and libraries including Jupyter Notebook, Python 3, Keras and TensorFlow. The trained model was distributed as an application known as APK, which will make it accessible as an Android application for personal diagnosis. It will be a web service for diagnosing Alzheimer's disease and determining the stage of the disease remotely.

Chapter 7 addresses a critical need for improved precision in the application of hyperthermia for clinical purposes. The developed approach advances our understanding of the physical mechanisms behind medical interventions and opens the door to more effective and personalized clinical hyperthermia, which could lead to improved patient outcomes and advancements in medical management.

Chapter 8 demonstrates the prospective synergy between physics-based knowledge and machine learning approaches, which contributes to improved understanding of temperature distribution in biological tissues and its implications for a variety of healthcare applications.

The objectives of Chap. 9 are to: (i) evaluate the performance of machine learning classification of brain tumors with WEKA and Python programming using MRI scan images and (ii) deploy the best model for the brain tumor classification in WEKA and PYTHON programming. By developing and implementing machine learning models capable of interpreting MRI scans, the method improves the speed and accuracy of diagnosis for brain tumors and provides better patient outcomes.

Chapter 10 applies neural radiance fields (NeRFs) to improve the rendering of magnetic resonance images to establish a strong framework for the three-dimensional rendering of MRI data, which will enhance the precision and depth of medical image interpretation. This will advance medical imaging technology and make a significant contribution to the visualization of anatomical structures with never-before-seen precision and details.

The aim of Chap. 11 is to develop computational magnetic resonance fingerprinting (MRF) methods based on Bloch NMR flow equations and artificial intelligence for differentiating intra-axial brain tumors. We calculate simultaneous measurement of multiple tissue properties in terms of T1 and T2 relaxation times for differentiating intra-axial brain tumors. Machine learning models for contralateral white matter, peritumoral white matter and solid tumors dataset are developed with graphic user interface (GUI) and shiny package for model implementation.

Chapter 12 explores the fundamental quantum physics based on the uncertainty principle, to initially describe the particle (electron or a biological cell) under study by a wave packet. Then, when the particle encounters a force (diseases such as cardiovascular diseases, diabetes, hepatitis and cancer, so its potential energy is no longer zero), the force modifies the wave packet. The chapter proposed first-order perturbation theory to find such propagation techniques and applied them appropriately to provide useful techniques for the design of low-field MRI that can be useful to find solutions to biological and medical problems which otherwise could not be easily solved.

The goal of Chap. 13 is to develop and implement a transfer learning approach for the artificial intelligence-based classification of brain tumors using MRI images by: (i) pre-processing magnetic resonance scans of brain tumors according to their class (glioma, meningioma or pituitary tumors), (ii) develop a transfer learning model capable of accurately classifying the various types of brain tumors and (iii) develop an easy-to-use web application/GUI based on the trained model.

Chapter 14 is a general conclusion of the book which amplifies the power of neural radiance fields (NeRFs) to completely transform the 3D rendering industry and provide more precise and thorough visualizations for applications such as medical imaging. NeRFs have the potential of greatly improving our understanding of complex 3D environments and open up new possibilities for a wide range of applications as they continue to develop and get better. NeRF techniques have great potential to advance the visualization and quantification of complex pathologies affecting important organs such as the brain and chest by allowing MRI's exceptional soft tissue perspectives to be unlocked.

Chapter 15 focuses attention on democratizing access of magnetic resonance imaging powered by quantum computing. Independent of artificial intelligence, quantum computing is going to be a big deal. However, coupled with AI, it will be a game changer. Quantum computing, an entirely new way of processing information, when combined with AI, has the potential, without exaggeration, to bring about a new computing revolution. The field of AI, particularly machine learning, is poised to experience a significant leap forward due to quantum computing. This advancement is largely attributed to quantum-specific algorithms tailored to enhance computational processes. A wide range of sectors in healthcare are expected to benefit from integrating AI with quantum computing. This synergy promises more efficient problem-solving and innovation within medical fields. The fusion of AI with quantum computing is not just about incremental changes but about paving the way for transformative advancements. This combination will revolutionize our approach to magnetic resonance imaging and its applications in various fields of medicine.

The book is written for students, scientists, engineers, medical personnel and researchers who are interested in developing novel concepts for innovative MRI for the management of tissue diseases at low cost yet with maximum efficiency and accuracy utilizing fundamental mathematical concepts, physics-based knowledge and artificial intelligence for digital magnetic resonance imaging.

This book opens the door to ultra-powerful digital molecular MRI powered by quantum computing that can perform calculations that would take supercomputers millions of years.

Minna, Nigeria

Bamidele O. Awojoyogbe
Michael O. Dada

Acknowledgements

The authors specifically appreciate the current Vice Chancellor, Prof. Faruk Adamu Kuta, Federal University of Technology, Minna, Nigeria, for his continuous encouragement to publish high-quality academic works with reputable and respected international publishers such as Springer Nature.

The Springer Nature exclusive training received by the authors during the book publishing process which significantly helped to turn manuscript into a book is gratefully acknowledged.

The encouraging discussions received by the authors from the International Society for Magnetic Resonance in Medicine (ISMRM), the education committee of the World Molecular Imaging Society (WMIS), the publication committee of International Organization of Medical Physics (IOMP) which stimulate the publication of this book is acknowledged. The encouragements of Dr. Aurea Martins Bach, Dr. Mohamed Tachrount and Prof. Karla Miller (Nuffield Department of Clinical Neurosciences, John Radcliffe Hospital, Oxford) through the WIN Global Scholars (WINGS) programme, are acknowledged.

The authors will continue to appreciate the students, scientists, engineers, medical personnel and researchers who are interested in developing innovative MRI from the novel scientific computing concepts presented in this book.

Contents

1 General Introduction 1
1.1 Introduction to Nuclear Magnetic Resonance (NMR) 2
1.2 Theory of Nuclear Magnetic Resonance 7
1.2.1 Nuclear Spin and Magnets 7
1.2.2 Values of Spin Angular Momentum 8
1.2.3 Spin Behaviour in a Magnetic Field 8
1.2.4 Resonance 9
1.2.5 Nuclear Shielding 9
1.2.6 Relaxation 10
1.3 Magnetic Resonance Imaging (MRI) 10
1.3.1 Three-Dimensional Magnetic Resonance Imaging 11
1.4 Diffusion Process in Magnetic Resonance 12
1.5 Image Processing with Functional MRI of the Brain Software Library (FSL) 12
1.6 Medical Image Denoising 13
1.7 Medical Image Feature Extraction 16
1.8 Image Segmentation 18
1.9 Combining Physics Principles with Neural Networks for Feature Extraction, Denoising, and Segmentation 19
1.9.1 Data Preparation 21
1.9.2 Skull Extraction with FSL's BET 21
1.9.3 Image Segmentation with FAST 24
1.9.4 Creating Labelled Image 24
1.9.5 Extraction of Slices from Volume Data 26
1.10 Conclusion 28
References 28

2 Physics Informed Neural Networks (PINNs) 33
2.1 Artificial Intelligence (AI) in Medicine 33
2.2 Neural Network in Scientific Computing 34
2.3 Neural Network Architecture in Physics-Informed Neural Network (PINNs) 35
2.4 Output and Input Layers 35
2.5 Activation Functions and Weight Initialization 35
2.6 Partial Differential Equations (PDEs) 35
2.7 Physics-Informed Neural Networks 37
2.7.1 Evolution of Physics-Informed Neural Networks 37
2.7.2 Components and Workings of Physics-Informed Neural Networks (PINNs) 37
2.7.3 Enhancing Molecular Imaging Precision Through Physics-Informed Modelling 38
2.7.4 Integration of Physics Principles in Neural Networks 38
2.7.5 Advantages and Challenges of Physics-Informed Neural Networks (PINNs) 38
2.7.6 Physics Informed Neural Network and Clinical Hyperthermia 39
2.8 The Neural Network 39
2.9 Machine Learning 40
2.10 Deep Learning (DL) 40
2.11 Convolutional Neural Network (CNN) 40
2.12 Machine Learning in Medicine 41
2.12.1 Supervised Learning 41
2.12.2 Unsupervised Learning 41
2.12.3 Reinforcement Learning 42
2.12.4 Deep Learning 42
2.12.5 Deep Neural Network 42
2.12.6 Activation Function 43
2.12.7 The Sigmoid Function 43
2.12.8 The Rectified Linear Units (ReLU) 43
2.12.9 Machine Learning Techniques for Healthcare Applications 44
2.12.10 Supervised Learning 44
2.12.11 Classification 44
2.12.12 Regression 45
2.12.13 Clustering 45
2.13 Conclusion 45
References 46

3 New Methodology and Modelling in Magnetic Resonance Imaging 49
3.1 Introduction 49
3.2 Computational Approach to Compressed Sensing MRI 50
3.2.1 Software for Programming and Data Analysis 50
3.2.2 Data for Model Analysis 54
3.3 Analytical Methods 56
3.3.1 Variable Fluid Flow Corrections to the Bloch NMR Flow Equation 56
3.3.2 Derivation of Diffusion Equation with Variable Diffusion Coefficient 59
3.3.3 Optical Bloch Equation (OBE) and Molecular Imaging 62
3.3.4 General Mathematical Solutions to Model Equations 74
3.3.5 Mathematical Solutions to Model Equations of Specific Imaging Problem 78
3.3.6 Mathematical Models for Theranostics in Molecular Imaging 106
3.3.7 Quantum Mechanical Model of Transverse Magnetisation and Transverse Polarisation for Molecular Imaging 116
3.4 Conclusion 138
References 140

4 Physics-Informed Neural Network for Addressing Spatial and Temporal Resolution Problems in Magnetic Resonance Molecular-Imaging of Adult Brain Tumours 147
4.1 Introduction 147
4.2 Spatial and Temporal Resolution 148
4.2.1 Importance of Spatial and Temporal Resolution 148
4.3 Magnetic Resonance Molecular Imaging 149
4.4 Artificial Intelligence in Medicine 149
4.4.1 Tkinter 150
4.4.2 Deep Learning 151
4.4.3 Physics-Informed Neural Network (PINN) Model Structure 151
4.4.4 Equation for Temporal Resolution 153
4.4.5 Equation for Spatial Resolution 153
4.4.6 Training the PINN 154
4.4.7 Evaluating the PINN 155
4.5 Transverse Magnetization Profiles 156
4.5.1 PINN Model Training Result for Glioblastoma Multiform (GBM) 157
4.5.2 PINN Model (Training Result) for Metastasis. 157

4.5.3 PINN Model (Training Result) for Lower Grade Glioma ... 157
4.5.4 Classification of Adult Brain Tumours ... 157
4.6 Conclusion ... 158
References ... 161

5 Machine Learning Model for Diagnosis of Pulmonary Arterial Hypertension and Severe Aortic-Valve Stenosis Using Magnetic Resonance Relaxometry Data ... 163
5.1 Introduction ... 163
5.2 Magnetic Resonance Relaxometry (MRR) ... 165
5.3 Relaxation Process ... 167
5.4 Visualization and Analysis of the Projection Data ... 168
5.5 Reading the Input File ... 169
5.5.1 Distribution of Classes ... 169
5.5.2 Bivariate Analysis ... 173
5.5.3 Multivariate Analysis ... 174
5.5.4 3D Visualization ... 174
5.5.5 Model Deployment ... 179
5.6 Conclusion ... 181
References ... 181

6 A Convolution Neural Network for Artificial Intelligence-Based Classification of Alzheimer's Diseases ... 183
6.1 Introduction ... 183
6.2 Signs and Symptoms of Alzheimer's Disease ... 184
6.3 Jupyter Notebook ... 185
6.4 Python Libraries ... 185
6.5 Data ... 185
6.6 Criteria for Exclusion from the Study ... 186
6.7 Definition of Moderate, Mild, Very and Non-demented ... 186
6.8 Spyder IDE ... 186
6.9 Machine Learning ... 187
6.10 Convolutional Neural Network ... 187
6.11 MaxPooling2D ... 187
6.12 Nonlinearity ... 188
6.13 Fully Connected Layer ... 188
6.14 Softmax ... 189
6.15 Evaluation of Result ... 193
6.15.1 Classification Report on Tested Data ... 193
6.15.2 Recall ... 194
6.15.3 F1-Score ... 194
6.15.4 Support ... 194

6.16 Confusion Matrix on the Test Data 195
6.17 Model Deployment 196
6.17.1 Deployment Process with Streamlit 196
6.17.2 Deployed Alzheimer's Disease Classification App 197
6.18 Conclusion 197
References 200

7 Physics-Informed Neural Networks for Nuclear Magnetic Resonance-Guided Clinical Hyperthermia 201
7.1 Introduction 201
7.2 Clinical Hyperthermia 202
7.3 The Use of Nuclear Magnetic Resonance 203
7.4 Python Modules/Packages 204
7.4.1 Jupyter Notebook 204
7.4.2 Google Colab 204
7.4.3 Deep Learning 205
7.5 Partial Differential Equation (PDE) 205
7.6 Physics-Informed Neural Network (PINN) 207
7.6.1 Tanh (Activation) Function 208
7.7 PINN Model (Training Result) for Kidney 208
7.7.1 Biothermal Profile for Kidney 210
7.8 PINN Model (Training Result) for Cortical Bone 210
7.8.1 Biothermal Profile for the Cortical Bone 211
7.8.2 PINN Model (Training Result) for Tumour 211
7.9 Predicted Biothermal Profile for Tumour 212
7.9.1 PINN Model (Training Result) for Liver 212
7.9.2 Predicted Biothermal Profile for Liver 212
7.9.3 PINN Model (Training Result) for Skeletal Muscle 213
7.9.4 Predicted Biothermal Profile for Skeletal Muscle 214
7.9.5 PINN Model (Training Result) for Adipose Tissue 214
7.9.6 Predicted Biothermal Profile for Adipose Tissue 214
7.10 Conclusion 215
References 216

8 Simulation of Temperature Distribution in Biological Tissues Using Physics-Informed Neural Networks 217
8.1 Introduction 217
8.2 The Pennes Bioheat Equation 218
8.3 Physics-Informed Neutral Network (PINN) 219
8.3.1 Activation Function 219
8.3.2 The Initialization or Constructor 220
8.3.3 Residual Function (F Method) 220
8.3.4 Closure Method (Closure) 220

8.3.5 Adam Optimizer 220
8.3.6 L-BFGS Optimizer 221
8.4 Tissue Parameters Used in the First Simulation 221
8.4.1 Tissue Parameter Used in the Second Simulation 221
8.5 PINN Training for the Kidney Tissue 222
8.5.1 Simulation Results for the Kidney 222
8.5.2 PINN Training for the Skin Tissue 223
8.5.3 Simulation Results for the Skin 224
8.6 Conclusion 224
References 228

9 Performance Evaluation of Machine Learning Classification of Brain Tumors with WEKA and Python Programming 229
9.1 Introduction 229
9.2 Image Filters 232
9.3 WEKA Software 232
9.4 Spyder 232
9.5 Data Collection 233
9.6 Data Preprocessing Both on WEKA and Python 233
9.7 Filters 234
9.8 Classifiers 234
9.9 Data Classification with CNN on WEKA and Python 235
9.10 Discussion 236
9.11 Data Filtering and Training in WEKA with Classifiers 236
9.12 Data Training with CNN Implementation on WEKA 237
9.13 Data Training and Filtering with Python Code 239
9.14 Data Training with CNN in Python Code 239
9.15 Model Deployment on Spyder 241
9.16 Conclusion 244
References 245

10 Neural Radiance Fields (NeRFs) Technique to Render 3D Reconstruction of Magnetic Resonance Images 247
10.1 Introduction 247
10.2 Magnetic Resonance Imaging (MRI) 249
10.2.1 Application of MRI 250
10.3 Virtual Studio Code 251
10.3.1 Python Programming Language 251
10.3.2 PyTorch 252
10.3.3 How NeRF Works 252
10.3.4 Training NeRFs 253
10.3.5 Volume Rendering 254
10.3.6 Medical Neural Radiance Fields (MedNeRF) 254
10.3.7 Data Preprocessing 255
10.4 Conclusion 256
References 257

11 Computational Molecular Magnetic Resonance Fingerprinting in Digital Health 259
11.1 Introduction 260
11.2 Computation Results of Magnetic Resonance Fingerprinting Dataset 262
11.3 Garson Features Result Analysis 263
11.4 SNNs Plots for CW, PW and ST Features 265
11.5 Perceptron Profiles for CW, PW and ST Features 267
11.6 Machine Learning Classification 269
11.7 Shiny App Deployment with Contralateral, Peritumoral White Matter and Solid Tumor 288
11.8 Conclusion 295
References 296

12 Fundamental Quantum Mechanical Dynamics in Computational Magnetic Resonance Imaging from Low Field to High Field 297
12.1 Introduction 298
12.1.1 Perturbation Theory for NMR Wave Equation 300
12.2 Discussion 303
12.3 Conclusion 305
References 307

13 Artificial Intelligence-Based Classification of Brain Tumors 309
13.1 Introduction 309
13.2 Python Modules/Packages 312
13.3 Streamlit 312
13.4 Jupyter Notebook 312
13.5 MRI Brain Tumor Dataset 312
13.5.1 Data Processing 313
13.5.2 Criteria for Exclusion from the Study 313
13.6 Proposed Model 314
13.6.1 Deep Learning 314
13.6.2 Convolutional Neural Network (CNN) 315
13.6.3 Global Average Pooling 315
13.6.4 Dropout Layer 315
13.6.5 Dense Layer 320
13.6.6 Softmax 320
13.7 EfficientNetB0 320
13.7.1 ImageNet 323
13.8 Transfer Learning Setting (Method) 323
13.8.1 Precision/Positive Predictive Value 323
13.8.2 Recall/Negative Predictive Value 324
13.9 CNN Model (Training and Testing Result) 324
13.9.1 Training Metrics Visualizations 324

13.9.2 Model Loss ... 324
13.9.3 Model Evaluation ... 324
13.10 Classification Report on Tested Data ... 326
13.10.1 Evaluation Metrics ... 326
13.10.2 F1-Score ... 328
13.10.3 Support ... 328
13.10.4 Accuracy ... 328
13.11 Confusion Matrix on the Test Data ... 329
13.12 Model Deployment Process with Streamlit ... 329
13.13 Model Deployment Process with DigitalOcean ... 329
13.14 Conclusion ... 333
References ... 334

14 General Conclusion ... 335
14.1 Introduction ... 335
14.2 Introduction to 3D Rendering Techniques ... 336
14.2.1 Surface Rendering ... 337
14.2.2 Volume Rendering ... 337
14.3 Overview of Neural Radiance Fields (NeRFs) ... 338
14.4 Applications of NeRFs in Medical Imaging ... 339
14.5 Challenges in Using NeRF in Medicine ... 340
14.6 Conclusion ... 341
References ... 342

15 Democratizing Access to Magnetic Resonance Imaging: Capacity Building in Quantum Computing ... 345
15.1 Introduction ... 345
15.2 Building Capacity for Digital Molecular Magnetic Resonance Imaging ... 346
15.3 Prospect of Quantum Artificial Intelligence (QAI) in Digital Molecular Magnetic Resonance Imaging ... 347
15.4 Conclusion and Expectation ... 347
References ... 348

List of Figures

Fig. 1.1	The downloaded dataset for the study	20
Fig. 1.2	Flowchart showing different stages executed in this study	20
Fig. 1.3	Proposed method flowchart showing traditional machine learning models used for image segmentation	20
Fig. 1.4	Coronal, sagittal and axial view of the dataset used for the study	21
Fig. 1.5	The main FSL user interface showing all the tools	22
Fig. 1.6	The BET user interface	23
Fig. 1.7	The FAST user interface	23
Fig. 1.8	Segmented image opened with FSLeyes	24
Fig. 1.9	Segments lablelled with unique colours	25
Fig. 1.10	Labelled image data	25
Fig. 1.11	Python function for extracting slices from 4D image data	26
Fig. 1.12	Extracted slices from 4D image data	27
Fig. 1.13	Python function for extracting slices from 3D image data	27
Fig. 1.14	Extracted slices from 3D image data	28
Fig. 4.1	PINN model structure	152
Fig. 4.2	Physics loss of the model	152
Fig. 4.3	Transverse magnetization profile for the three types of adult brain tumours (**a**) Glioblastoma Multiform (**b**) for Metastasis (**c**) for Lower Grade Glioma	156
Fig. 4.4	Training Progress of PINN Model for Glioblastoma Multiforme	158
Fig. 4.5	Training progress of PINN model for metastasis	158
Fig. 4.6	Training progress of PINN model for low grade glioma	159
Fig. 4.7	Brain tumour classification deployment	160
Fig. 5.1	A cross-section generated dataset for Pulmonary Arterial hypertension (PAH), severe Aortic Stenosis (AS) and healthy patients	168
Fig. 5.2	R-Studio screen	169
Fig. 5.3	Visualized class distribution	170

Fig. 5.4	Box plot	171
Fig. 5.5	Violin plot	172
Fig. 5.6	Bivariate analysis	173
Fig. 5.7	Multivariate analysis	174
Fig. 5.8	3D visualization	175
Fig. 5.9	Checking correlations	176
Fig. 5.10	Confusion matrix	176
Fig. 5.11	Different values for N-neighbors	177
Fig. 5.12	Final result for different models	177
Fig. 5.13	Deployment on Django	180
Fig. 6.1	CNN model architecture	190
Fig. 6.2	Epoch training networks	191
Fig. 6.3	The model accuracy	191
Fig. 6.4	The model AUC	192
Fig. 6.5	The model loss	192
Fig. 6.6	Model evaluation	193
Fig. 6.7	Other classification metrics result	195
Fig. 6.8	Confusion matrix	196
Fig. 6.9	Welcome page of the web app	198
Fig. 6.10	A test page detecting that there is no presence of disease	199
Fig. 6.11	A section of the web app. showing the Alzheimer disease classification web app	199
Fig. 7.1	Architecture of deep neural network	206
Fig. 7.2	Neural Network and partial differential equation	207
Fig. 7.3	Schematic of PINNs (Moseley 2021)	208
Fig. 7.5	Training losses for the kidney	210
Fig. 7.6	Predicted biothermal profile of the kidney	210
Fig. 7.7	Training losses for cortical bone	211
Fig. 7.8	Predicted biothermal profile of the cortical bone tissue	211
Fig. 7.9	Training losses for tumour	212
Fig. 7.10	Predicted biothermal profile of tumour	212
Fig. 7.11	Training losses for liver	213
Fig. 7.12	Predicted biothermal profile of the tissue liver	213
Fig. 7.13	Training losses for skeletal muscle	213
Fig. 7.14	Predicted biothermal profile of the skeletal muscle	214
Fig. 7.15	Training losses for adipose tissue	214
Fig. 7.16	Showing the biothermal profile of adipose tissue	215
Fig. 8.1	Architecture of PINN (Pioch et al. 2023)	219
Fig. 8.2	PINN model training for the kidney	222
Fig. 8.3	Simulation of the 1D-biothermal profile (kidney)	223
Fig. 8.4	2-D representation of the biothermal PINN profile (kidney)	223
Fig. 8.5	Temperature distribution curve in the kidney	225
Fig. 8.6	PINN model training for the skin	225
Fig. 8.7	Simulation after 0.07 s of the skin tissue in 1-D	226
Fig. 8.8	Simulation after 0.74 s of the skin tissue in 1-D	226

Fig. 8.9 Simulation of the biothermal profile for the skin in 2D 227
Fig. 8.10 2D curve of the temperature distribution in the skin 227
Fig. 9.1 Flowchart describing the method used in this study 233
Fig. 9.2 Preprocessor output image 235
Fig. 9.3 Code for the model 236
Fig. 9.4 Result of the support vector machine after training 237
Fig. 9.5 Result of k-nearest neighbor after training 238
Fig. 9.6 Result of decision tree after training 238
Fig. 9.7 Error test shown upon training with logistic regression 239
Fig. 9.8 **a** Convolutional neural network on WEKA. **b** Result after training with CNN on WEKA 240
Fig. 9.9 Logistic regression code and output 241
Fig. 9.10 K-nearest neighbour code and output 241
Fig. 9.11 Decision tree code and output 242
Fig. 9.12 Support vector machine code and output 242
Fig. 9.13 CNN implementation depicting the absence of brain tumor ... 243
Fig. 9.14 CNN implementation depicting the presence of brain tumor 243
Fig. 9.15 Model deployment with Spyder 244
Fig. 9.16 Result after model deployment 244
Fig. 10.1 3D rendering of a brain MRI 249
Fig. 10.2 Functional magnetic resonance images 250
Fig. 10.3 NeRFs process (Mildenhall et al. 2021) 253
Fig. 10.4 NeRFs training process (Mildenhall et al. 2021) 253
Fig. 10.5 Overview of MedNeRF 255
Fig. 11.1 Plot of linear regression for contralateral white matter (CW) features 269
Fig. 11.2 CW chart with Garson model 270
Fig. 11.3 Plot of linear regression for peritumoral white matter (PW) features 270
Fig. 11.4 PW chart with Garson model 271
Fig. 11.5 Plot of linear regression with solid tumor (ST) features 271
Fig. 11.6 ST chart with Garson model 272
Fig. 11.7 SNNs plot for contralateral white matter (CW) features showing **a** iterative fit error and **b** iterative test error 272
Fig. 11.8 Plot of receiver operating characteristics for **a** training and **b** testing CW data 273
Fig. 11.9 SNNs plot for peritumoral white matter (PW) features showing **a** iterative fit error and **b** iterative test error 273
Fig. 11.10 Plot of receiver operating characteristics for **a** training and **b** testing PW data 274
Fig. 11.11 SNNs plot for solid tumor (ST) features showing **a** iterative fit error and **b** iterative test error 274
Fig. 11.12 Plots of receiver operating characteristics for **a** training and **b** testing ST data 275

Fig. 11.13 Scatterplot of T1 relaxation time against T_0 relaxation rate for contralateral white matter (CW) according to the tumor class ... 275
Fig. 11.14 Optimal weights for the perceptron training model for CW ... 276
Fig. 11.15 Scatterplot of T1 relaxation time against T_0 relaxation rate for peritumoral white matter (PW) according to the tumor class ... 276
Fig. 11.16 Optimal weights for the perceptron training model for PW ... 277
Fig. 11.17 Scatterplot of T1 relaxation time and T_0 relaxation rate for solid tumor (ST) according to tumor class ... 277
Fig. 11.18 Optimal weights for the perceptron training model for ST ... 278
Fig. 11.19 Box plots for each attribute distributions on one image for CW ... 278
Fig. 11.20 Correlation plots for the datapoint attribute features for CW ... 278
Fig. 11.21 Box and whisker plots for each attribute features for CW ... 279
Fig. 11.22 Density plots for each attribute by class value for CW ... 279
Fig. 11.23 Machine learning models tested on CW dataset ... 280
Fig. 11.24 Dot plot comparison of accuracy of models tested on CW dataset ... 280
Fig. 11.25 Naive Bayes (NB) models performance with CW dataset ... 281
Fig. 11.26 Shows box plot for each attribute distribtions on one image for PW ... 281
Fig. 11.27 Correlation plots for the datapoint attribute features for PW ... 282
Fig. 11.28 Box and whisker plots for each attribute features for PW ... 282
Fig. 11.29 Density plots for each attribute by class value for PW ... 282
Fig. 11.30 Machine learning models tested on PW dataset ... 283
Fig. 11.31 Dot plot comparison of accuracy of models tested on PW dataset ... 283
Fig. 11.32 SVM model performance with PW dataset ... 284
Fig. 11.33 Box plots for each attribute distributions on one image for ST ... 284
Fig. 11.34 Box plots for the datapoint attribute features for ST ... 285
Fig. 11.35 Box and whisker plots for each attribute features for ST ... 285
Fig. 11.36 Density plots for each attribute by class value for ST ... 285
Fig. 11.37 Several models tested on the ST dataset ... 286
Fig. 11.38 Dot plots comparing accuracy of models for ST dataset ... 286
Fig. 11.39 Glmnet model performance with ST dataset ... 288
Fig. 11.40 App deployment with contralateral white matter ... 289
Fig. 11.41 Contralateral white matter dataset for app deployment ... 289
Fig. 11.42 Contralateral white matter tumor predictions column (1–10) ... 290
Fig. 11.43 Contralateral white matter tumor predictions column (101–110) ... 290

Fig. 11.44 Contralateral white matter tumor predictions column (201–210) ... 290
Fig. 11.45 App deployment with peritumoral white matter ... 291
Fig. 11.46 Peritumoral white matter dataset for app deployment ... 291
Fig. 11.47 Peritumoral white matter tumor predictions column (1–10) ... 292
Fig. 11.48 Peritumoral white matter tumor predictions column (101–110) ... 292
Fig. 11.49 Peritumoral white matter tumor predictions column (201–210) ... 292
Fig. 11.50 App deployment with solid tumor ... 293
Fig. 11.51 Solid tumor dataset for app deployment ... 293
Fig. 11.52 Solid tumor white matter tumor predictions column (1–10) ... 294
Fig. 11.53 Solid tumor predictions column (101–110) ... 294
Fig. 11.54 Solid tumor predictions column (201–210) ... 294
Fig. 13.1 Dataset sample ... 313
Fig. 13.2 Architecture of deep neural network ... 314
Fig. 13.3 Convolutional neural network (CNN) ... 315
Fig. 13.4 CNN model training ... 316
Fig. 13.5 Imagenet weights ... 320
Fig. 13.6 EfficientNetB0 Architecture ... 321
Fig. 13.7 Transfer learning ... 323
Fig. 13.8 CNN model training process ... 325
Fig. 13.9 Model accuracy ... 326
Fig. 13.10 Model loss ... 327
Fig. 13.11 Model result on the validation dataset ... 327
Fig. 13.12 Confusion matrix of the model according to tumor class ... 328
Fig. 13.13 Brain tumor app welcome page ... 330
Fig. 13.14 Brain tumor app prediction page ... 330
Fig. 13.15 Brain tumor app info page ... 331
Fig. 13.16 MR scan testing ... 331
Fig. 13.17 Glioma tumor detection ... 332
Fig. 13.18 DigitalOcean deployment page ... 332
Fig. 13.19 Deployment successful message ... 333
Fig. 14.1 Surface rendering of human head CT data, corresponding to (left) skin and (right) skull ... 337
Fig. 14.2 Example of volume rendering ... 338
Fig. 14.3 NeRF rendering of a 3D scene from 100 input view ... 339

List of Tables

Table 1.1 FSL tools and their functionalities 13
Table 4.1 Relaxometric data for solid tumour at 3.0 T (Badve et al. 2017) 151
Table 5.1 Reading the impute file 170
Table 5.2 PAH data of the first 25 entries 178
Table 5.3 AS data of the first 25 entries 179
Table 5.4 Control data of the first 25 entries 180
Table 6.1 Classification report table 193
Table 7.1 Table of values for Tissues used for the modeling (Awojoyogbe and Dada 2018) 209
Table 8.1 Approximated thermal properties for the kidney tissue (IT'IS Foundation 2023) 221
Table 8.2 Approximated thermal properties for the skin tissue (IT'IS Foundation 2023) 221
Table 11.1 Computed data for contralateral white matter 264
Table 11.2 Computed data for peritumoral white matter 266
Table 11.3 Computed data for solid tumor 268
Table 11.4 Machine learning classification with contralateral white matter CW (summarised attribute distribution values for CW) 287
Table 11.5 Machine learning classification with peritumoral white matter PW (summarised attribute distribution values for PW) 287
Table 11.6 Machine learning classification with solid tumor ST (summarised attribute distribution values for ST) 287
Table 13.1 A description of the brain data set 313
Table 13.2 The specifications of the layers in the proposed learning model 321
Table 13.3 Classification report 327

Chapter 1
General Introduction

Abstract Digital health refers to using various technologies, including telehealth services, health information technology, and mobile health (mHealth), to improve communication between doctors and patients throughout healthcare systems. The breadth of digital health offers consumers greater access to and ownership over their information and allows doctors to comprehensively understand their patients' health. Digital health technology has become more significant over time, particularly in light of the COVID-19 pandemic. Modern digital health solution improves relationships between patients and their doctors, expands preventive disease strategies, and shifts healthcare toward value-based treatment, increasing access to health information for patients and providers. Communication between systems and devices is possible. Medical professionals can better understand their patients' health status as data can now be safely transferred between platforms. The tremendous transformation that will take healthcare to the next level will be led by cloud-based digital health solutions that exchange data. By enabling people to track their health and giving doctors easy access to patient data, digital health is predicted to stop the rising rates of various diseases. Digital health includes mobile technologies, telemedicine collaborations, wearables and sensors, big data, and artificial intelligence, amongst other interventions. Digital health provides an opportunity to address health system challenges if designed with the consideration of the end-user, proper evidence, and a goal for large-scale roll-out. It is, however, important to remember that digital health interventions can enhance the functions of the health systems but cannot replace the fundamental health systems building blocks of (i) service delivery, (ii) health workforce, (iii) health information systems, (iv) access to essential medicines, vaccines, and technology, (v) financing, and (vi) leadership/governance. Although digital health is still relatively new, consumers can take control of their health as healthcare services continue to shift to a digital paradigm. As a result, people of all generations will be better informed about their health and able to decide what kind of care they need.

Keywords Nuclear magnetic resonance · Nuclear spin and magnets · Resonance · Relaxation

B. O. Awojoyogbe and M. O. Dada, *Digital Molecular Magnetic Resonance Imaging*, Series in BioEngineering, https://doi.org/10.1007/978-981-97-6370-2_1

1.1 Introduction to Nuclear Magnetic Resonance (NMR)

Nuclear magnetic resonance (NMR) refers to a physical principle—response of nuclei to a magnetic field. Many nuclei have a magnetic moment—they behave like spinning bar magnets. These spinning magnetic nuclei can interact with externally applied magnetic fields, producing measurable signals. It is the effect produced when a radiofrequency field is imposed at right angles to a (usually much larger) static magnetic field B_0 to perturb the orientation of nuclear magnetic moments generated by spinning electrically charged atomic nuclei. When the spinning nuclei interact with the very large (10,000–50,000 gauss) static magnetic field, characteristic spectral shifts and fine structure are produced that reflect the molecular or chemical environment seen by the nucleus.

Nuclear magnetic resonance is a physical phenomenon which is based upon the magnetic property of an atom's nucleus. All nuclei that contain odd numbers of nucleons and some that contain even numbers of nucleons have an intrinsic magnetic moment. Hydrogen nuclei, fluorine, carbon-13 and oxygen-17 all have distinctive magnetic properties that make them suitable for NMR studies. A comprehensive description of the fundamental physics of magnetic resonance imaging has been given by Dada and Awojoyogbe (2021).

The technique of the nuclear magnetic response requires that a substance be placed in a strong magnetic field. This strong magnetic field affects the spin of the atomic nuclei of certain isotopes of common elements, such as hydrogen molecules. Atomic nuclei have an angular momentum arising from their inherent property of rotation, or spin. An electric charge in such motion creates a magnetic field with strength and direction represented by a magnetic vector denoted by μ. Each nucleus of nonzero spin, therefore, has a magnetic moment, or dipole, associated with it. Only nuclei with an odd number of nucleons (protons or neutrons) exhibit a net spin, and therefore, lend themselves to NMR spectroscopy. In general, the magnetic dipoles of the nuclei with spin will be points in random directions. When placed in a magnetic field, however, these magnetic dipoles will orient themselves with the field's lines of force. For nuclei of ½ spin, such as protons (hydrogen nuclei), the only allowed orientation of the dipoles are parallel or anti parallel to the field. The energy associated with the magnetic dipoles depends, therefore, on its magnetic moment, the strength of the external magnetic field and the angle between the direction of the moment of the magnetic particle and the direction of the external field.

Technically speaking, NMR measurements consist of a series of manipulation of hydrogen protons (and some other NMR—sensitive nuclei) in fluid molecules. As earlier mentioned, protons have a magnetic moment and behave like small bar magnets, so their orientations may be controlled by magnetic fields. They also spin, which make them act like gyroscopes. A measurement sequence starts with proton alignment followed by spin tipping, precession, and repeated dephasing and refocusing. Transverse relaxation and longitudinal relaxation limit how long a measurement must last. Only after the completion of all steps—which takes a few seconds—can the measurement be repeated.

Proton alignment—Hydrogen protons are aligned by the application of a large constant magnetic field, B_0 (Katti et al. 2011). Alignment takes a few seconds and the protons will remain aligned unless disturbed using powerful permanent magnets. The protons precess about an axis parallel to the B_0 direction—the net magnetization being the sum of all the precessing protons.

Spin tipping—The next step is to tip the aligned protons by transmitting an oscillating magnetic field, B_1, perpendicular to the direction of B_0. For effective spin tipping:

$$f = \gamma B_0$$

where f is the frequency of B_1 - called Larmor frequency—and γ is a constant called the gyromagnetic ratio. For example, the Larmor frequency for hydrogen nuclei in a field of 0.55 T is about 2.3 MHz.

The angle through which spins are tipped is controlled by the strength of B_1 and the length of the time it is switched on. For example, to tip spins by 90^0, a B_1 field of 0.0004 T is switched on for 16 microseconds (μs).

Precession and dephasing—when protons are tipped 90^0 from the B_0 direction, they precess in the plane perpendicular to B_0. In this respect, they act like gyroscopes in a gravitational field.

At first, all the protons precess in unison. While doing so, they generate a small magnetic field at the Larmor frequency which is detected by the antenna and forms the basis of NMR measurements. However, the magnetic field, B_0, is not perfectly homogeneous, causing the protons to precess at slightly different frequencies. Gradually, they lose synchronization—they diphase—causing the antenna signal to decay. The decaying signal is called free induction decay (FID) and the decay time is called T_2^* (usually shorter than T_2).

Refocusing spin echoes—The dephasing caused by the inhomogeneity of B_0 is reversible. The runners start off in unison, but after several laps they become dispersed around the track—due to their slightly different speeds. Now if the starter gives another signal by firing a 180^0 pulse, the runners turn around and race in the opposite direction. The fastest runners have the greatest distance to travel back to the start. However, if the conditions remain the same—which is never the case—the runners arrive back at the same time.

Similarly, the hydrogen protons—precessing at slightly different Larmor frequencies—can be refocused when a 180^0 pulse is transmitted. The 180^0 pulse is the same strength as a 90^0 pulse, but switched on for twice as long. As the protons rephrase, they generate a signal in the antenna—a spin echo.

Of course, the spin echo quickly decays again. However, the 180^0 pulses can be applied repeatedly—typically several hundred times in a single NMR measurement. The usual procedure is to apply pulses in an evenly spaced train, as close together as possible. The entire pulse sequence, a 90^0 pulse followed by a long series of 180^0 pulses, is called a CPMG sequence after its inventors, Carr, Purcell, Meiboom and Gill.

Each NMR measurement comprises a sequence of transverse magnetic pulses transmitted by an antenna-called a CPMG (Carr-Purcell-Meiboom-Gill) sequence.

Each CPMG sequence starts with a pulse that tips hydrogen protons 90^0 and is followed by several hundred pulses that refocus the protons by tipping them 180^0 (top). After each pulse, the antenna becomes a receiver that records the signal amplitude (middle). The fast decay of each echo—called free induction decay—is caused by variations in the static magnetic field, B_0. The decay of each of the echo amplitude is caused by molecular interactions and has a characteristic time constant T_2—transverse relaxation time.

Transverse relaxation, T_2—The CPMG pulse sequence compensates for dephasing caused by B_0 field imperfections. However, molecular processes also cause dephasing, but this is irreversible. Irreversible dephasing is monitored by measuring the decaying amplitude of the spin echo in the CPMG echo train. The characteristic echo amplitude decay is called the transverse relaxation (spin—lattice relaxation) time, T_2, because dephasing occurs in the plane transverse to the static field B_0.

As the protons precess about the static magnetic field, they gradually lose synchronization. This causes the magnetic field in the transverse plane to decay. Dephasing is caused by inhomogeneities in the static magnetic field and by molecular interactions.

Longitudinal relaxation, T_1—After a period of several times T_2, the protons completely lose coherence, and no further refocusing is possible. After the CPMG pulse sequence is delivered, the protons return to their equilibrium direction parallel to B_0. This process occurs with a different time constant—longitudinal relaxation (spin—spin relaxation), T_1. The next spin tipping measurement is not started until the protons have returned to their equilibrium position in the constant B_0 field. T_1 and T_2 both arise from molecular processes and this is the reason why NMR is invaluable in probing molecular processes.

When the CPMG pulse sequence ends, the protons gradually relax back towards the static magnetic field. They do so with characteristic time constant T_1, the longitudinal relaxation time.

The magnetic vector μ of a spinning (Slichter 2013) charged particle lies along the axis of rotation. The surrounding magnetic field, symbolized by the vector **H**, exerts a torque that tends to bring $\boldsymbol{\mu}$ and **H** into alignment. However, this torque also interacts with the angular momentum vector; the effect of this interaction is to cause the spin axis to describe a cone about the direction of the magnetic field. This phenomenon is called the Larmor precession, named after Sir Joseph Larmor, the Irish Physicist, who was the first to calculate the rate at which energy is radiated by an accelerated electron and the first to explain the splitting of spectrum lines by a magnetic field.

When the natural frequency of the precessing nuclear magnets corresponds to the frequency of a weak external radio wave striking the material, energy is absorbed from the radio wave. This selective absorption, called resonance, may be produced either by tuning the natural frequency of the nuclear magnets to that of a weak radio wave of fixed frequency or by tuning the frequency of the weak radio wave to that of nuclear magnets determined by the strong constant external magnetic field. This motion of the magnetization vector of uncoupled spins is easily expressed in terms of the Bloch equations, the solution of which we attempt to find in this study.

Nuclear magnetic resonance is inherently a three-dimensional phenomenon. The spatial resolution of a three-dimensional set of data is usually equal in all three directions. At the core of the NMR, a magnetic resonance imaging (MRI) makes use of the fields dependency on the precession frequency by superimposing a magnetic field gradient onto the static polarizing field $H_0 = H_z$ to spatially encode information into the RF (radio frequency) signal. With this three-dimensional data in hand, surfaces can be detected mathematically. A computer translates these signals into highly detailed cross sectional images. The images are essentially maps of the locations of hydrogen in the body.

The sensitivity of nuclear magnetic resonance to molecular structure has made it a valuable research tool in organic chemistry, enabling chemists to determine hydrogen locations in crystals, something that cannot be done using X-ray diffraction. Nuclear magnetic resonance has also been used to study electron densities, chemical bonding, the compositions of mixtures, and to make purity determinations.

The basic requirements for NMR spectroscopy are that the magnetic field be homogenous over the volume of the sample, that there be a radio frequency field rotating in a plane perpendicular to the static field, and that there be a means of detecting the interaction of the frequency field with the sample.

Techniques that have been developed for the observation of NMR signals fall into two categories: pulsed and continuous wave. In the case of the pulsed methods, an applied rotating (or alternating) magnetic field with a frequency at or near the Larmor frequency (that is, frequency of precession) of the nucleus to be studied is directed at right angle to the static field. If the rotating field is applied at exact resonance, the nuclei precess about that field as though there was no static field. Continuous wave methods are either high resolution or broad line. Broad line widths are produced by most oriented molecules exhibiting strong magnetic dipolar interactions, so broad line spectroscopy does not permit measurements of chemical shifts and spin–spin coupling. High resolution on the other hand, has been used to identify molecules, to measure the electronic effects, to determine structure, to study reaction intermediates, and to follow the motion of molecules or groups of atoms within molecules. For high resolution studies, the magnetic field must be uniform to 1 part in 108 for a 100 MHz instrument, if a resolution of 7 MHz is to be obtained. In the case of broad line studies, 5 parts in 106 may be adequate.

Nuclear magnetic resonance has been used to study the physics and chemistry of solids, including metals, semiconductors, magnetic solids and organic materials. Physical phenomena studied include conduction-electron paramagnetism; spin waves and magnetic fluctuations in ordered magnetic materials; metal molecular transitions; charge density wave phenomena; spin-freezing in spin glasses and frequency shift and spin–lattice relaxation effects. At low temperatures, NMR has been used to make temperature measurements and to study the super fluid phases of ^{3}He.

In the fields of organic chemistry and materials science, NMR has been used to study polymers, amorphous systems and complex molecular solids. In many of these systems, the NMR line widths of the nuclei are dominated by dipolar fields arising from neighbouring magnetic moments. These systems exhibit complex NMR spectra due to shifts in nuclear magnetic resonance frequencies.

In the case of complex molecules in liquid environments, the molecules undergo a tumbling motion, producing very sharp NMR spectra. The technique used to study these systems is known as Fourier transforms NMR Spectroscopy.

Specifically, nuclear magnetic response is extremely useful for analysing samples non-destructively. Radio waves and static magnetic fields easily penetrate many types of matter and anything that is not inherently ferromagnetic. For example, various expensive biological samples, such as nucleic acids, including RNA (Ribonucleic Acid) and DNA (Deoxyribonucleic Acid), or proteins, can be studied using nuclear magnetic resonance for weeks or months before using destructive biochemical experiments. This also makes nuclear magnetic resonance a good source for analysing dangerous samples.

Nuclear magnetic resonance has been used in data acquisition in petroleum industry for petroleum and natural gas exploration and recovery. A borehole is drilled into rock and sedimentary strata into which nuclear magnetic resonance logging equipment is lowered. Nuclear magnetic resonance analysis of these boreholes is used to measure rock porosity, estimate permeability from pore size distribution and identify pore fluids (water, oil and gas).

Nuclear magnetic resonance has also entered the arena of real-time process control and process optimization in oil refineries and petrochemical plants. Two different types of NMR analysis are utilized to provide real time analysis of feeds and products in order to control and optimize unit operations. Time-domain NMR (TD-NMR) spectrometers operating at low field (2–20 MHz for ^{1}H) yield free induction decay (FID) data that can be used to determine absolute hydrogen content values, rheological information, and component composition. These spectrometers are used in mining, polymer production, cosmetics and food manufacturing as well as coal analysis. High resolution FT-NMR spectrometers operating in the 60 MHz range with shielded permanent magnet systems yield high resolution ^{1}H NMR spectra of refinery and petrochemical streams. The variation observed in these spectra with changing physical and chemical properties is modeled utilizing chemo metrics to yield predictions on unknown samples. The prediction results are provided to control systems through analogue or digital outputs from the spectrometer.

Nuclear magnetic response has been adapted to medical studies in the form of magnetic responses imaging (MRI), which we have mentioned earlier. MRI is a technique that is capable of producing high quality images of the internal human body without requiring invasive surgery. MRI started as a tomographic imaging technique because it produced an image that corresponds to a thin slice of the human body, MRI later evolved into a three-dimensional imaging technique.

The complex structural organization of the human tissues can be depicted in vivo in great detail with advanced diffusion magnetic resonance (MR) imaging schemes. Diffusion MR imaging techniques are increasingly varied, from the simple stand most commonly used technique—the mapping of apparent diffusion coefficient values—to the more complex, such as diffusion tensor imaging, q-ball imaging, diffusion spectrum imaging, and tractography. The type of structural information obtained differs according to the technique used. To fully understand how diffusion MR imaging works, it is helpful to be familiar with the physical principles of water diffusion in the brain and the conceptual basis of each imaging technique. In this study, we attempt to show the relationship between diffusion as visualized in NMR and the basic medical imaging principles.

1.2 Theory of Nuclear Magnetic Resonance

At this point, we shall give a very concise description of the principles of nuclear magnetic resonance. This will reveal the basic physics behind the concept of nuclear magnetic resonance, and subsequently, of diffusion in nuclear magnetic resonance. Although the thrust of this study is applying NMR diffusion in most advanced medical imaging methods, we may need to follow the principles step by step.

1.2.1 Nuclear Spin and Magnets

Electrons, neutrons and photons all have an intrinsic property called spin. This spin is defined by the fourth quantum number for any given wave function obtained by solving the relativistic form of the Schrödinger equation. It represents a general property which can be described using the properties of electrons. Electrons in atoms move around the nucleus, generating a magnetic field. This generated field has an angular momentum associated with it. It so turns out that there is also an angular momentum with the electron particle itself, denoting the spin, and this gives rise to the spin quantum number, m_s.

Spin angular momentum is quantized and can take different integer or half integer values, depending on what system is under study. If we solve the relativistic Schrödinger equation for the electron, we get the values + ½ and – ½. Since the Pauli exclusive principle states that no two fermions can have the same quantum number, then, only two electrons, paired antiparallel (one with positive spin and one negative with negative spin) can appear in a single atomic orbital.

Like electrons, protons and neutrons also have a spin angular momentum which can take values of + ½ and – ½. In the atomic nucleus, protons can pair with other antiparallel protons much in the same way that electrons pair in a chemical bond. Neutrons do the same. Paired particles, with one positive and one negative spin thus have a net spin of zero "0". We can see that a nucleus with unpaired protons and neutrons will have an overall spin, with the number unpaired contributing ½ to the overall nuclear spin quantum number, ***I***. When this is larger than zero, a nucleus will have a spin angular momentum and an associated magnetic moment, μ, which is dependent on the direction of the spin. It is this magnetic moment that we manipulate in modern NMR experiments. It is worth noting here that nucleus can have more than one unpaired proton and one unpaired neutron, much in the same way that electronic structures in transition metals can have many unpaired spins. For example ^{27}Al has an overall spin $I = 5/2$.

1.2.2 Values of Spin Angular Momentum

The spin angular momentum of a nucleus can take values from $+ I$ to $- I$ in integral steps. This value is known as the magnetic quantum number, m. for any given nucleus (Dada and Awojoyogbe 2021), there is a total $(2\,I + 1)$ angular momentum states. Spin angular momentum is a vector quantity, the z-component of which is quantized and is given as:

$$I_z = \frac{mh}{2\pi} \quad \text{(where h is the Planck constant)}$$

The resultant magnetic moment of the nucleus is intrinsically connected with its spin angular momentum. This magnetic moment is intrinsically related to I with a proportionality constant, γ called the gyromagnetic ratio by the relation $\mu = \gamma I$.

1.2.3 Spin Behaviour in a Magnetic Field

Considering the case of nuclei which have a spin of a half, like ^{1}H, ^{13}Cr or ^{19}F, the nucleus thus has two possible magnetic moments it could take, often referred to as up or down, + ½ or -½, which are also called the spin states α and β. The energies of each state are degenerate (that is, they are the same). The effect is that the number of atoms (their population) in the up or α-state is the same as the number of atoms in the β-state.

If a nucleus is placed in a magnetic field, the angular momentum axis coincides with the field direction. The resultant magnetic momentum, space quantized from the angular momentum axis, will no longer have the same energy since one state has z-component aligned with an external field and one lower in energy (positive I values) and the other opposes the external field and is higher in energy. This causes a population bias toward the lower energy states.

The energy of a magnetic moment μ when in a magnetic field B_0 (the zero subscript is used to distinguish this magnetic field from any other applied filed) is the negative scalar product of the vectors

$$E = -\mu_z B_0$$

And since we have already defined $\mu = \gamma I_z$, it follows that

$$E = -\frac{mh\gamma B_0}{2\pi} \quad \text{(where } m \text{ is the magnetic quantum number)}$$

1.2.4 Resonance

The energy gap between our α and β states is $\frac{h\gamma B_0}{2\pi}$. Resonance occurs between the states; therefore, equalizing populations, if a radio frequency is applied with the same energy difference ΔE between the spin states. The energy of a photon is $E = h\upsilon$; where υ is the frequency of the photon.

$$\Delta E = \frac{h\gamma B_0}{2\pi}$$

Thus, the frequency of electromagnetic radiation that will produce resonance of a specific nucleus in a field B is

$$\upsilon = \frac{\gamma B_0}{2\pi}$$

It is this frequency that we are concerned with, and detect in NMR. It is this frequency that describes the sample we are observing. But more importantly, it is this resonance that gives rise to the nuclear magnetic spectrum. Nuclei within the body are always in transit, mostly by diffusion; hence, the frequency as stated above when detected carry information about the diffusion process that goes on within the region under investigation.

1.2.5 Nuclear Shielding

It would seem, from the above equation, that all nuclei of the same nuclide (Dada and Awojoyogbe 2021), which have the same gyromagnetic ratio (γ) resonate at the same frequency. This is not the case since the gyromagnetic ratio of a given nuclide does not change; we conclude that the effect of the external magnetic field is different for different nuclei, and also for molecules freely diffusing, and those bound within body cavities. Local effects of other nuclei, especially spin-active nuclei, and local electron effects shield each nucleus differently from the main external field.

It has been stated that the energy of a spin state is defined by $E = -\mu_z B_0$. It can be seen that by shielding the strength of the magnetic field, the experienced effect, or effective magnetic field at the nucleus is lower: $B_{effective} < B_0$.

Thus, the energy gap is different, and hence the frequency required for resonance, deviates from the expected value. These differences due to nuclear shielding give rise to many peak frequencies in a nuclear magnetic resonance spectrum. It can be seen now why nuclear magnetic resonance is a direct probe of chemical structure.

It is possible for the shielding to change as the orientation of the molecule changes. This is called chemical shift anisotropy, and can even be used in some types of experiment, especially if the sample is solid, external fields of the crystal structure interfere with the nuclear field too much for the spectrum to be of any use.

1.2.6 Relaxation

The process called population relaxation refers to nuclei that return to the thermodynamic state in the magnet. This process is also called T_1 relaxation, where T_1 refers to the mean time required for an individual nucleus to return to its equilibrium state. Once the population is relaxed, it can be probed again, since it is in the initial state (Dada and Awojoyogbe 2021).

The precessing nuclei can also fall out of alignment with each other (returning the net magnetization vector to a non-precessing field) and stop producing signal. This is called T_2 relaxation. It is possible for the nuclei to be in this state and not have the population difference required to give a net magnetization vector at its thermodynamic state. Because of this, T_1 is always large (slower) than T_2. This happens because some of the spins were flipped by the pulse and will remain so until they have undergone population relaxation.

In practice, the T_2 time is the life time of the observed NMR signal, the free induction decay (FID) in the NMR spectrum (the Fourier transform of the free induction decay), the T_2 time defines the width of the NMR signal. Thus, a nucleus having a large T_2 time gives rise to a sharp signal, whereas nuclei with shorter T_2 give rise to more broad signals. The length of T_1 and T_2 is closely related to molecular motion. This molecular motion is then what is studied, analysed and adapted in the various applications of nuclear magnetic resonance. The molecular motion is guided by a set of equations known as the Bloch NMR equations, the solution of which this study attempts to find. It would be necessary at this point to introduce Magnetic Resonance Imaging (MRI) because of its great relevance in this study.

1.3 Magnetic Resonance Imaging (MRI)

Magnetic resonance imaging (MRI), formerly referred to as magnetic resonance tomography (MRI) is a method used to visualize the inside of living organisms as well as to detect the composition of geological structures. It is primarily used to demonstrate pathological alterations of living tissues and is a commonly used form of medical imaging.

Magnetic resonance imaging was developed from the knowledge gained in the study of nuclear magnetic resonance. The original name of the medical technology is nuclear magnetic resonance imaging (NMRI), but the word nuclear is almost universally dropped. This is done to avoid the negative connotations of the word nuclear, and to prevent patients from associating the examination with radiation exposure, which is not one of the safety concerns for MRI.

Like NMR, magnetic resonance imaging is based on the absorption and emission of radio frequency energy under the influence of a magnetic field. On a molecular level, the human body consists largely of hydrogen-rich fat and water molecules, and approximately 63% of the human body contains hydrogen atoms, which are constantly in motion. The hydrogen nuclei emit signals that can be observed by magnetic resonance imaging. When exposed to a magnetic field, the spin of the protons in the hydrogen nuclei of water which ordinarily assume random orientations, line up (align with the magnetic field). Although short pulses of radio wave briefly disturb this spin alignment, the spins promptly re-align in the direction of the magnetic field. In the process of realigning, small signals are produced that can be picked up by sensitive scanners. The signals are then compiled by a computer and an image is formed.

Because water is the major component of soft tissues, MRI creates excellent images of soft tissue and organs, but poor images of dense structures like bone due to proton flow restriction. This study shall address each of these processes. MRI thus complements X-ray imaging, which by contrast is fine for dense structures but useless for soft tissues.

1.3.1 Three-Dimensional Magnetic Resonance Imaging

In clinical practice, MRI is used to distinguish pathologic tissue (such as a brain tumor) from normal tissue. One advantage of an MRI scan is that it is harmless to the patient. It uses strong magnetic fields and non-ionizing radiation in the radio frequency range. This is unlike the CT scans and traditional X-rays which involve doses of ionizing radiation and may increase the chance of malignancy, especially in a fetus (Amouzeshi and Zarghani 2018).

While CT provides good spatial resolution (the ability to distinguish two structures which are at an arbitrarily small distance from each other as separate), MRI provides comparable resolution with far better contrast resolution (that is, the ability to distinguish differences between two arbitrarily similar but not identical tissues). The basis of this ability is the complex library of pulse sequences that the modern medical MRI scanner includes, each of which is optimized to provide image contrast based on the chemical sensitivity of MRI. For example, with particular values of the echo time (TE) and the repetition time (TR), which are the basic parameters of image acquisition, a sequence will take on the property of T_2—weighting. On T_2—weighted scan, fat-, water-, and fluid-containing tissues are bright (most modern T_2 sequences are actually fast T_2 sequences). Damaged tissue tends to develop edema (an excessive accumulation of fluid in the cells, tissue spaces or body cavities due to a disturbance in the fluid exchange mechanism or diffusion) which makes a T_2-weighted sequence sensitive for pathology, and generally able to distinguish pathologic tissue from normal tissue. With an additional radio frequency pulse and additional manipulation of the magnetic gradient, a T_2-weighted sequence can be converted to a FLAIR (Fluid Light Attenuation Inversion Recovery) sequence in which free water (which diffuses quite faster) is now dark, but edematous tissues (tissues that have developed edema) remain bright. This sequence in particular is currently the most sensitive way to evaluate the brain for demyelinating diseases (diseases associated with the destruction or removal of myelin from nerves), such as multiple sclerosis. The typical MRI examination consists of 5–20 sequences (Brown and Semelka 2011), each of which is chosen to provide a particular type of information about the subject tissues. This information is then synthesized by the interpreting physician.

1.4 Diffusion Process in Magnetic Resonance

The molecular movements that are seen in fluid spins are frequently random and haphazard enough that these processes directly impact NMR signal attenuation. Nuclear spins are continually in motion while traversing relatively tiny distances, so the spins are viewed as particles diffusing over small distances (Cowan 1997). This results in the establishment of irreversible dampening of the size of the nuclear magnetization and an eventual erasure of the information included in the spatial frequency. The consequence of this knowledge is that molecule diffusion is directly imprinted on the NMR signal, given that the information collected could be understood suitably. To account for the migration of spins in magnetic resonance, a diffusion factor is typically added to the Bloch NMR equation. This modification to the basic Bloch NMR equation is known as the Bloch-Torrey equation (Cowan 1997; Torrey 1956). The Bloch-Torrey equation generalises the Bloch NMR equations to incorporate factors that characterise magnetization transfer owing to molecular diffusion (Torrey 1956).

1.5 Image Processing with Functional MRI of the Brain Software Library (FSL)

FSL is a software library (developed by the Analysis Group, FMRIB, Oxford, UK) containing image analysis and statistical tools for functional, structural and diffusion MRI brain imaging data. It is a comprehensive library of analysis tools for fMRI, MRI and diffusion brain imaging data. It runs on macOS, Linux, and Windows via the Windows Subsystem for Linux. Its tools can be run both from the command line and as GUIs. The tools of FSL can be summarized as in Table 1.1 (Smith 2002; Jenkinson et al. 2002; Smith et al. 2006).

Table 1.1 FSL tools and their functionalities

Tool	Functionality
Functional MRI FEAT	Model-based fMRI analysis with straightforward but powerful GUI: data preprocessing (including slice timing correction, MCFLIRT motion correction and PRELUDE + FUGUE EPI unwarping); FILM GLM timeseries analysis with prewhitening; registration to structural and/or standard space images; and fully generalised mixed-effects group analysis using advanced Bayesian techniques
MELODIC	Model-free FMRI analysis using Probabilistic Independent Component Analysis (PICA). MELODIC automatically estimates the number of interesting noise and signal sources in the data and because of the associated "noise model", is able to assign significances ("*p*-values") to the output spatial maps
FLOBS	Generation of optimal HRF basis functions and Bayesian activation estimation
SMM	Spatial mixture modelling-alternative hypothesis testing using histogram mixture modelling with spatial regularisation of the voxel classification into activation and non-activation
Structural MRI BET/BET2	Brain Extraction Tool (BET) segments brain from non-brain in structural and functional data, and models skull and scalp surfaces
SUSAN	Nonlinear noise reduction
FAST	FMRIB's automated segmentation tool is for brain segmentation (into different tissue types) and bias field correction
FLIRT	FMRIB's linear image registration tool is used for linear inter- and intra-modal registration
FUGUE	Unwarps geometric distortion in EPI images using B0 field maps
SIENA	Structural brain change analysis, for estimating brain atrophy
Diffusion MRI FDT	FMRIB's diffusion toolbox contains tools for low-level diffusion parameter reconstruction and probabilistic tractography
TBSS	Tract-Based Spatial Statistics (part of FDT)—voxel wise analysis of multi-subject diffusion data
Other tools Inference	Various inference/thresholding tools, including Randomise (permutation-based inference tool for nonparametric statistical thresholding), cluster (cluster-based thresholding using GRF theory for inference), FDR (false discovery rate inference) and Glm (a GUI for creating model design matrices)
FSLeyes	Interactive display tool for 3D and 4D data
AVWUTILS	Miscellaneous utils for converting and processing images

1.6 Medical Image Denoising

Denoising is a key topic in image processing. It attempts to improve the quality of an image by eliminating the parasitic information that randomly adds to the features of the scene. Guedj and Rengot (2020), in their work, stated that this noise may be attributed to image capture conditions (lack of light, blurring, erroneous tuning of field depth) or to the camera itself (increase of sensor temperature, data transfer mistake, approximations made during digitization). Therefore, the difficulty lies in eliminating the noise from the image while keeping its structure. Fan et al. (2019) and Goyal et al. (2020) stated that many techniques of denoising have already been established in the last few decades; although excellent performance has been achieved, denoised images still tend to be overly smooth (certain details are lost) and blurred (edges are less sharp). Seeking to enhance the performance of these algorithms is a highly active research area.

Medical image denoising is regarded as a method of increasing the perceptual quality of degraded, noisy images collected using specialist medical image acquisition instruments. Regrettably, such imaging systems are prone to capturing noise despite the rise in imaging technology (Chen et al. 2018). However, the presence of noise in the images has a starling effect on medical image analysis as well as might convolute the decision-making manoeuvre of an expert (Xing et al. 2020). Hence, denoising is considered conventional, but hard medical image analysis work. Typically, medical image suffers from various noises due to the different medical imaging modality employed in the capture of patient images (Guo et al. 2019; Kazeminia et al. 2020). Some of these noises include Gaussian noise, Poisson noise, impulse noise, and quantization noise. While the development of any sort of noise in diverse photographs is a fully random stochastic process, the additive white Gaussian noise is the most frequent noise in images since it happens during capture and transmission via analogous circuitry, which is very much the fundamental function necessary for computerised operation. Besides the fact that Gaussian noise impacts the whole image in a uniform way, it may be claimed that Gaussian noise influences practically each pixel of an image, which substantially limits the degree of image interpretation.

Goyal et al. (2020), in their work, showed that impulse noise, on the other hand, influences a restricted number of pixels in the dark and white modes, and this still provides some space for common interpretation of the image data. The predominance of various forms of noise, such as quantization noise, impulse noise, speckle noise, and Poisson, is a nearly uncommon occurrence due to defective manufacturing, a high bit error rate, and insufficient photon count during image capture (Majeski 2023). However, the influence of Gaussian noise is most significant as far as the interpretation of feature details is concerned, and this becomes more crucial in the case of medical and remote sensing images.

Gaussian noise has a probability density function similar to the normal distribution. A random value determined by the normal distribution is applied to each clean pixel. The noise samples generated are independent of each other, and for every pixel, the variance and the mean are the same. Gaussian noise is the most common noise that may develop owing to the thermally agitated electrons in an electrical wire (Goyal et al. 2020; Gupta 1975). It may also be characterised as the noise that has a Gaussian amplitude distribution, and this can be technically defined as:

$$P_f = \frac{1}{\sigma\sqrt{2\pi}} e^{-\frac{1}{2}\left(\frac{x-\mu}{\sigma}\right)^2} \tag{1.1}$$

where μ is the mean (the position of the peak in the bell-shaped curve for a Gaussian distribution) and is the standard deviation. The distribution with a zero mean and a variance equal to one is the standard normal distribution. In statistical and functional analysis, the first approximation for any real-valued random variables that is uniformly distributed around a single mean is considered normally distributed (Karunasingha 2022). This According to the central limit theorem, the mean of an extended collection of random variables non-dependently removed from an identical distribution is roughly distributed normally, independent of the shape or type of the original distribution.

A large number of unique approaches have been presented in the recent past. According to their optimisation methodologies, the medical image denoising works may be split into two broad categories: (i) classical techniques; and (ii) learning-based approaches.

Classical approaches: The filter-based denoising technique has dominated the classical medical image denoising for a long period. Most of the current filter-based efforts focused on developing a low-pass filter, where the filter seeks to remove precipitate peaks from a noisy image based on the local estimate (Beck et al. 2022). In the respective research performed by Kulkarni and Madathil (2018), Majeeth and Babu (2019), Sagheer and George (2020) and Chandra and Verma (2020), they highlighted the numerous variants of filter-based approaches, Gaussian averaging filters, median filters, mean filters, diffusion filters which are used widely for removing noise from specific types of medical imaging modalities such as ultrasound images, magnetic resonance images (MRI), computed tomography (CT) images. Despite the extensive adoption of such filter-based approaches, such methods tend to smoothen the provided images while eliminating the noise.

Many subsequent efforts enhanced medical image denoising by employing adaptive filters to solve the deficiencies of previously described static filter-based techniques. The non-local denoising approach, like Block Matching 3D (BM3D) (Wang et al. 2019), is one of the ideal instances of adaptive filter-based methods. Similarly, non-local means filters-based methods by Arora et al. (2023), was used to denoise speckle noise while retaining the edges and important features of the image, for MR image denoising, Optimised Bayesian Non-Local Mean (OBNLM) (Masse et al. 2018).Their enhancement includes decreasing numerous estimates of an identical pixel using a masking approach and altering the spatial size of the comparable patch search region, they provide multiple studies and explain ideal choices for their parameters, providing a gain in computing time of 50% with an optimised masking approach. Sharif et al. (2020), in their study, modified non-local-based (MNL) for ultrasound image denoising. Moreover, bio-inspired bilateral filter and CT image denoising are representative of the adaptive filter-based techniques for learning residual noise from a wide breadth of data samples. In addition, network design uses the feature correlation known as the attention mechanism and blends it with spatially refined residual characteristics. However, their suggested technique can manage real-world image noise and may increase the performance of numerous medical image processing tasks without introducing any visually distracting artefacts.

Another well-known classical medical imaging denoising genre is referred to as multi-scale analysis-based approaches. This genre can be seen in the study by Sharif et al. 2020, where the representative techniques aim to process noisy images at multiple image resolutions. In recent years, a considerable amount of work has exploited such denoising approaches and applied the time–frequency analysis as seen in the study conducted by Sagheer and George (2020). Nevertheless, in several circumstances, multi-scale medical image denoising illustrates inadequacies in describing the distribution of noisy inputs on distinct scales. The drawbacks of the multi-scale approaches have been solved using nonlinear estimators (Tian and Chen 2011). They used a maximum a posteriori (MAP) estimation-based image despeckling strategy by putting a non-parametric statistical model into a Bayesian inference framework. In creating their wavelet coefficient distributions, the suggested statistical model outperforms traditional approaches. Furthermore, as shown by their studies, the suggested image despeckling technique outperforms standard image despeckling approaches. However, the performance of this conventional image-denoising category is still far from expectations.

Learning-Based Methods: The learning-based image-denoising approach has recently begun to attract attention in medical image denoising. In a recent work, a feedforward autoencoder, as seen in Gondara (2016) work, has been utilised to learn medical image denoising. Gondara (2016) showed that denoising autoencoders developed using convolutional layers may be applied for successful denoising of medical photos even with a small sample set. To boost denoising efficiency, heterogeneous photos may be blended to increase sample size. However, the following work (Jifara et al. 2019) on medical image denoising adopted the idea from Zhang et al. (2017) and increased the effectiveness of their technique by applying residual learning batch normalization as regularization in the deep model. Incorporating batch normalisation with residual learning enables better model learning accuracy and training time. In another study on medical image denoising conducted by Liu et al. (2019), they used a different network construction technique and employed a genetic algorithm (GA) to search the hyper-parameters of their deep network. Despite the good performance at specified noise levels as well as on specific datasets, none of the available publications extend their strategy to multidisciplinary modalities which Sharif et al. (2020) clearly stated.

1.7 Medical Image Feature Extraction

Feature extraction in medical imaging is a key operation that includes finding and extracting important patterns from medical images, including MRIs. Extracting this critical information from medical photos is essential for further healthcare analysis, diagnosis, and decision-making. It plays a crucial function in image processing, boosting image quality by lowering dimensionality (Kumar et al. 2020). Extracting features increases the accuracy of trained models by separating critical information, thereby, eliminating redundant input, as found by Kumar et al. (2020). Strategies for extracting features include developing new combinations from the original collection of characteristics (Liu et al. 1993). They employed optimum discriminant criteria to extract algebraic characteristics. For the provided training image samples, a set of appropriate discriminant projection vectors is constructed according to a generalized Fisher criteria function. The findings revealed that their given technique has high recognition performance and they highlighted that the mechanism is crucial in multimedia systems. Image processing is vital for guaranteeing image accuracy as pointed out in the work of Russ (1990), and feature extraction comprises a key part of image processing (Mutlag et al. 2020).

Humans can develop tales based on photographs and their previous knowledge; however, the issue arises: can a computer system do the same? The answer is positive. By applying models to teach the system feature extraction, computers may construct tales analogous to humans. To train a computer system, knowing the needed characteristics, extraction methodologies, and available approaches is crucial. Feature extraction tries to retrieve useful information from medical images, such as pixels, image texture, and dimensions, crucial for comprehending the characteristics of an image.

While people learn key qualities visually, quantifying features like giving a number value to ground-glass texture presents a hurdle. Image characteristics need to endure fluctuations in noise, intensity, and rotation angles often encountered in medical imaging data. While multiple characteristics may be calculated from an image, having an excess could lead to overfitting rather than appropriate decision-making (Erickson et al. 2017). Selecting a subset of characteristics for the best predictions is called feature selection (Bolón-Canedo and Alonso-Betanzos 2019). Correlation analysis between characteristics is one way to reduce the number of features without losing crucial information (Jiang et al. 2020). Their research findings on benchmark datasets demonstrate that the proposed FSLCLC outperforms the state-of-the-art multi-label feature selection techniques. However, analysing a characteristic in isolation could be dangerous, as indicated by Erickson et al. (2017). Feature extraction has applications in radiomics, computer-aided diagnosis (CAD), image segmentation, and registration.

The extraction of medical image characteristics is a vital step in leveraging the huge information contained in medical images (Sarvamangala and Kulkarni 2022). These traits provide the foundation for additional investigations, assisting in diagnosis, therapeutic planning, and understanding illness development. Advancements in computer technologies, notably deep learning, has contributed to the extraction of strong and distinguishing information from medical images. This improvement leads to more accurate and precise medical image processing and interpretation (Razzak et al. 2018). Overcoming issues linked to image variability while retaining the reliability and interpretability of derived characteristics remains a focus for researchers in the medical imaging area.

1.8 Image Segmentation

Image segmentation is a core approach in computer vision that includes splitting an image into several segments to simplify its representation and ease the extraction of useful information. This method plays a vital role in different applications such as object recognition, medical imaging, scene interpretation, and image analysis (Minaee et al. 2021; Liu et al. 2021). It is a sophisticated and multidimensional subject in computer vision that incorporates multiple methodologies, tactics, and issues.

Medical image segmentation is a vital activity in the field of medical imaging that includes splitting an image into distinct sections or structures to identify and characterise certain anatomical or pathological areas of interest. Das et al. (2021) noted that segmentation serves a key role in different medical applications, assisting in diagnosis, treatment planning, and tracking the course of illnesses. In their work, they used hybrid EF-based blood-cell segmentation method which has the potential to identify a range of haematological diseases. Contributions from their work are: (1) a novel least-squares (LS)-based geometric EF approach; (2) more precise seed-point detection based on BO-FRS; and (3) enhanced segmentation performance through the use of a hybridised version of geometric and algebraic EF techniques that retain the advantages of both approaches. The importance of medical image segmentation might be noticed in disease diagnosis and localization, treatment planning and guidance, and disease progression monitoring.

According to these various research studies by Zhang (1996), Anjna and Kaur (2017) and Abdulateef and Salman (2021), the techniques and methods typically used in image segmentation include: thresholding and region-based methods, edge-based methods, clustering and classification, model-based segmentation; and deep learning-based approaches: convolutional neural networks (CNNs) and deep learning architectures, notably U-Net, SegNet, or Mask R-CNN. These have demonstrated substantial success in medical image segmentation owing to their capacity to learn hierarchical features. As with every machine learning approach applied to medical images, there are various hurdles that might restrict image segmentation: anatomical variability, image noise and artefacts and manual annotation and ground truth. Recent studies on the partitioning of medical images have included tasks such as skin lesion delineation (Jafari et al. 2016). In Jafari et al. (2016), the CNN generates a label for every pixel by fusing local and global contextual data, resulting in a segmentation mask that displays the lesion area. There were several post-processing steps to further enhance this mask. Their experimental findings demonstrate that the suggested technique can achieve higher segmentation accuracy than some current state-of-the-art methods. In a study on identification of brain tumours carried out by Kadkhodaei et al. (2016), preprocessing processes employed included image improvement and normalisation to the same scale, followed by 3D super-voxel segmentation based on intensities. They go on to offer a unique feature that makes use of an algorithm for saliency identification. The borders of the original image are matched with the saliency map, using an edge-aware filtering strategy, which enhances the tumour boundaries. Supervoxels are utilised to produce a collection of strong textural qualities for the aim of diagnosing cancers in brain imaging. According to experimental data, our recommended solution performs better in terms of dice score than a comparable state-of-the-art algorithm. In the observation of heart ventricles by Nasr-Esfahani et al. (2018), they used an automated approach for cardiac MR image segmentation of the left ventricle, which utilise the region of interest as an input for a fully convolutional network after first automatically extracting it. In their recommended post-processing phase, they did thresholding on the fully convolutional network's output map and chose regions based on their roundness on the York dataset of heart photos, their employed approach acquired a Dice score of 87.24%. Rafiei et al. (2018) performed an assessment of liver conditions while utilising the proposed 3D to 2D fully convolution network (3D-2D-FCN) to segment livers effectively. The conditional random field on the border of the organ is utilised to enhance the segmented mask. They segmented a target liver with a dice score of 93.52% in less than a minute.

1.9 Combining Physics Principles with Neural Networks for Feature Extraction, Denoising, and Segmentation

Anaconda is a prominent open-source version of the Python and R programming languages for scientific computing, data research, and machine learning. It comes with a broad collection of pre-installed packages and tools, making it simple to build up a solid development environment. Anaconda contains Spyder, Jupyter Notebooks, and other integrated development environment which is extensively used for interactive data exploration and visualization. It can be downloaded from https://www.anaconda.com/products/distribution.

The MRI datasets used in the study were obtained from W-ResearchWorks Archive which is a digital library of Washington education platform and was downloaded from the link (https://digital.lib.washington.edu/researchworks/handle/1773/33311?show=full). In this dataset, diffusion MRI at multiple b-values (200, 400, 1000, 2000, and 3000) were collected on a 3 T Siemens Prisma MRI instrument in a single healthy individual.

Data were acquired on a Siemens 3 T PRISMA MRI (maximum gradient strength of 80 mT/m) using a 64-channel coil, at CENIR, ICM (Paris). The multiband epi2ddiff sequence from Minneapolis (http://www.cmrr.umn.edu/multiband/) was used with a multiband factor of 3, TR = 3 s, TE = 74 ms, flip angle = 72, voxel size = 2 mm, isotropic FOV = 208 × 208 × 144, bandwidth = 2090, monopolar acquisition scheme with δ of 15.7 ms and Δ of 36.5 ms (spoiler b-value = 2.6 s/mm^2). For b = 200, 400, 8 directions were used for each b-value. For other b-values, 90 directions for each b-value were chosen o ensure that there is different direction for each b-value. A b = 0 image was interleaved every 8 or 9 directions. Figure 1.1 is the compressed form of the downloaded data needed for this study.

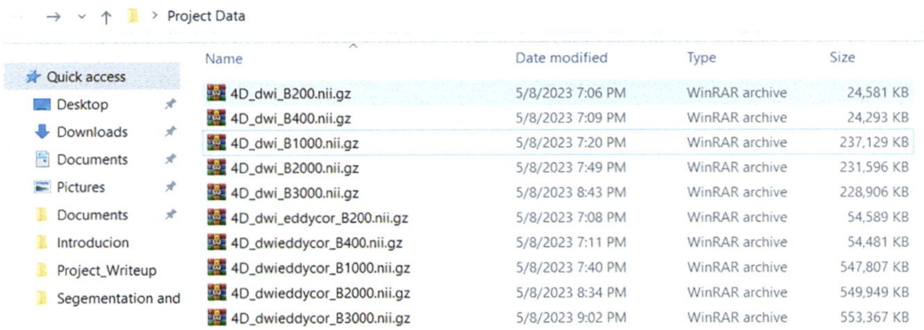

Fig. 1.1 The downloaded dataset for the study

The flow charts showing the method used in this study are shown in Figs. 1.2 and 1.3, respectively.

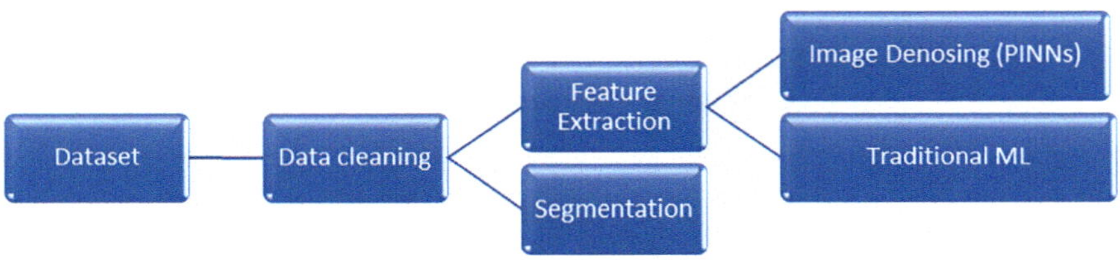

Fig. 1.2 Flowchart showing different stages executed in this study

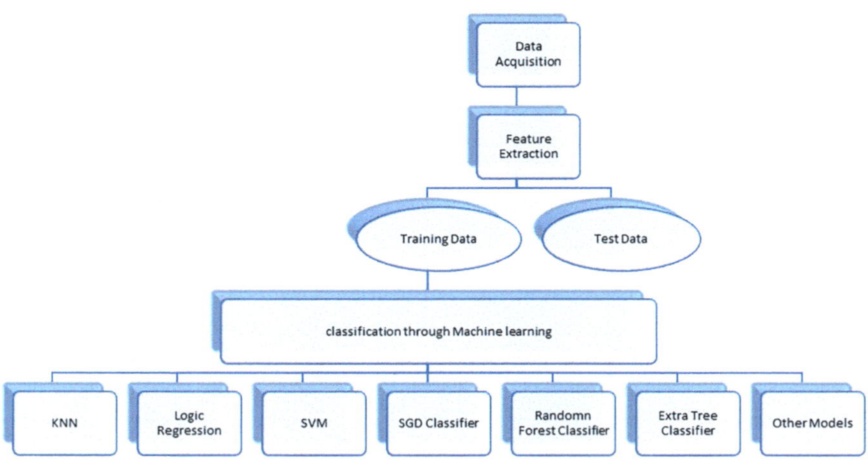

Fig. 1.3 Proposed method flowchart showing traditional machine learning models used for image segmentation

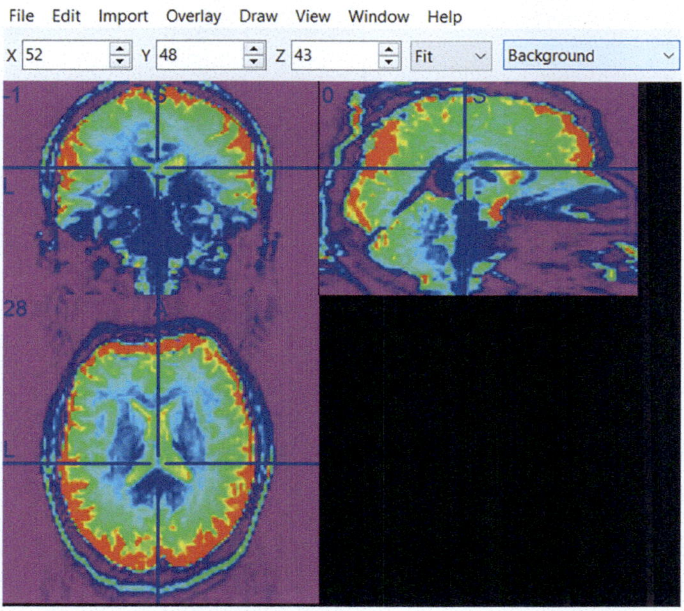

Fig. 1.4 Coronal, sagittal and axial view of the dataset used for the study

For image segmentation, traditional machine learning models have been used for training the crafted features. This is illustrated in Fig. 1.3.

The detailed implementation of the steps summarised in Figs. 1.2 and 1.3 are presented below:

1.9.1 Data Preparation

The dataset was first viewed using MRIcron as seen in Fig. 1.4, and then decompressed using some set of code in image of Fig. 1.5 into a folder which was set as the working directory of the Jupyter notebook. Figure 1.6 is the extracted image.

It is important to note that each image data is in the Neuroimaging Informatics Technology Initiative (NIfTI) format. This comprises of various image slices from the MRI scanner.

1.9.2 Skull Extraction with FSL's BET

From Fig. 1.4, it would be observed that the image data contains contributions from the skull, and as a preliminary step in processing, skull stripping needs be performed

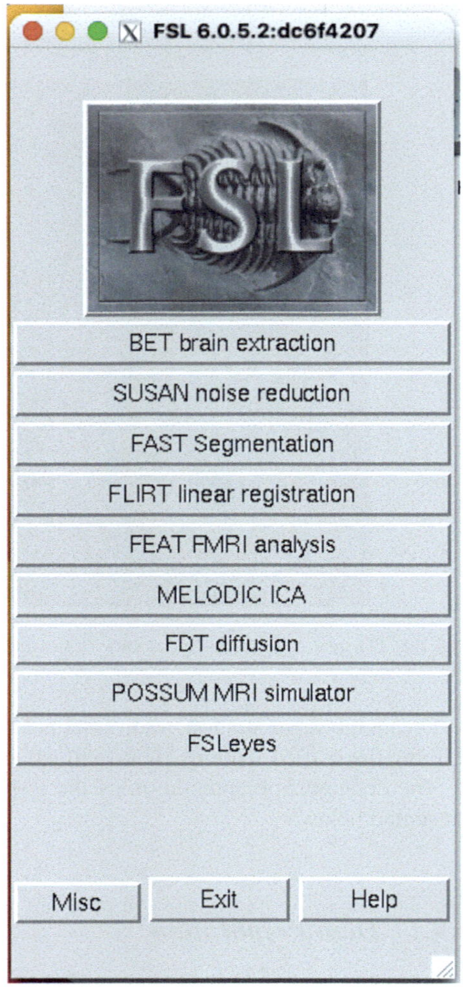

Fig. 1.5 The main FSL user interface showing all the tools

to increase the speed and accuracy of diagnosis in manifold. The essence of this is simply to remove non-cerebral tissues such as skull, scalp, and dura from the image data. The main FSL interface is shown in Fig. 1.7.

Upon clicking on the BET brain extraction tab, the BET dialog box, as shown in Fig. 1.8, is obtained. In the "Input image" field, the image to be processed can be selected and the path to the output was set at the "Output image" field.

As shown in Fig. 1.6, the fractional intensity threshold was set to 0.5, threshold gradient was set to 0 while the coordinates for centre of initial brain surface was set to (X = 0, Y = 0, Z = 0). In the Advanced options, the BET tool was set to output the brain extracted image and the mask image.

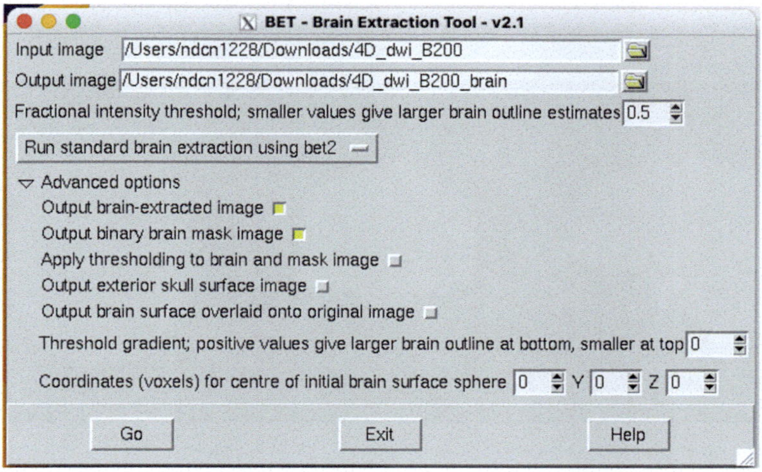

Fig. 1.6 The BET user interface

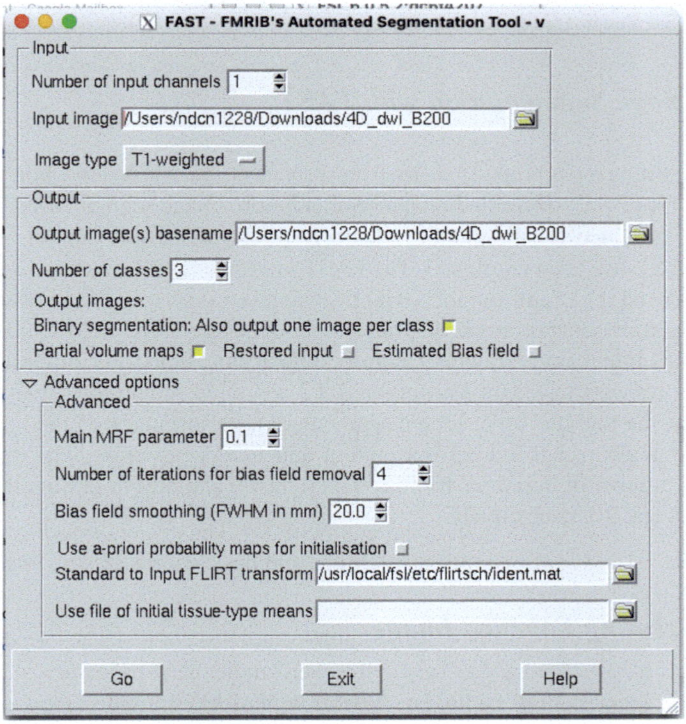

Fig. 1.7 The FAST user interface

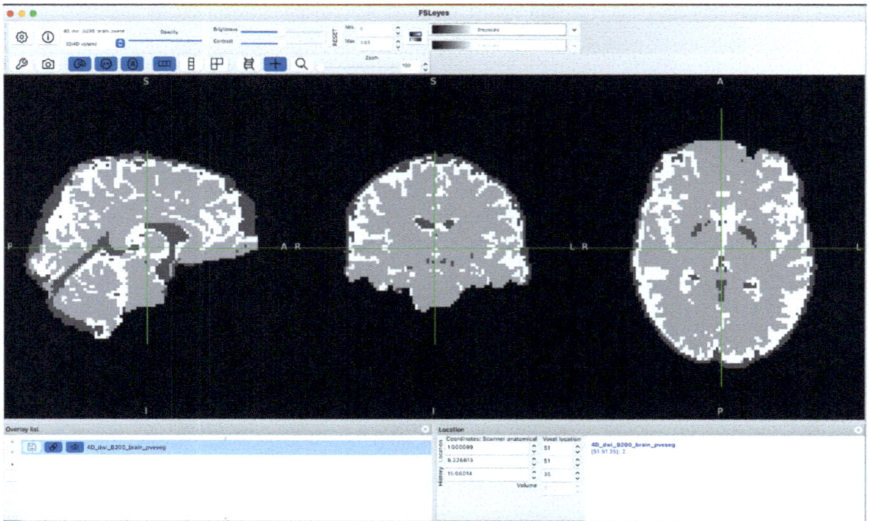

Fig. 1.8 Segmented image opened with FSLeyes

1.9.3 Image Segmentation with FAST

Machine learning models need to learn from a dataset to get a task done. For this study, especially for physics-informed image segmentation, image segmentation examples need to be provided for the model to learn from. To create this dataset, the FAST module in FSL has been employed. The FAST interface in FSL is shown in Fig. 1.7.

As shown in Fig. 1.7, the number of input channels was set to 1, the input and output image paths were set appropriately. The number of classes were set to 3 and expected to represent white matter, gray matter, and cerebrospinal fluid. It is important to note that these tools were tested on T1-weighted, T2-weighted and proton density images. Fortunately, the tool had an excellent segmentation performance on the image data. The tool was also set to output one output and masks per class. The main MRF parameter, number of iterations for bias field removal and bias field smoothing were set to 0.1, 4 and 20, respectively.

1.9.4 Creating Labelled Image

After image segmentation had been completed, a labelled image data was created and the segmented image was opened in FSLeyes, as shown in Fig. 1.8, with the image segments listed as 0, 1, and 2, representing white matter, gray matter, and cerebrospinal fluid.

The next step was to change the colour function to "HSV" as shown in Fig. 1.9.

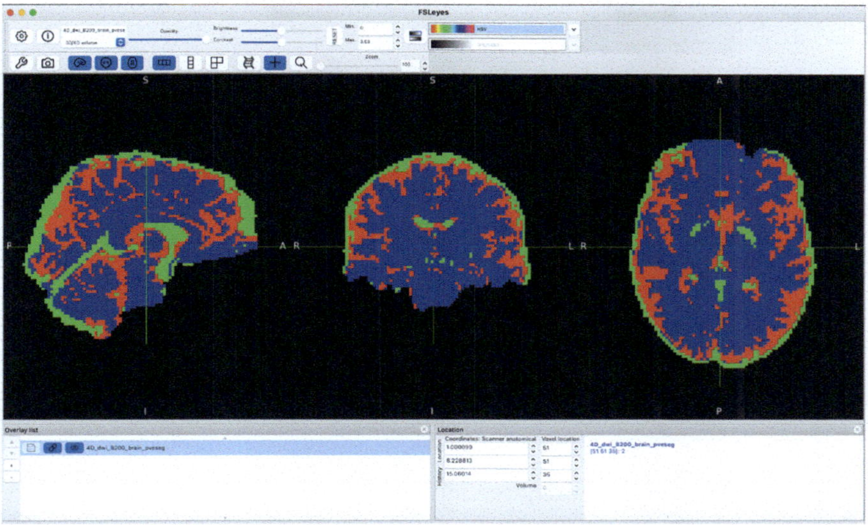

Fig. 1.9 Segments lablelled with unique colours

In the Toggle canvases, the Sagittal and coronal orthogonal planes were switched off (leaving only the transverse plane) while the Toggle cursor/labels was also switched off. Hence, the labelled image produced is shown in Fig. 1.10 and was exported as a single slice in. TIF format.

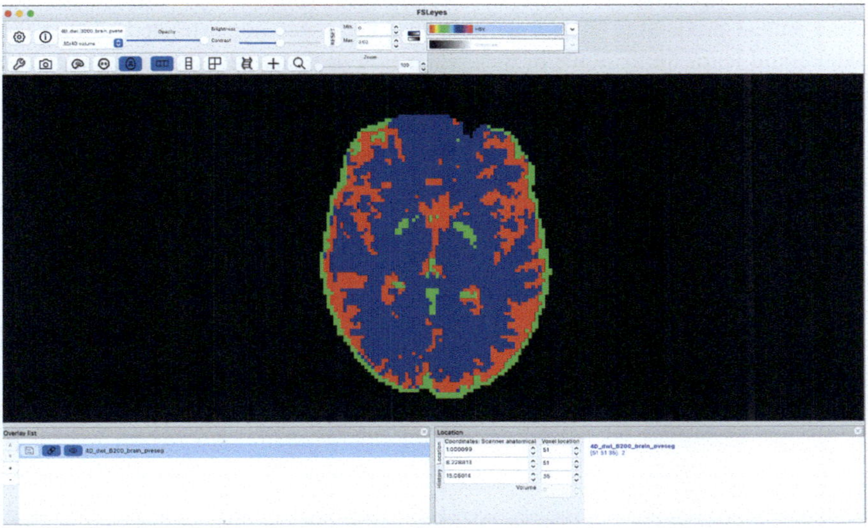

Fig. 1.10 Labelled image data

```python
def convert_dwi_to_images(input_path, output_dir):
    # Load the DWI NIfTI file
    dwi_img = nib.load(input_path)
    dwi_data = dwi_img.get_fdata()

    # Iterate over each slice in the 4th dimension
    for slice_index in range(dwi_data.shape[-1]):
        # Extract the 3D slice
        slice_data = dwi_data[..., slice_index]

        # Iterate over each slice in the 4th dimension
        for slices in range(slice_data.shape[-1]):
            # Extract the 2D slice
            slice_data_2d = slice_data[..., slices]

            # Normalize the slice data to [0, 255]
            slice_data_2d = (slice_data_2d - np.min(slice_data_2d)) / (np.max(slice_data_2d) - np.min(slice_data_2d))
            slice_data_2d = (slice_data_2d * 255).astype(np.uint8)

            # Create a PIL image from the slice data
            slice_image = Image.fromarray(slice_data_2d)

            # Rotate the image 90 degrees clockwise
            rotated_image = slice_image.rotate(90, expand=True)

            # Save the slice image as TIFF
            output_path = os.path.join(output_dir, f"slice_{slice_index}{slices}.tif")
            rotated_image.save(output_path)

            print(f"Saved slice {slice_index}{slices} as {output_path}")

# Example usage
input_file = "4D_dwi_B200.nii.gz"
output_directory = "b200"

convert_dwi_to_images(input_file, output_directory)
```

Fig. 1.11 Python function for extracting slices from 4D image data

1.9.5 Extraction of Slices from Volume Data

For NIfTI files, images and other data are stored in a 3D format. This is specifically designed to overcome spatial orientation challenges of other medical image file formats. The slices are 2D layers with which these images have been made. For further image processing, it is important that the slices are extracted so that the model could learn the features in these images. For this study, the image data has been extracted into the Tag Image File Format (TIF) because of its high-quality graphics. Python functions were written for this purpose for 4D nifti files and 3D nifti files. The image data with $b = 200$ s/mm^2 has been used as the reference data. These Python functions are shown in Figs. 1.11 and 1.13, respectively.

Figures 1.12 and 1.14 show the images that were extracted using the Python functions. It is important to note that the original 4D image data loses few features as image processing is executed from one stage to another. This is why the Python function to process 3D nifti images was necessary.

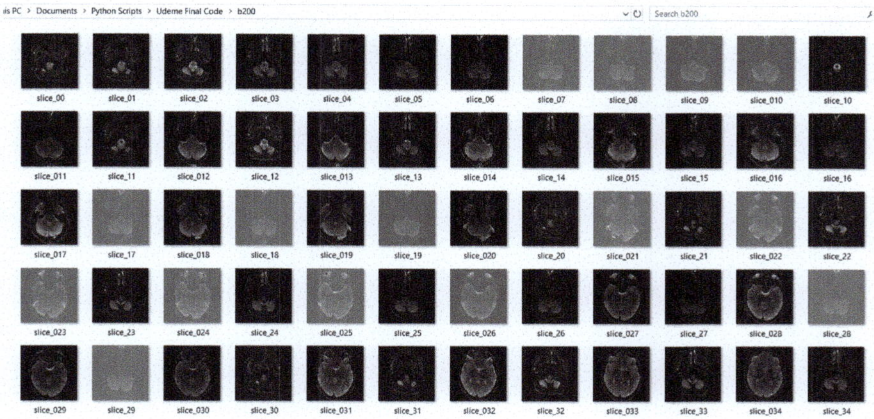

Fig. 1.12 Extracted slices from 4D image data

```
def convert_dwi_to_images3D(input_path, output_dir):
    # Load the DWI NIfTI file
    dwi_img = nib.load(input_path)
    slice_data = dwi_img.get_fdata()

    slice_data = (slice_data - np.min(slice_data)) / (np.max(slice_data) - np.min(slice_data))

        # Iterate over each slice in the 4th dimension
    for slices in range(slice_data.shape[-1]):
        # Extract the 2D slice
        slice_data_2d = slice_data[:,:, slices]


        # Normalize the slice data to [0, 255]
        #slice_data_2d = (slice_data_2d - np.min(slice_data_2d)) / (np.max(slice_data_2d) - np.min(slice_data_2d))
        slice_data_2d = (slice_data_2d * 255).astype(np.uint8)

        # Create a PIL image from the slice data
        slice_image = Image.fromarray(slice_data_2d)

        # Rotate the image 90 degrees clockwise
        rotated_image = slice_image.rotate(90, expand=False)

        # Save the slice image as JPEG
        output_path = os.path.join(output_dir, f"img_{slices}.tif")
        rotated_image.save(output_path)

        print(f"Saved slice {slices} as {output_path}")


# Usage
input_file = "4D_dwi_B200_brain.nii.gz"
output_directory = "python_ext"

convert_dwi_to_images3D(input_file, output_directory)
```

Fig. 1.13 Python function for extracting slices from 3D image data

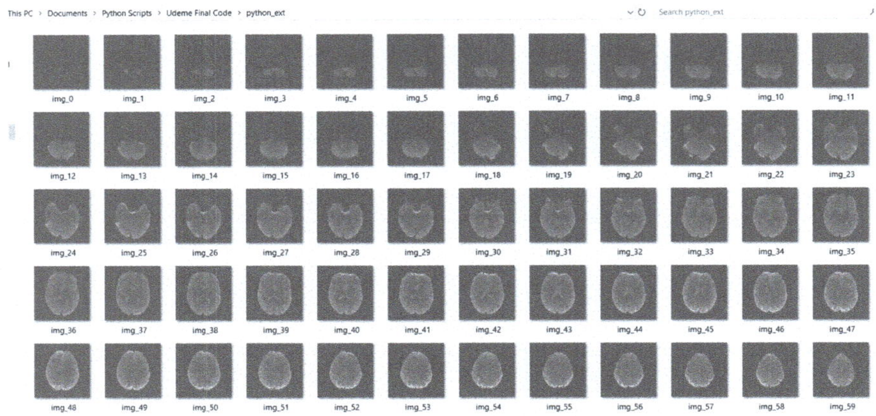

Fig. 1.14 Extracted slices from 3D image data

1.10 Conclusion

We have introduced physics-informed neural network (PINNs) algorithms particularly in instances where conventional, exclusively data-driven algorithms may fail owing to noise or a lack of annotated data. Since medical imaging data are expensive or complicated, the incorporation of physics-based constraints in neural network topologies has the potential to reduce dependency on vast amounts of labelled training data which could be beneficial. Combining physics principles with neural networks may pave the way to new insights and breakthroughs in fundamental medical health knowledge (Anazodo et al. 2023), showing relationships and patterns in medical data that would not be evident using traditional methodologies.

Acknowledgements The authors acknowledge the motivated research efforts of Udeme Iniobong Nicholas (Department of Physics, Federal University of Technology, Minna, Nigeria) as a promising young researcher. The authors would like to appreciate Prof. Karla Miller (Nuffield Department of Clinical Neurosciences, John Radcliffe Hospital, Oxford) for providing tools for image processing.

References

Abdulateef, S.K., Salman, M.D.: A comprehensive review of image segmentation techniques. Iraqi J. Electr. & Electron. Eng. **17**(2) (2021)

Amouzeshi, A., Zarghani, H.: A narrative review on the Radiation Risk of Medical Imaging for traumatic pregnant patient and the fetal risks. J. Surg. Trauma **6**(3), 77–85 (2018)

Anazodo, U.C., Ng, J.J., Ehiogu, B., Obungoloch, J., Fatade, A., Mutsaerts, H.J., Secca, M.F., Diop, M., Opadele, A., Alexander, D.C., Dada, M.O., Ogbole, G., Nunes, R., Figueiredo, P., Figini, M., Aribisala, B., Awojoyogbe, B.O., Aduluwa, H., Sprenger, C., Wagner, R., Olakunle, A., Romeo, D., Sun, Y., Fezeu, F., Orunmuyi, A.T., Geethanath, S., Gulani, V., Nganga, E.C., Adeleke, S., Ntobeuko, N., Minja, F.J., Webb, A.G., Asllani, I., Dako, F. & Consortium for Advancement of

MRI Education and Research in Africa (CAMERA): A framework for advancing sustainable magnetic resonance imaging access in Africa. NMR Biomed. **36**(3), e4846 (2023)
Anjna, E., Kaur, E.R.: Review of image segmentation technique. Int. J. Adv. Res. Comput. Sci. **8**(4), 36–39 (2017)
Arora, P., Singh, P., Girdhar, A., Vijayvergiya, R.: Performance analysis of various denoising filters on intravascular ultrasound coronary artery images. Int. J. Imaging Syst. Technol. **33**(3), 965–984 (2023)
Beck, I., Angot, H., Baccarini, A., Dada, L., Quéléver, L., Jokinen, T., Laurila, T., Lampimäki, M., Bukowiecki, N., Boyer, M., Gong, X., Gysel-Beer, M., Petäjä, T., Wang, J., Schmale, J.: Automated identification of local contamination in remote atmospheric composition time series. Atmos. Meas. Tech. **15**(14), 4195–4224 (2022)
Bolón-Canedo, V., Alonso-Betanzos, A.: Ensembles for feature selection: a review and future trends. Inf. Fusion **52**, 1–12 (2019)
Brown, M.A., Semelka, R.C.: MRI: Basic Principles and Applications. John Wiley & Sons (2011)
Chandra, T.B., Verma, K.: Analysis of quantum noise-reducing filters on chest X-ray images: A review. Measurement **153**, 107426 (2020)
Chen, C., Chen, Q., Xu, J., Koltun, V.: Learning to see in the dark. In: Proceedings of the IEEE Conference on Computer Vision and Pattern Recognition, pp. 3291–3300 (2018)
Cowan, B.P.: Nuclear Magnetic Resonance and Relaxation, 1st edn. Cambridge University Press (1997)
Dada, M.O., Awojoyogbe, B.O.: Fundamental physics of nuclear magnetic resonance. In: Computational Molecular Magnetic Resonance Imaging for Neuro-oncology. Biological and Medical Physics, Biomedical Engineering. Springer, Cham (2021). https://doi.org/10.1007/978-3-030-76728-0_2
Das, P.K., Meher, S., Panda, R., Abraham, A.: An efficient blood-cell segmentation for the detection of hematological disorders. IEEE Trans. Cybern. **52**(10), 10615–10626 (2021)
Erickson, B.J., Korfiatis, P., Akkus, Z., Kline, T.L.: Machine learning for medical imaging. Radiographics **37**(2), 505–515 (2017)
Fan, L., Zhang, F., Fan, H., Zhang, C.: Brief review of image denoising techniques. Vis. Comput. Ind. Biomed. Art **2**, 1–12 (2019)
Gondara, L.: Medical image denoising using convolutional denoising autoencoders. In: 2016 IEEE 16th International Conference on Data Mining Workshops (ICDMW), pp. 241–246. IEEE (2016)
Goyal, B., Dogra, A., Agrawal, S., Sohi, B.S., Sharma, A.: Image denoising review: From classical to state-of-the-art approaches. Inf. Fusion **55**, 220–244 (2020)
Guedj, B., Rengot, J.: Non-linear aggregation of filters to improve image denoising. In: Intelligent Computing: Proceedings of the 2020 Computing Conference, vol. 2, pp. 314–327. Springer International Publishing (2020)
Guo, Z., Li, X., Huang, H., Guo, N., Li, Q.: Deep learning-based image segmentation on multimodal medical imaging. IEEE Trans. Radiat. Plasma Med. Sci. **3**(2), 162–169 (2019)
Gupta, M.S.: Applications of electrical noise. Proc. IEEE **63**(7), 996–1010 (1975)
Jafari, M.H., Karimi, N., Nasr-Esfahani, E., Samavi, S., Soroushmehr, S. M.R., Ward, K., Najarian, K.: Skin lesion segmentation in clinical images using deep learning. In: 2016 23rd International Conference on Pattern Recognition (ICPR), pp. 337–342. IEEE (2016)
Jenkinson, M., Bannister, P., Brady, M., Smith, S.: Improved optimization for the robust and accurate linear registration and motion correction of brain images. Neuroimage **17**(2), 825–841 (2002)
Jiang, L., Yu, G., Guo, M., Wang, J.: Feature selection with missing labels based on label compression and local feature correlation. Neurocomputing **395**, 95–106 (2020)
Jifara, W., Jiang, F., Rho, S., Cheng, M., Liu, S.: Medical image denoising using convolutional neural network: a residual learning approach. J. Supercomput. **75**, 704–718 (2019)
Kadkhodaei, M., Samavi, S., Karimi, N., Mohaghegh, H., Soroushmehr, S.M.R., Ward, K., All, A., Najarian, K.: Automatic segmentation of multimodal brain tumor images based on classification of super-voxels. In: 2016 38th Annual International Conference of the IEEE Engineering in Medicine and Biology Society (EMBC), pp. 5945–5948. IEEE (2016)

Karunasingha, D.S.K.: Root mean square error or mean absolute error? Use their ratio as well. Inf. Sci. **585**, 609–629 (2022)
Katti, G., Ara, S.A., Shireen, A.: Magnetic resonance imaging (MRI)–a review. Int. J. Dent. Clin. **3**(1), 65–70 (2011)
Kazeminia, S., Baur, C., Kuijper, A., van Ginneken, B., Navab, N., Albarqouni, S., Mukhopadhyay, A.: GANs for medical image analysis. Artif. Intell. Med. **109**, 101938 (2020)
Kulkarni, P., Madathil, D.: A review on echocardiographic image speckle reduction filters. Biomed. Res. **29**(12), 2582–2589 (2018)
Kumar, K.K., Chaduvula, K., Markapudi, B.: A Detailed Survey on feature extraction techniques in image processing for medical image analysis. Eur. J. Mol. Clin. Med. **7**(10), 2020 (2020)
Liu, K., Cheng, Y.Q., Yang, J.Y.: Algebraic feature extraction for image recognition based on an optimal discriminant criterion. Pattern Recogn. **26**(6), 903–911 (1993)
Liu, P., El Basha, M.D., Li, Y., Xiao, Y., Sanelli, P.C., Fang, R.: Deep evolutionary networks with expedited genetic algorithms for medical image denoising. Med. Image Anal. **54**, 306–315 (2019)
Liu, X., Song, L., Liu, S., Zhang, Y.: A review of deep-learning-based medical image segmentation methods. Sustainability **13**(3), 1224 (2021)
Majeeth, S.S., Babu, C.N.K.: Gaussian noise removal in an image using fast guided filter and its method noise thresholding in medical healthcare application. J. Med. Syst. **43**, 1–9 (2019)
Majeski, J.B.: Co-Registered Speckle Contrast Optical Tomography and Frequency Domain-Diffuse Optical Tomography for Monitoring of Hemodynamics in the Fifth Metatarsal (Doctoral dissertation, University of Rochester) (2023)
Masse, A., Lefevre, S., Binet, R., Artigues, S., Blanchet, G., Baillarin, S.: Denoising very high resolution optical remote sensing images: Application and optimization of nonlocal bayes method. IEEE J. Sel. Top. Appl. Earth Obs. Remote. Sens. **11**(3), 691–700 (2018)
Minaee, S., Boykov, Y., Porikli, F., Plaza, A., Kehtarnavaz, N., Terzopoulos, D.: Image segmentation using deep learning: a survey. IEEE Trans. Pattern Anal. Mach. Intell. **44**(7), 3523–3542 (2021)
Mutlag, W.K., Ali, S.K., Aydam, Z.M., Taher, B.H.: Feature extraction methods: a review. In: Journal of Physics: Conference Series, vol. 1591, no. 1, p. 012028. IOP Publishing (2020)
Nasr-Esfahani, M., Mohrekesh, M., Akbari, M., Soroushmehr, S.R., Nasr-Esfahani, E., Karimi, N., Samavi, S., Najarian, K.: Left ventricle segmentation in cardiac MR images using fully convolutional network. In: 2018 40th Annual International Conference of the IEEE Engineering in Medicine and Biology Society (EMBC), pp. 1275–1278. IEEE (2018)
Rafiei, S., Nasr-Esfahani, E., Najarian, K., Karimi, N., Samavi, S., Soroushmehr, S.R.: Liver segmentation in CT images using three dimensional to two dimensional fully convolutional network. In: 2018 25th IEEE International Conference on Image Processing (ICIP), pp. 2067–2071. IEEE (2018)
Razzak, M.I., Naz, S., Zaib, A.: Deep learning for medical image processing: overview, challenges and the future. Calcifications BioApps Autom. Decis. Mak. 323–350 (2018)
Sagheer, S.V.M., George, S.N.: A review on medical image denoising algorithms. Biomed. Signal Process. Control **61**, 102036 (2020)
Sarvamangala, D.R., Kulkarni, R.V.: Convolutional neural networks in medical image understanding: a survey. Evol. Intel. **15**(1), 1–22 (2022)
Sharif, S.M.A., Naqvi, R.A., Biswas, M.: Learning medical image denoising with deep dynamic residual attention network. Mathematics **8**(12), 2192 (2020)
Slichter, C.P.: Principles of Magnetic Resonance, vol. 1. Springer Science & Business Media (2013)
Smith, S.M. (2002). Fast robust automated brain extraction. Hum. Brain Mapp. **17**(3), 143–155.6
Smith, S.M., Jenkinson, M., Johansen-Berg, H., Rueckert, D., Nichols, T.E., Mackay, C.E., Watkins, K.E., Ciccarelli, O., Cader, M.Z., Matthews, P.M., Behrens, T.E.: Tract-based spatial statistics: voxelwise analysis of multi-subject diffusion data. Neuroimage **31**(4), 1487–1505 (2006)
Tian, J., Chen, L.: Image despeckling using a non-parametric statistical model of wavelet coefficients. Biomed. Signal Process. Control **6**(4), 432–437 (2011)
Torrey, H.C.: Bloch equations with diffusion terms. Phys. Rev. **104**(3), 563 (1956)

Wang, H., Cao, S., Jiang, K., Wang, H., Zhang, Q.: Seismic data denoising for complex structure using BM3D and local similarity. J. Appl. Geophys. **170**, 103759 (2019)
Xing, L., Giger, M.L., Min, J.K. (eds.): Artificial Intelligence in Medicine: Technical Basis and Clinical Applications. Academic Press (2020)
Zhang, K., Zuo, W., Chen, Y., Meng, D., Zhang, L.: Beyond a Gaussian denoiser: residual learning of deep CNN for image denoising. IEEE Trans. Image Process. **26**(7), 3142–3155 (2017)
Zhang, Y.J.: A survey on evaluation methods for image segmentation. Pattern Recogn. **29**(8), 1335–1346 (1996)

Chapter 2
Physics Informed Neural Networks (PINNs)

Abstract Physics Informed Neural Networks (PINNs) are a class of machine learning models that train neural networks based on physical constraints or laws. In order to detect patterns, conventional neural networks only use the input–output pairs that are supplied to them during training. They are determined by data. Conversely, PINNs combine data-driven learning with well-established physical principles in an effort to increase learning efficiency and generalisation. Because they can solve real-world problems with noisy data and often incomplete physics, physics-informed neural networks, or PINNs, are gaining popularity in many engineering domains. In PINNs, automatic differentiation is used to evaluate differential operators without discretization errors, and a multitask learning problem is defined to fit simultaneously observed data while adhering to the underlying laws of physics. This book has mostly focused on modifying the PINN through the use of different activation functions, neural network architectures, gradient optimisation techniques, and loss function structures. Numerous other applications have been demonstrated for PINNs, even though they have proven to be more useful in some scenarios when compared to more conventional numerical techniques such as the Finite Element Method (FEM). Still, there is potential for improvement, especially in terms of open theoretical problems.

Keywords Physics-informed neural network · Partial differential equations · Physics principles in neural networks · Machine learning in medicine · Deep learning

2.1 Artificial Intelligence (AI) in Medicine

Over the past ten years, there has been a lot of interest in applying artificial intelligence (AI) to the medical field because of its attractive automation and efficiency features. In the field of disease management, artificial intelligence (AI) has become a powerful force that is changing important aspects such as prognosis, therapy, and diagnosis. Its power stems from machine learning algorithms that are highly proficient in analysing

B. O. Awojoyogbe and M. O. Dada, *Digital Molecular Magnetic Resonance Imaging*, Series in BioEngineering, https://doi.org/10.1007/978-981-97-6370-2_2

vast datasets, identifying intricate patterns, and identifying minute correlations. By quickly processing and analysing medical data, AI increases the accuracy of illness diagnosis and enables healthcare professionals to make more accurate decisions more quickly. AI's ability to tailor interventions to individual patient profiles enhances therapeutic approaches by maximizing them for better results.

Artificial Intelligence (AI) models the neural network of the human brain by using multidimensional information to learn how to analyse data and use this knowledge for decision making (Xu et al. 2023). As artificial intelligence (AI) continues to improve, the job of physicians is set to change. AI is projected to boost doctors' capacities rather than replace them, allowing them to focus on higher-level decision-making, improved patient interaction, and fostering interdisciplinary collaboration. AI systems with data-driven insights and pattern recognition can help with everyday chores, diagnostics, and information processing. This change frees up clinicians' time for individualized patient care, empathy-driven encounters, and negotiating complex medical scenarios requiring human intuition and knowledge. The collaboration of AI and clinicians has the potential to improve healthcare delivery, improve diagnostic accuracy, and enable novel treatment techniques. The emerging landscape anticipates a harmonic relationship in which artificial intelligence serves as a valued ally, empowering healthcare personnel.

2.2 Neural Network in Scientific Computing

The use of neural networks in scientific computing has advanced significantly, showcasing their ability to represent complex relationships and resolve challenging issues. Neural networks were traditionally linked to applications like natural language processing and image recognition. But by approximating functions and patterns present in experimental or simulated data, researchers quickly realized their potential to address scientific problems.

The capacity of neural networks to capture non-linear dependencies and adapt to high-dimensional data has been a defining feature of their evolution in scientific applications. This flexibility is particularly useful when working with complex physical phenomena that might not fit neatly into linear models. Neural networks have proven to be more effective than traditional methods in a variety of tasks, including simulating fluid dynamics and predicting molecular structures. This is especially true in situations where the underlying relationships are dynamic and complex (Syed et al. 2020). The use of neural networks in scientific research has increased as a result of developments in deep learning architectures, such as convolutional neural networks (CNNs) and recurrent neural networks (RNNs) (Schmidhuber 2015). These architectures are ideal for applications ranging from material science to climate modeling because they perform exceptionally well in tasks involving temporal and spatial dependencies.

2.3 Neural Network Architecture in Physics-Informed Neural Network (PINNs)

Feed forward neural network architecture with input, hidden, and output layers is frequently used by PINNs. The complexity of the problem determines how many neurons are used in these layers, and the architecture can be modified to strike a balance between computational efficiency and model expressiveness. The adaptability of PINN architectures is further enhanced by taking into account specialised layers, such as those intended to manage temporal or spatial dependencies (Ayzek 2023).

2.4 Output and Input Layers

Usually, the parameters or features pertinent to the physical problem at hand are accommodated in the input layer of a PINN. Time, space, and other pertinent variables might be among them. Whether the task involves forecasting temperature distributions, fluid velocities, or other physical phenomena, the output layer's structure is determined by the nature of the problem and relates to the quantity of interest.

2.5 Activation Functions and Weight Initialization

The neural network gains non-linearity from activation functions such as hyperbolic tangent or sigmoid, which allows it to model intricate relationships. Effective learning and convergence depend on the careful selection of activation functions. Furthermore, the stability and effectiveness of the learning process depend greatly on the initialisation of weights, which establish the strength of connections between neurons (Antoniadis 2023).

2.6 Partial Differential Equations (PDEs)

In order to fully understand the concept of Physics-Informed Neural Networks (PINNs), one must have a basic understanding of partial differential equations (PDEs). PDEs provide the mathematical foundation for a wide range of phenomena, including heat transfer and fluid dynamics. PDEs are especially useful for modeling physical processes that change over space and time because, in contrast to ordinary differential equations, they take into account variations in multiple dimensions (Polyanin et al. 2008). The heat, wave, and Navier–Stokes equations are examples of common PDEs that are designed to clarify particular physical phenomena. The intrinsic capability of PINNs to incorporate the governing PDEs straight into the neural network architecture sets them apart. This is accomplished during the training phase by expressing the PDEs as constraints in the loss function (Li et al. 2023). By doing this, PINNs effectively uphold the system's physical laws, enabling precise prediction-making even in scenarios with sparse data.

Partial Differential Equations (PDEs) are mathematical equations that involve numerous independent variables and their partial derivatives. They are used in engineering, physics, biology, and other scientific fields to describe the behavior of physical systems, natural occurrences, and diverse processes. PDEs are critical for modeling systems that change in space and time. The structural linkages between numerical techniques for PDEs and neural architectures are being investigated by researchers in order to transfer the rich set of mathematical foundations from the domain of PDEs to neural networks. The goal is to create new mathematically sound neural building blocks that may be utilised to tackle a wide range of problems (Alt et al. 2022). Deep neural networks can be used to solve PDEs and PDEs can be used to improve deep neural networks.

For the past 20 years, researchers have used mathematical and numerical techniques to mimic the different biological processes that affect heat distribution. There is now a definite preference for mathematical modeling, driven by practical considerations. Although experimental methods are useful, their application is constrained by financial resources, logistical difficulties, and time restraints. The efficiency and cost-effectiveness of mathematical modeling have clearly shifted in response to the time-consuming, expensive, and complex nature of experiments. This tactical decision emphasises a methodological change by highlighting computational methods as crucial tools for investigating and understanding the thermal dynamics that define biological systems. This persistent focus on temperature dynamics highlights how important it is for a variety of medical interventions and helps to progress in preservation and treatment techniques.

A commonly used mathematical model for predicting the temperature distribution inside biological tissues is the Pennes Bioheat Equation which helps to explain the intricate interaction between thermal and physiological factors that drive heat propagation. It is a fundamental tool in many medical fields, providing insights into temperature dynamics that are essential for therapeutic uses such as cryosurgery, thermal-based interventions, and the treatment of hyperthermia. The Pennes Bioheat Equation is still essential for improving therapeutic approaches and deepening our knowledge of heat transfer in biological contexts. It takes into account factors such as blood perfusion, metabolic heat generation, and thermal conductivity. Pennes overlooked the direction of blood flow when determining the blood perfusion effect and metabolic heat production within the heat transfer of living tissue. According to the Pennes model, tissue and the venous circulation are in an equilibrium state.

The definition of Pennes's bio-heat equation in the presence of an RF heating source is as follows:

$$\rho_{\mathrm{t}} c_{\mathrm{t}}(\partial T/\partial t) = k\partial^2 T/\partial x^2 + \mathrm{pSAR} + W_{\mathrm{b}}\rho b c_{\mathrm{b}}(T - T_b) \tag{2.1}$$

where W_0 is the volumetric perfusion rate, cb is the specific heat of blood, Tb is the supplying arterial blood temperature, k is the thermal conductivity of tissue, SAR is the local specific absorption rate, and ρ is the tissue density and ρb is the density of blood (Awojoyogbe and Dada 2018).

2.7 Physics-Informed Neural Networks

Physics–Informed Neural Networks (PINNs) are a scientific machine learning technique used to solve problems involving Partial Differential Equations (PDEs). PINNs approximate PDE solutions by training a neural network to minimise a loss function; it includes terms reflecting the initial and boundary conditions along the space–time domain's boundary and the PDE residual at selected points in the domain (called collocation point) (Cuomo et al. 2022).

2.7.1 Evolution of Physics-Informed Neural Networks

PINNs have evolved from earlier attempts in the late 1990s to more recent advancements by Raissi et al. (2019). This evolution has been driven by improved computing power, algorithmic enhancements, and the availability of open-source tools and methodologies. Despite significant progress, there are still theoretical challenges and unresolved issues in PINNs, suggesting areas for further research and development.

2.7.2 Components and Workings of Physics-Informed Neural Networks (PINNs)

i. **Neural Network Architecture**:

PINNs employ standard feedforward neural networks (NNs) with the basic physical laws encoded into the NN using automatic differentiation trained to minimise a loss function that encompasses terms representing the initial and boundary conditions across the space-time domain and the residual of the PDE at specific points within the domain (collocation points) (Anitescu et al. 2023).

ii. **Methodology**:

PINN training does not necessarily require labeled data from simulations or experiments; it is essentially an unsupervised strategy. Instead of directly solving the governing equations, PINNs convert the problem into an optimisation task, where the goal is to optimise the loss function. This function includes the mathematical model and a residual term from the governing equation, acting as a penalty to constrain acceptable solutions (Anitescu et al. 2023).

iii. **Incorporating Physics into Learning**:

PINNs differentiate themselves by incorporating prior knowledge of physical laws directly into the learning process. This contrasts with traditional machine learning approaches that solely rely on fitting a neural network to a set of state-value pairs (Anitescu et al. 2023).

2.7.3 *Enhancing Molecular Imaging Precision Through Physics-Informed Modelling*

PINNs can be used in Magnetic Resonance Molecular Imaging (MRMI) to better understand the interactions between molecular characteristics and imaging parameters. This knowledge might help extract more accurate molecular information while taking acquisition variances into consideration (Monakhova 2022).

2.7.4 *Integration of Physics Principles in Neural Networks*

Physics principles can be incorporated into the design of neural networks to ensure that the networks produce physically plausible results. For example, the signal decay in MRI can be modelled using exponential functions, and this knowledge can be used to constrain the outputs of the neural network. Tissue interactions, such as the different relaxation times of various tissues in the body, can also be incorporated into the neural network design. This can help the network to better distinguish between different types of tissues and enhance the contrast in the resulting images. PINNs are a type of deep learning model that incorporates known physical laws described by differential equations into the training process. This means they are not just learning from the data, but also from the underlying physics of the problem.

2.7.5 *Advantages and Challenges of Physics-Informed Neural Networks (PINNs)*

Physics-Informed Neural Networks (PINNs) have shown promise in addressing resolution problems in medical imaging. Here are some of the benefits and limitations:

Benefits:

a. They are frequently employed to address a variety of scientific computer problems due to their superior approximation and generalisation capabilities (Lawal et al. 2022).
b. PINNs have contributed to improvements in many areas of computer science and engineering due to their simplicity (Lawal et al. 2022).

Limitations:

a. Conventional neural networks, including PINNs, can struggle when data is sparse, which is often the case in many scientific and engineering domains (Hua et al. 2023).
b. While PINNs show promise, the accuracy and computational performance can still be a limiting factor for their direct use (Hua et al. 2023).

2.7.6 Physics Informed Neural Network and Clinical Hyperthermia

In the field of Nuclear Magnetic Resonance (NMR)-guided Clinical Hyperthermia, this integration offers a particularly promising prospect of a potent solution for addressing complex problems. The ground breaking paper "Physics-Informed Neural Networks: A Deep Learning Framework for Solving Forward and Inverse Problems Involving Nonlinear Partial Differential Equations" Raissi et al. (2017) presents the concept of PINN, which provides a novel approach that unites data-driven learning with the complex physical dynamics involved in hyperthermic treatment planning.

A PINN's architecture is especially important when discussing NMR-guided Clinical Hyperthermia because of its two distinct parts: the data-driven portion that adjusts to observed NMR data, and the physics-informed portion that incorporates established physical laws governing hyperthermia. With this two-component structure, PINN can faithfully adhere to the governing physics of the hyperthermic treatment while simultaneously leveraging the power of deep learning. In practical terms, PINN finds great use in situations unique to NMR-guided Clinical Hyperthermia, where there may be a lack of data or where the dynamic nature of temperature distribution in tissues is difficult for traditional methods to capture. Its efficacy is most evident in cases where treatment optimisation requires a deep comprehension of the underlying physics, but where some aspects of the hyperthermic system are unknown or difficult to explicitly model.

2.8 The Neural Network

Deep learning, which uses multi-layered Artificial Neural Networks (ANNs) as its essential building block, can be based on neural network techniques. Deep learning has been recognised as one of the dominant tools due to its ability to handle massive amounts of data, and the literature has shown how widely used it is. Deeper hidden layers that are focused have recently been introduced, and they outperform conventional techniques in many fields of study, particularly image and pattern classifications. Among the most popular deep neural networks is the CNN.

2.9 Machine Learning

Machine learning is an artificial intelligence (AI) application that allows computers to learn from their experiences without the need for explicit programming. Machine learning improves pattern recognition accuracy by utilising data and algorithms. The effectiveness and efficiency of a machine learning solution are often determined by the nature and qualities of the data as well as the performance of the learning algorithms. In the field of machine learning algorithms, classification analysis, regression, data clustering, and dimensionality reduction, association rule learning, or reinforcement learning approaches exist that can be used to efficiently develop and construct data-driven systems effectively (Sarker 2021).

2.10 Deep Learning (DL)

Deep learning is a subset of machine learning that employs multilayer artificial neural networks to learn complex patterns and correlations within data. Deep Learning (DL) does not need any human-designed rules to function; instead, it leverages a massive amount of data to map given input to specified labels. DL is built with multiple layers of algorithms (artificial neural networks, or ANNs), each of which delivers a unique interpretation of the data provided to it (Alzubaidi et al. 2021). Deep learning is a vital technology in self-driving cars, voice recognition, brain tumour segmentation, and virtual assistants like Alexa, Siri, and Google Assistant. To be effective in the clinical domain, deep learning algorithms and models must be enhanced in terms of data integration, interpretability, security, and temporal modelling.

2.11 Convolutional Neural Network (CNN)

Convolutional Neural Network (CNN) is a type of neural network approach that is commonly used for addressing complex issues. CNN is made up of four layers, a convolution layer, a pooling layer, an activation function, and a fully connected layer. The convolution layer produces feature maps by applying filters to the incoming data. The pooling layer decreases the feature maps' size and complexity. Non-linearity is introduced into the network by the activation function. The fully connected layer, also known as the dense layer, uses the retrieved characteristics to conduct categorisation.

The pooling layer decreases the feature maps' size and complexity, non-linearity is introduced into the network by the activation function. The dense layer performs classification based on the extracted features (Indolia et al. 2018).

2.12 Machine Learning in Medicine

Several machine-learning models for diagnostic disease prediction have evolved, covering problems such as Alzheimer's disease, cardiovascular ailments, and a multi-class model capable of predicting up to 39 diseases by examining 86 laboratory tests (Zlobina et al. 2022). These improvements demonstrate machine learning's adaptability in healthcare, allowing for more accurate and timely forecasts for a variety of medical issues. The combination of many models improves disease prediction precision and scope, providing valuable tools for early detection and proactive healthcare treatment. An example of machine learning algorithms include regularised General Linear Model regression (GLMs), Support Vector Machines (SVMs) with a radial basis function kernel, and single-layer Artificial Neural Networks. These are all utilised in medicine (Sidey-Gibbons and Sidey-Gibbons 2019).

2.12.1 Supervised Learning

Supervised learning is a machine learning paradigm in which the algorithm is trained on a labelled dataset of input–output pairs. The algorithm is fed a training set that comprises inputs and correct outputs, allowing the model to learn over time. The algorithm evaluates its accuracy using the loss function and adjusts until the error is sufficiently minimised. There are two types of supervised learning: classification and regression. One of the most common examples of supervised learning in healthcare is the automated interpretation of an electrocardiogram (EKG), where the recognition of patterns is performed to choose from a limited number of diagnoses (i.e. a classification task), and in radiology, automated detection of a lung nodule from a chest X-ray would also represent supervised learning.

2.12.2 Unsupervised Learning

Unsupervised learning is a paradigm in which algorithms investigate and identify inherent structures in data that lack explicit labels. Unsupervised learning extracts useful information from unstructured datasets. In healthcare and medicine, one of application of unsupervised learning is the use of clustering algorithms which can be applied to identify different diseases. Clustering has been shown to be a valuable technique in the medical sector for detecting patterns and structure in labelled and unlabelled information. In neurological illness research, many clustering algorithms, such as k-means, hierarchical clustering, and DBSCAN, have been used, most notably in Alzheimer's disease (AD) datasets (Alashwal et al. 2019). These techniques aid in the identification of patterns and subgroups within complex data; hence, improving the understanding of illness heterogeneity and possibly individualised treatment strategies for affected individuals.

2.12.3 Reinforcement Learning

Reinforcement learning (RL), a subtype of machine learning, allows AI systems to learn by trial and error, modifying their behaviour depending on feedback from actions. RL algorithms receive signals in the form of rewards for positive outcomes and penalties for negative ones through a continuous loop of decision-making, execution, and outcome evaluation. The system's goal is to maximise the total reward, which drives the AI to develop optimal tactics and behaviours in dynamic context. RL can solve sequential decision-making problems with sampling, evaluative, and delayed data at the same time. It is an excellent choice for building powerful solutions in the healthcare area such as to improve the treatment of chronic diseases such as cancer, HIV, and epilepsy.

2.12.4 Deep Learning

Deep learning has received a lot of attention in the medical industry; it involves training algorithms to learn from huge amounts of data, such as medical imaging, electronic health records, and genetic data, in order to generate predictions or detect patterns, making it valuable for early disease identification and individualised medication. Researchers have used deep learning to gain new insights into the intricacies of biological systems, allowing for a more sophisticated understanding of the data's relationships and patterns. Deep learning's ability to detect high-level features in big and complex biological datasets makes it a great tool for expanding our understanding of high-throughput biology and expediting discoveries in genomics and molecular biology.

2.12.5 Deep Neural Network

A deep neural network (DNN) is a particular kind of artificial neural network with numerous layers between the input and output layers (deep architecture). Each layer is made up of interconnected nodes (neurons) that process and change the data input. The depth of a DNN allows it to learn hierarchical representations and detailed patterns in data autonomously, distinguishing it from simpler structures. This hierarchical feature extraction enables DNNs to collect subtle information at many levels of abstraction, allowing them to be used for diagnosis, treatment planning, and patient care by revolutionising the distinct features of medical image processing (Mall et al. 2023). The training procedure entails modifying the weights and biases of the network using algorithms such as back-propagation to improve the network's capacity to reduce the difference between expected and actual outputs. DNNs, which are widely used in a variety of fields, show deep learning's transformative potential, ushering in a new era of artificial intelligence in which automated learning from data delivers extraordinary insights and capabilities.

2.12.6 Activation Function

Deep learning relies heavily on activation functions. They add non-linearity to the neural network, allowing it to learn more complicated data patterns. Without activation functions, the network could only learn linear patterns and could not learn more complicated patterns. Examples of activation functions include Sigmoid function (Logistic function), Tanh activation function: hyperbolic tangent (Tanh), ReLU (Rectified Linear Units) activation function, and Exponential Linear Units (ELU).

2.12.7 The Sigmoid Function

The Sigmoid function, also known as the Logistic function, squashes the input values to a range between 0 and 1, making them suitable for binary classification problems where the output needs to represent probabilities. Sigmoid function is a widely used activation function. The Tanh (hyperbolic tangent) activation function is similar to the Sigmoid in that it maps input values to a range between -1 and 1, but it produces a more symmetric output and avoids the vanishing gradient problem.

2.12.8 The Rectified Linear Units (ReLU)

Rectified Linear Units (ReLU) have gained widespread popularity due to their simplicity and effectiveness. The ReLU activation function outputs the input directly if it is positive; otherwise, it outputs zero. ReLU helps overcome the vanishing gradient problem associated with other activation functions, facilitating the training of deeper networks.

2.12.9 *Machine Learning Techniques for Healthcare Applications*

Different machine-learning techniques are used for a variety of healthcare applications, including breast cancer detection, tumours detection, cervical cancer detection, and cognitive health assessment. The following machine learning techniques RF, SVM, AdaBoost1, and RUSBoost are used to localise brain tumour on FLAIR scans MRI. In automatic brain tumour classification system, the K-nearest neighbour algorithm is used to categorise MRI images as abnormal or normal. The segmentation of tumour regions uses the fuzz C-means clustering method. The experiment uses the two datasets MICCAI and BRATS, which evaluate the outcomes with 96.5% accuracy, 100% sensitivity, and 93% specificity. Zhou et al. (2020) present a model that has been manually optimised by a machine learning expert and compare its performance to that of the automated machine learning method Tree-Based Pipeline optimizing tool. The proposed model was applied to the MRI scans of 288 patients, and the results revealed that the best AUC value and accuracy were 0.94% and 0.88%, respectively.

2.12.10 *Supervised Learning*

Through the analysis of outputs of interest selected by a (typically human) supervisor, computer programmes are taught associations between inputs and outputs in data. This process is known as supervised learning. It is possible to predict future examples using associations that have been learned based on the data already available. One of the most developed areas of machine learning is supervised learning, with numerous examples in and outside of the healthcare industry (Panch et al. 2018). One method used to address this project's problem statement is classification, which is a typical implementation of the supervised learning task.

2.12.11 *Classification*

In machine learning, classification is regarded as a supervised learning technique that also refers to the problem of predictive modelling, in which a class label is predicted for a given 19 examples. A function (f) is mathematically mapped from input variables (X) to output variables (Y) as targets, labels, or categories. It can be used with structured or unstructured data to determine the class of a set of data points. A classification issue can arise, for instance, when spam is detected and classified as "spam" or "not spam" by email service providers.

2.12.12 *Regression*

A number of machine learning techniques used in regression analysis enable the prediction of a continuous (y) result variable based on the value of one or more (x) predictor variables (Han et al. 2011). The major distinction between classification and regression is that the former predicts labels for specific classes while the latter facilitates the prediction of a continuous quantity. The two categories of machine learning algorithms frequently have some overlaps. In many different industries today, regression models are frequently used for financial forecasting or prediction, cost estimation, trend analysis, marketing, time series estimation, drug response modeling, and many other purposes. Regression algorithms of the linear, polynomial, lasso and ridge varieties are among the more well-known ones.

2.12.13 *Clustering*

Unlabelled data are grouped using the data mining technique of clustering according to their similarities or differences. Algorithms called clustering are used to group raw, unclassified data objects into groups that can be visualised as patterns or structures in the data. It does this by categorising a collection of objects so that those belonging to the same category, known as a cluster, are in some ways more similar to one another than those belonging to other groups (Han et al. 2011). Several types of clustering algorithms, including exclusive, overlapping, hierarchical, and probabilistic algorithms, can be distinguished.

2.13 Conclusion

Although neural networks clearly have applications in scientific computing, there are obstacles in the way of their widespread use. Research is still being conducted in areas like interpretability, the requirement for big labelled datasets, and generalisation to novel scenarios. However, the accomplishments to date highlight how revolutionary neural networks can be in expanding scientific knowledge and problem-solving skills.

Acknowledgements The authors acknowledge the motivated research efforts of Oyedim Victor Ejike, as a promising young researcher.

References

Alashwal, H., El Halaby, M., Crouse, J.J., Abdalla, A., Moustafa, A.A.: The application of unsupervised clustering methods to Alzheimer's disease. Front. Comput. Neurosci. **13** (2019). https://doi.org/10.3389/fncom.2019.00031

Alt, T., Schrader, K., Augustin, M., Peter, P., Weickert, J.: Connections between numerical algorithms for PDEs and neural networks. J. Math. Imaging Vis. **65**(1), 185–208 (2022). https://doi.org/10.1007/s10851-022-01106-x

Alzubaidi, L., Zhang, J., Humaidi, A.J., Al-Dujaili, A., Duan, Y., Al-Shamma, O., Santamaría, J., Fadhel, M.A., Al-Amidie, M., Farhan, L.: Review of deep learning: concepts, CNN architectures, challenges, applications, future directions. J. Big Data **8**(1) (2021). https://doi.org/10.1186/s40537-021-00444-8

Anitescu, C., İsmail Ateş, B., Rabczuk, T.: Physics-informed neural networks: theory and applications. In: Machine Learning in Modeling and Simulation: Methods and Applications, pp. 179–218. Springer International Publishing, Cham (2023)

Antoniadis, P., Antoniadis, P.: Activation Functions: Sigmoid vs Tanh | Baeldung on Computer Science. Baeldung on Computer Science (2023). https://www.baeldung.com/cs/sigmoid-vs-tanh-functions

Awojoyogbe, B.O., Dada, M.O.: Computational design of an RF controlled theranostic model for evaluation of tissue biothermal response. J. Med. Biol. Eng. **38** 993–1013 (2018)

AyZek.: Decoding the Architecture: Determining Layers and Neurons in Deep Neural Networks (DNNs). Medium. Retrieved December 2, 2023 (2023), from: https://blog.gopenai.com/decoding-the-architecture-determining-layers-and-neurons-in-deep-neural-networks-dnns-c2ac349989c2

Cuomo, S., Di Cola, V.S., Giampaolo, F., Rozza, G., Raissi, M., Piccialli, F.: Scientific machine learning through physics–informed neural networks: Where we are and what's next. J. Sci. Comput. **92**(3), 88 (2022)

Han, B., Zhu, H., Ding, Y.: Bottom-up saliency based on weighted sparse coding residual. In: Proceedings of the 19th ACM International Conference on Multimedia, pp. 1117–1120 (2011)

Hua, J., Li, Y., Liu, C., Wan, P., Liu, X.: Physics-informed neural networks with weighted losses by uncertainty evaluation for accurate and stable prediction of manufacturing systems. IEEE Trans. Neural Netw. Learn. Syst. (2023)

Indolia, S., Goswami, A.K., Mishra, S., Asopa, P.: Conceptual understanding of convolutional neural network- a deep learning approach. Procedia Comput. Sci. **132**, 679–688 (2018). https://doi.org/10.1016/j.procs.2018.05.069

Lawal, Z.K., Yassin, H., Lai, D.T.C., Che Idris, A.: Physics-informed neural network (PINN) evolution and beyond: a systematic literature review and bibliometric analysis. Big Data Cogn. Comput. **6**(4), 140 (2022)

Li, Q., Geng, J., Evje, S., Rong, C.: Solving nonlinear conservation laws of partial differential equations using graph neural networks. In: Proceedings of the Northern Lights Deep Learning Workshop, vol. 4 (2023). https://doi.org/10.7557/18.6808

Mall, P.K., Singh, P.K., Srivastav, S., Narayan, V., Paprzycki, M., Jaworska, T., Ganzha, M.: A comprehensive review of deep neural networks for medical image processing: recent developments and future opportunities. Healthc. Anal. **4**, 100216 (2023). https://doi.org/10.1016/j.health.2023.100216

Monakhova, K.: Physics-Informed Machine Learning for Computational Imaging. University of California, Berkeley (2022). http://www2.eecs.berkeley.edu/Pubs/TechRpts/2022/EECS-2022-177.pdf. Retrieved Jan 18 2024

Panch, T., Szolovits, P., Atun, R.: Artificial intelligence, machine learning and health systems. J. Glob. Health **8**(2) (2018)

Polyanin, A.D., Schiesser, W.E., Zhurov, A.I.: Partial Differential Equation. Scholarpedia (2008). https://doi.org/10.4249/scholarpedia.4605

Raissi, M., Perdikaris, P., Karniadakis, G.E.: Physics informed deep learning (part i): Data-driven solutions of nonlinear partial differential equations (2017). arXiv preprint arXiv:1711.10561

Raissi, M., Perdikaris, P., Karniadakis, G.: February). Physics-informed neural networks: a deep learning framework for solving forward and inverse problems involving nonlinear partial differential equations. J. Comput. Phys. **378**, 686–707 (2019). https://doi.org/10.1016/j.jcp.2018.10.045

Ronaghan, S.: Deep Learning: Common Architectures—Stacey Ronaghan—Medium. Medium (2018). https://srnghn.medium.com/deep-learning-common-architectures-6071d47cb383

Sarker, I.H.: Deep learning: a comprehensive overview on techniques, taxonomy, applications and research directions. SN Comput. Sci. **2**(6) (2021). https://doi.org/10.1007/s42979-021-00815-1

Schmidhuber, J.: Deep learning in neural networks: an overview. Neural Netw. **61**, 85–117 (2015). https://doi.org/10.1016/j.neunet.2014.09.003

Sidey-Gibbons, J.A.M., Sidey-Gibbons, C.J.: Machine learning in medicine: a practical introduction. BMC Med. Res. Methodol. **19**(1) (2019). https://doi.org/10.1186/s12874-019-0681-4

Syed D.A., Rajul A.: The Evolution and Core Concepts of Deep Learning & Neural Networks (n.d.). www.Analyticsvidhya.Com. Retrieved November 28, 2020, from: https://www.analyticsvidhya.com/blog/2016/08/evolution-core-concepts-deep-learning-neural-networks/

Xu, N., Yang, D., Arikawa, K., Bai, C.: Application of artificial intelligence in modern medicine. Clin. EHealth **6**, 130–137 (2023). https://doi.org/10.1016/j.ceh.2023.09.001

Zlobina, K., Jafari, M., Rolandi, M., Gomez, M.: The role of machine learning in advancing precision medicine with feedback control. Cell Rep. Phys. Sci. **3**(11), 101149 (2022). https://doi.org/10.1016/j.xcrp.2022.101149

Zhou, H., Hu, R., Tang, O., Hu, C., Tang, L., Chang et al.: Automatic machine learning to differentiate pediatric posterior fossa tumors on routine MR imaging. Am. J. Neuroradiol. **41**(7), 1279–1285 (2020)

Chapter 3
New Methodology and Modelling in Magnetic Resonance Imaging

Abstract The partial differential equations employed in this chapter are those based on the Bloch NMR flow equations and NMR diffusion equation. Each physical problems solved in this chapter are described by a Bloch related equation based on close observation of how the system usually behaves in the presence of NMR magnetic fields. A few numbers of realistic assumptions have been made in order to obtain a solution to each differential equation that has been developed for each problem identified. Justification for the choice of solution may not be presented in detail. However, basic information is appropriately provided. The solutions are obtained based on the following two properties (i) ability of the solution (describing the NMR transverse magnetisation) to demonstrate significant tissue contrast based on the unique relaxation times (ii) ability of the solution to give a significant value of NMR Signal when the spatial resolution is very small. Meanwhile, it may not be quite easy to achieve these criteria in all situations; however, at least one of these criteria was met in the model developed. In terms of methodology employed, the first two steps of developing a model (setting up a model in terms of differential equations and seeking for their mathematical/computational solutions) is adequately captured.

3.1 Introduction

The anatomy component of Magnetic Resonance (MR) image is derivable from the Bloch NMR model obtained from the Bloch NMR flow equations (Awojoyogbe 2002, 2003, 2004; Awojoyogbe et al. 2011; Dada and Awojoyogbe 2021):

$$v^2\frac{\partial^2 M_y}{\partial x^2} + 2v\frac{\partial^2 M_y}{\partial x \partial t} + T_o\frac{\partial M_y}{\partial t} + vT_o\frac{\partial M_y}{\partial x} + \frac{\partial^2 M_y}{\partial t^2} + (T_g + \gamma^2 B_1^2)M_y = F_o \gamma B_1, \tag{3.1}$$

where $T_o = \frac{1}{T_1} + \frac{1}{T_2}$, $T_g = \frac{1}{T_1 T_2}$ and $F_o = \frac{M_o}{T_1}$.

Equation (3.1) is a general second order differential equation with constant coefficient (which is the Awojoyogbe version of the Bloch NMR flow equation) derived

B. O. Awojoyogbe and M. O. Dada, *Digital Molecular Magnetic Resonance Imaging*, Series in BioEngineering, https://doi.org/10.1007/978-981-97-6370-2_3

from the Bloch NMR equation and can be applied to any fluid flow problem. At any given time t, we can obtain information about the system, provided that appropriate boundary conditions are applied. From Eq. (3.1), we have obtained the wave equation and solve them in terms of NMR parameters by the application of appropriate initial or boundary conditions. Hence, we could get very important information about the dynamics of any system. It should be noted however that the term $F_o \gamma B_1(x, t)$ is the forcing function or disturbance ($F_o = \frac{M_o}{T_1}$) where M_o is the equilibrium Magnetization. If the function is zero, there is no disturbance and we have a freely vibrating system; else, the system is undergoing a forced vibration.

Equation (3.1) is applicable under the following conditions: (i) the selected slice must be homogeneous with one compartment (either of tissue or of fluid compartment). (ii) Rate of fluid influx and efflux should be very small and equal (such as in some slow blood flow junctions) or static fluid (e.g. edema). (iii) The size of the chemical substance to be investigated must be significantly very small compared to the tissue compartment in which they are found. (iv) the geometry of the compartment should be defined and could be approximated to common geometrical system. Assuming that the excitation coil is large enough to cover the slice under investigation, the amount of radiofrequency (RF) power received by any of the fluid particles should be the same irrespective of the position of the particle spin. In this chapter, a general second order differential equation with variable coefficients in the form of Eq. (3.1) will be derived.

3.2 Computational Approach to Compressed Sensing MRI

In this book, the differential equations which have been set up to address specific problems are solved analytically so that data interpretation is very flexible and better model control is achieved. However, few cases exist where analytical solutions are very difficult to obtain and consequently, numerical methods have to be used. As stated in Chapter one, the general task in this book is to develop techniques with which magnetic resonance molecular imaging could be implemented with few data. This is like a theoretical and computational approach to compressed sensing MRI. Various computer programs that allow users to enter few experimental data to obtain MR molecular images have been developed. Wolfram Mathematica v.9 and Python 3.4 have been employed for computer programming while SigmaPlot 12 and Wolfram Mathematica v.9 have been used for data analysis in this study.

3.2.1 Software for Programming and Data Analysis

The factors considered in the choice of software for this study are flexibility, data manipulation and unique function representation. These soft ware are mathematically robust and allow the use of a large number of functions whose representation are less complex. That is, irrespective of the complexity and functions obtained in the solutions to the model differential equations, direct representations are possible without the use of complicated matrices. Overviews of the software that have been used in this study are given as follows.

Wolfram Mathematica

Mathematica is a symbolic mathematical and computational programdue Stephen Wolfram, which is employed in computations involving mathematical, engineering, and computing field's problems (https://en.wikipedia.org/wiki/Mathematica). The program uses the Wolfram Language and few of the features of the computer program are:

i. Libraries containing special mathematical functions
ii. Impressive tools for data and matrix manipulation
iii. Evaluation, visualisation and animation of two-dimensional and three-dimensional data.
iv. Evaluation of solvers for different systems and types of equations.
v. Symbolic and numerical tools for continuous and discrete calculus
vi. Robust libraries for statistical computations.
vii. Contains computational support for various data types.
viii. Calculations and simulations on randomprocesses and queues
ix. Machine learning tools for sounds, images and data
x. Two and higher dimensional geometrical computations
xi. Two and three dimensional adaptive mesh generation
xii. Support for functional, procedural and object orientedprogrammings.
xiii. Tools for inclusion of graphic user interfaces toprograms, image processingand image recognition.
xiv. Ability to connect to other programming languages such as forrtran, Java, .NET, C++, OpenCL, etc.

The program interface of Mathematica consist oftwo different ends, parts, the kernel and the front end. The kernel interprets expressions (Wolfram Language code) and returns result expressions. The front end that is designed by Theodore Gray, provides a GUI which allows the creation and editing ofNotebook documents containing program code with prettyprinting, formatted text together with resultsincluding typeset mathematics, graphics, GUI components, tables, and sounds. All contents andformatting can be generated algorithmically or interactively edited. There is support for most standard word processingcapabilities and this includes a spellchecker.

Recently, the capabilities for highperformancecomputing have been extended with theintroduction of packed arrays (version 4, 1999) and sparse matrices (version 5, 2003), and byadopting the GNU Multi-Precision library to evaluate high-precision arithmetic. Version 5.2 (2005) introduced automatic multithreading when computations are performed on multi-core computers. This release included CPU specific optimized libraries and Mathematica is supported by third party specialist acceleration hardware such as ClearSpeed. In 2002, grid Mathematica was introduced to allow user level parallel programming on heterogeneous clusters and multiprocessor systems. In 2008, parallel computing technology was included in all Mathematica licenses with support for grid technology such as Windows HPC Server 2008, Microsoft Compute Cluster Server and Sun Grid. Support for CUDA and Open CL GPU hardware was added in 2010. Also, since version 8 it can generate C code, which is automatically compiled by a system C compiler, such as the Intel C++ Compiler or Visual Studio 2010.

There are several ways to deploy applications written in Wolfram Mathematica. These are (https://en.wikipedia.org/wiki/Mathematica):

i. Mathematica Player Pro is a runtime version of Mathematica that will run any Mathematicaapplication but does not allow editing or creation of the code.
ii. A free of charge version of Wolfram CDF Player has been provided for running Mathematica programs that have been saved in the Computable Document Format (CDF). It can also view standardMathematica files, but not run them. It includes plugins for common web browsers on Windows and Macintosh.
iii. webMathematica allows a web browser to act as a front end to a remote Mathematica server. It isdesigned to allow a user written application to be remotely accessed via a browser on anyplatform. It may not be used to give full access to Mathematica.
iv. Wolfram Language code can be converted to C code or to an automatically generated DLL.
v. Wolfram Language code can be run on a Wolfram cloud service.

Communication with other applications occurs through a protocol called MathLink. This allows communication between theMathematica kernel and frontend, and provides a general interface between the kernel and otherapplications. Wolfram Research freely distributes a developer kit for linking applications written in theC programming language to the Mathematica kernel through *MathLink*. Using *J/Link*, a Javaprogram can ask Mathematica to perform computations. Likewise, a Mathematica program can load Javaclasses, manipulate Java objects and perform method calls. Similar functionality is achieved with *.NET/Link*, but with .NET programs instead of Java programs. Other languages that connect toMathematica include Haskell, AppleScript, Racket, Visual Basic, Python andClojure (https://en.wikipedia.org/wiki/Mathematica).

Links are available to many mathematical software packages including OpenOffice.org Calc, Microsoft Excel, MATLAB, R, Sage, SINGULAR, Wolfram SystemModeler, and Origin. Mathematical equations can be exchanged with other computational or typesettingsoftware via MathML.Communication with SQL databases is achieved through built-in support for JDBC. Mathematica canalso install web services from a WSDL description. It can access HDFS data via Hadoop.Mathematica can capture real-timedata via a link to LabVIEW, from financial data feeds anddirectly from hardware devices via GPIB (IEEE 488), USB and serial interfaces.

Python Programming

Python is a widely used general-purpose, high-level programming language whose design philosophy emphasizes code readability, and its syntax allows programmers to express concepts in fewer lines of code than would be possible in languages such as C++ or Java. The language provides constructs intended to enable clear programs on both a small and large scale (https://en.wikipedia.org/wiki/Mathematica). Python supports multiple programming paradigms such as object-oriented, imperative and functional programming or procedural styles. It features a dynamic type system and automatic memory management and has a large and comprehensive standard library.

Python interpreters are available for installation on many operating systems, allowing Python code execution on a wide variety of systems. Using third-party tools, such as Py2exe or Pyinstaller, Python code can be packaged into stand-alone executable programs for some of the most popular operating systems, allowing the distribution of Python-based software for use on those environments without requiring the installation of a Python interpreter. These executable programs are developed and applied to specific problems in this study. CPython is the reference implementation of Python and it is free and open-source software and has a community-based development model. CPython is managed by the non-profit Python Software Foundation.

In terms of program features, Python is a multi-paradigm programming language: object-oriented programming and structured programming are fully supported, and there are a number of language features which support functional programming and aspect-oriented programming. Many other paradigms are supported using extensions, including design by contract and logic programming. Python uses dynamic typing and a combination of reference counting and a cycle-detecting garbage collector for memory management. An important feature of Python is dynamic name resolution (late binding), which binds method and variable names during program execution.

In summary, the core philosophy of the Python programming is to make programs that are beautiful, explicit, readable and simple with some element of reasonable complexity (https://en.wikipedia.org/wiki/Python_(programming_language)).

Rather than requiring all desired functionality to be built into the language's core, Python was designed to be highly extensible. Python can also be embedded in existing applications that need a programmable interface.

SigmaPlot

SigmaPlot is a software package for scientific graphing and data analysis. It runs on Microsoft Windows. The software can read multiple formats (such as Microsoft Excel spreadsheets), and can also perform mathematical transforms and statistical analyses.

SigmaPlot was developed by Jandel Corporation for Windows 3.1× and maintained by them until version 4.0. In 1996, Jandel Corporation merged into SPSS Inc (https://en.wikipedia.org/wiki/SigmaPlot). SigmaPlot was maintained by SPSS through version 8. As of version 9, it is owned and maintained by SYSTAT Software. The features of this software package are:

i. Graphing Features
ii. Statistical Analysis
iii. Global Curve Fitting
iv. Dynamic Curve Fitting
v. Advisory Statistical Analysis
vi. Standard Curves Macro
vii. Improved 3D Graph Customization and Data sharing
viii. PreformattedWorksheets
ix. Import/Export Functionality
x. Automation of Routine and Complex Tasks.

3.2.2 Data for Model Analysis

It has been noted earlier on that, the models developed in the study are executed with few basic data. These data are experimental parameters that have been mined from literature for building a medical scenario that is as close to reality as possible. The data for this study are classified as follows:

i. NMR Relaxometric Data
ii. Physical Data
iii. Chemical Data.

NMR Relaxometric Data

These are parameters which are dependent on NMR relaxation in experimental settings. They are the most important in the various data that is required to execute the developed models. Since they are dependent on molecular dynamics, these parameters have lent themselves for molecular imaging due to their ability to display contrasts between normal different tissues; and between normal and disease-affected tissues. These parameters are as follows.

T_1: The relaxation of the longitudinal magnetisation M_z characterized by the longitudinal or spin–lattice relaxation time T_1 (Haacke et al. 1999; Brix et al. 2008).

T_2: The relaxation of the transverse magnetisation$M_y$characterized by the transverse or spin–spin relaxation time T_2 (Cowan 1997; Haacke et al. 1999; Brix et al. 2008).

r_1 and r_2: Both of these parameters are known as relaxivity. It is the ability of the unpaired electrons in a substance to alter the relaxation time of adjacent protons (Aime et al. 2005; Brix et al. 2008). They have formed the basis for quantifying MR contrast enhancement in contrast agents. Relaxivity depends on extrinsic factors likeapplied field, temperature and the molecular properties of the contrast agent (Dumas et al. 2010; Jacques et al. 2010). Experimentally, they are determined by finding the slope of the linear regression generated from a plot of the measured relaxation rate (1/Ti, where i = 1, 2) versus the concentration of MRI contrast agents (Rohrer et al. 2005).

Physical Data

The physical parameters are physical conditions under which magnetic resonance measurements are made. There are lots of physical parameters which are used in MR measurements but for the purpose of this study, the major ones are outlined as follows.

TR: This is called repetition time. It is the delay time after which MR sequence is repeated; that is the time between two consecutive RF excitations (Haacke et al. 1999; Brix et al. 2008).
TE: This is the time of observation of the free induction decay (FID). That is, the time required for the spin echo to decay (Haacke et al. 1999). This parameter is known as echo time.
G: Spatial information can be generated in MRI by space-dependent magnetic fields additionally applied along the three space coordinates (Cowan 1997; Brix et al. 2008). These spatial magnetic fields or magnetic field gradients-are small as compared with the main external field B_0 and are generated by special coils mounted in the bore of the magnet. G describes the magnitude or steepness of the orthogonal gradient fields.
δ or τ: This is called pulse duration. It is the length of time for which the RF pulse is applied (Brix et al. 2008).
B_0: This is the flux density of the external magnetic field which forces the magnetic moments μ of the tissue to align along the magnetic field (Brix et al. 2008).

Chemical Data

These parameters are dependent on the chemical features of the biological system that is being considered. Some of these parameters are functions of the composition of the tissues that is to be imaged.

γ: This is known as the gyromagnetic ratio and it is the ratio of an object's magnetic dipole moment to its angular momentum (Haacke et al. 1999). It is a specific property of electrons ($\gamma_e = 28.0249540$ [GHz T^{-1}]) and protons (and also neutrons, $\gamma_p = 42.5774825$ [MHz T^{-1}]). Because of the abundance of the protons in nature (Cowan 1997), it is generally used for nuclear resonance. Hence, the value of $\gamma_p = \gamma$ shall be used throughout this study to represent the gyomagnetic ratio.
D: This represents the tissue diffusion coefficient. It is the proportionality constant between the diffusion flux and the gradient in the concentration of the diffusing species (Basser and Jones 2002).

3.3 Analytical Methods

For this study, differential equations which describe a molecular imaging problem are used. In cases where the suitable differential equations have been derived in earlier studies (as stated in Chapter two), the equations were selected based on the physical prevailing processes around the modelling problems in this study.

However, there are modifications that need to be made to the existing differential equations so that they could take into consideration new observations and move the modelling much closer to real-life situations in clinical settings. Secondly, according to the points made earlier, this study intends to develop new approach to the Optical Bloch Equation (OBE). This is aimed at demonstrating the application of OBE in clinical scenarios, especially in the diagnosis of diseases at cellular level and possible treatment technique. Therefore, the modifications to the earlier modelling equations and new ones that are required to achieve the objectives of the study are presented.

3.3.1 Variable Fluid Flow Corrections to the Bloch NMR Flow Equation

In earlier studies, constant velocity approximations have been used. It could be observed as well that this led to the constant diffusion coefficient. However, in real biological systems, molecular dynamics problems generally involve non-uniform velocity fields and variable diffusion coefficients (Li et al. 1992; Dahlem and Müller 2004; Rungseevijitprapa et al. 2009) and since this study deals with molecular imaging, the sizes of the flowing molecular species are small enough such that the effect of surrounding geometric structure is no longer negligible. Hence, this challenge demands a modification of the Bloch NMR flow equation and derivation of the coupled differential equations for flow problems with spatial dependence.

3.3.1.1 NMR Flow with Variable Fluid Velocity

In biological systems, fluid flow is not always as we have described so far. Fluid velocity shows complex behaviour such they change their behaviour from point to point (Price 2009; Nelson and Cox 2008; Awojoyogbe and Dada 2013). This is especially very conspicuous in blood flow (Dada et al. 2013a, b) and hence, it is very crucial for to outline how the Bloch NMR flow formulation will look like when the fluid velocity has spatial dependence.

The General Bloch NMR Flow Equation for Variable Fluid Velocity

If it is assumed that the variable fluid velocity in the NMR system is dependent only on x, Eq. (2.193) may now be written as follows:

$$\vec{v}\cdot\nabla M_x = v(x)\frac{\partial M_x}{\partial x}, \vec{v}\cdot\nabla M_y = v(x)\frac{\partial M_y}{\partial x}, \vec{v}\cdot\nabla M_z = v(x)\frac{\partial M_z}{\partial x}, \tag{3.2}$$

The earlier assumption made in (Dada and Awojoyogbe 2021) may now be modified as follows:

$$\vec{v} = \hat{i}v(x) + \hat{j}v(x) + \hat{k}v(x), \tag{3.3}$$

then we obtain:

$$\frac{dM_x}{dt} = \frac{\partial M_x}{\partial t} + v(x)\frac{\partial M_x}{\partial x} = -\frac{M_x}{T_2}, \tag{3.4}$$

$$\frac{dM_y}{dt} = \frac{\partial M_y}{\partial t} + v(x)\frac{\partial M_y}{\partial x} = \gamma M_z B_1 - \frac{M_y}{T_2}, \tag{3.5}$$

$$\frac{dM_z}{dt} = \frac{\partial M_z}{\partial t} + v(x)\frac{\partial M_z}{\partial x} = -\gamma M_y B_1 + \frac{(M_0 - M_z)}{T_1}, \tag{3.6}$$

or

$$\frac{\partial M_x}{\partial t} + v(x)\frac{\partial M_x}{\partial x} = -\frac{M_x}{T_2}, \tag{3.7}$$

$$\frac{\partial M_y}{\partial t} + v(x)\frac{\partial M_y}{\partial x} = \gamma M_z B_1 - \frac{M_y}{T_2}, \tag{3.8}$$

$$\frac{\partial M_z}{\partial t} + v(x)\frac{\partial M_z}{\partial x} = -\gamma M_y B_1 + \frac{(M_o - M_z)}{T_1}. \tag{3.9}$$

From Eq. (3.9), the following expression could be written:

$$\left(v(x)\frac{\partial}{\partial x} + \frac{\partial}{\partial t} + \frac{1}{T_1}\right)M_z = -\gamma M_y B_1 + \frac{M_o}{T_1},$$

$$M_z = \left(-\gamma M_y B_1 + \frac{M_o}{T_1}\right)\frac{1}{\left(v(x)\frac{\partial}{\partial x} + \frac{\partial}{\partial t} + \frac{1}{T_1}\right)}. \tag{3.10}$$

Substituting for Eq. (3.10) in Eq. (3.8) gives:

$$\frac{\partial M_y}{\partial t} + v(x)\frac{\partial M_y}{\partial x} = \gamma\left(-\gamma M_y B_1 + \frac{M_o}{T_1}\right)\frac{B_1}{\left(v(x)\frac{\partial}{\partial x} + \frac{\partial}{\partial t} + \frac{1}{T_1}\right)} - \frac{M_y}{T_2},$$

$$\left(v(x)\frac{\partial}{\partial x} + \frac{\partial}{\partial t} + \frac{1}{T_1}\right)\left\{v(x)\frac{\partial M_y}{\partial x} + \frac{\partial M_y}{\partial t} + \frac{M_y}{T_2}\right\}$$

$$
\begin{aligned}
&= \gamma\left(-\gamma M_y B_1 + \frac{M_o}{T_1}\right) B_1, v^2(x)\frac{\partial^2 M_y}{\partial x^2} + v(x)\frac{dv}{dx}\frac{\partial M_y}{\partial x} \\
&+ v(x)\frac{\partial^2 M_y}{\partial x \partial t} + \frac{v(x)}{T_1}\frac{\partial M_y}{\partial x} + v(x)\frac{\partial^2 M_y}{\partial x \partial t} + \frac{\partial^2 M_y}{\partial t^2} \\
&+ \frac{1}{T_1}\frac{\partial M_y}{\partial t} + \frac{v(x)}{T_2}\frac{\partial M_y}{\partial x} \\
&+ \frac{1}{T_2}\frac{\partial M_y}{\partial t} + \frac{1}{T_1 T_2} M_y \\
&= -\gamma^2 B_1^2 M_y + \frac{\gamma B_1 M_o}{T_1}, \\
&v^2(x)\frac{\partial M_y}{\partial x^2} + v(x)\frac{dv(x)}{dx}\frac{\partial M_y}{\partial x} + 2v\frac{\partial^2 M_y}{\partial x \partial t} + v(x)\left(\frac{1}{T_1} + \frac{1}{T_2}\right)\frac{\partial M_y}{\partial x} \\
&+ \left(\frac{1}{T_1} + \frac{1}{T_2}\right)\frac{\partial M_y}{\partial t} + \frac{\partial^2 M_y}{\partial t^2} + \left(\frac{1}{T_1 T_2} + \gamma^2 B_1^2\right) M_y = \frac{\gamma B_1 M_o}{T_1}.
\end{aligned}
$$

This expression may be written as:

$$
\begin{aligned}
&v^2(x)\frac{\partial^2 M_y}{\partial x^2} + 2v(x)\frac{\partial^2 M_y}{\partial x \partial t} + v(x)\left(T_o + \frac{dv(x)}{dx}\right)\frac{\partial M_y}{\partial x} \\
&+ T_o\frac{\partial M_y}{\partial t} + \frac{\partial^2 M_y}{\partial t^2} + (T_g + \gamma^2 B_1^2) M_y = F_o \gamma B_1, \qquad (3.11)
\end{aligned}
$$

where

$T_o = \frac{1}{T_1} + \frac{1}{T_2}$, $T_g = \frac{1}{T_1 T_2}$ and $F_o = \frac{M_o}{T_1}$.

Equation (3.11) is a general second order differential equation which can be applied to NMR flow system with spatial fluid velocity dependence.

The Time-Independent Bloch NMR Flow Equation for Variable Fluid Velocity

Within steady molecular fluid flow situations and spatial fluid velocity (Dada and Awojoyogbe 2021), the Bloch equation could be modified as follows:

$$
v(x)\frac{dM_x}{dx} = -\frac{M_x}{T_2}, \qquad (3.12)
$$

$$
v(x)\frac{dM_y}{dx} = \gamma M_z B_1 - \frac{M_y}{T_2}, \qquad (3.13)
$$

$$
v(x)\frac{dM_z}{dx} = -\gamma M_z B_1 + \frac{(M_0 - M_z)}{T_1}. \qquad (3.14)
$$

From Eq. (3.14), the following equation is obtained:

$$M_z = \left(-\gamma M_y B_1 + \frac{M_o}{T_1}\right)\frac{1}{\left(v(x)\frac{d}{dx} + \frac{1}{T_1}\right)}. \tag{3.15}$$

Substituting for Eq. (3.15) into Eq. (3.13), we have:

$$v(x)\frac{dM_y}{dx} = \gamma\left(-\gamma M_y B_1 + \frac{M_o}{T_1}\right)\frac{1}{\left(v(x)\frac{d}{dx} + \frac{1}{T_1}\right)}B_1 - \frac{M_y}{T_2},$$

$$v(x)\frac{dM_y}{dx} + \frac{M_y}{T_2} = \left(-\gamma^2 M_y B_1^2 + \frac{\gamma B_1 M_o}{T_1}\right)\frac{1}{\left(v(x)\frac{d}{dx} + \frac{1}{T_1}\right)},$$

$$\left(v(x)\frac{d}{dx} + \frac{1}{T_1}\right)v(x)\frac{dM_y}{dx} + \left(v(x)\frac{d}{dx} + \frac{1}{T_1}\right)\frac{M_y}{T_2}$$
$$= -\gamma^2 M_y B_1^2(x) + \frac{\gamma B_1 M_o}{T_1},$$

$$v^2(x)\frac{d^2M_y}{dx^2} + v(x)\frac{dv(x)}{dx}\frac{dM_y}{dx} + v\left(\frac{1}{T_1} + \frac{1}{T_2}\right)\frac{dM_y}{dx}$$
$$+ \left(\frac{1}{T_1T_2} + \gamma^2 B_1^2(x)\right)M_y = \frac{\gamma B_1(x)M_o}{T_1},$$

$$v^2(x)\frac{d^2M_y}{dx^2} + v(x)\left(T_o + \frac{dv(x)}{dx}\right)\frac{dM_y}{dx}$$
$$+ (T_g + \gamma^2 B_1^2(x))M_y = \frac{M_o}{T_1}\gamma B_1(x). \tag{3.16}$$

Equation (3.16) is a variable time independent Bloch NMR flow equation.

3.3.2 *Derivation of Diffusion Equation with Variable Diffusion Coefficient*

Biological systems usually have compartments with fluids of different hydrodynamic properties occupying the compartments (Nelson and Cox 2008) such that the diffusion coefficient varies in different locations. As noted earlier on, this is consistent with fluid flow in real biological systems (Li et al. 1992; Dahlem and Müller 2004; Rungseevijitprapa et al. 2009). Therefore, quantifying the spatial dependence may hold important information about the system and associated molecular dynamics at any instant. Meanwhile, this problem has been stated in earlier studies (Awojoyogbe et al. 2011; Gupta et al. 2014) but at this point, it is important to state the conditions under which the variable diffusion equation holds. Consequently, assuming:

$$\frac{\partial^2 M_y}{\partial t^2} + 2v(x)\frac{\partial^2 M_y}{\partial x \partial t} = 0. \tag{3.17}$$

Then, the following differential equation is obtained:

$$v^2(x)\frac{\partial^2 M_y}{\partial x^2} + v(x)\left(T_o + \frac{dv(x)}{dx}\right)\frac{\partial M_y}{\partial x} + T_o\frac{\partial M_y}{\partial t} + (T_g + \gamma^2 B_1^2)M_y = F_o\gamma B_1. \tag{3.18}$$

If the applied RF field is then designed such that

$$\gamma B_1 = \sqrt{\frac{v(x)}{M_y}\left(T_o + \frac{dv(x)}{dx}\right)\frac{\partial M_y}{\partial x} - T_g}. \tag{3.19}$$

If the velocity gradient changes very slowly compared to the relaxation rate of molecular dynamics, we may assume that

$$T_o \ll \frac{dv(x)}{dx}. \tag{3.20}$$

Equations (3.19) and (3.18) then become:

$$\gamma B_1(x) = \sqrt{\frac{v(x)}{M_y}\frac{dv(x)}{dx}\frac{\partial M_y}{\partial x} - T_g},$$

$$v^2(x)\frac{\partial^2 M_y}{\partial x^2} + 2v(x)\frac{dv(x)}{dx}\frac{\partial M_y}{\partial x} + T_o\frac{\partial M_y}{\partial t} = F_o\gamma B_1(x), \tag{3.21}$$

or

$$\frac{\partial}{\partial x}\left(v^2(x)\frac{\partial M_y}{\partial x}\right) + \frac{1}{T_0}\frac{\partial M_y}{\partial t} = F_0\gamma B_1. \tag{3.22}$$

This equation may be written as

$$\frac{\partial M_y}{\partial t} = \frac{\partial}{\partial x}\left(-\frac{v^2(x)}{T_0}\frac{\partial M_y}{\partial x}\right) + F_0\gamma B_1. \tag{3.23}$$

If the following term holds

$$D(x) = -\frac{v^2(x)}{T_0}. \tag{3.24}$$

Provided that D(x) represents the spatial diffusion coefficient, we finally have:

$$\frac{\partial M_y}{\partial t} = \frac{\partial}{\partial x}\left(D(x)\frac{\partial M_y}{\partial x}\right) + \frac{F_o}{T_o}\gamma B_1(\vec{r}, t). \tag{3.25}$$

Equation (3.25) is the 1D NMR diffusion equation with variable coefficient.

3.3.2.1 One-Dimensional Diffusion–Advection with Variable Diffusion Coefficient

If the fluid velocity and the diffusion coefficient D are both spatially dependent on x and making an assumption (Eq. 3.11) of the following form:

$$\frac{\partial^2 M_y}{\partial t^2} + 2v(x)\frac{\partial^2 M_y}{\partial x \partial t} = 0. \tag{3.26}$$

The equation becomes

$$v^2(x)\frac{\partial^2 M_y}{\partial x^2} + v(x)\left(T_o + \frac{dv(x)}{dx}\right)\frac{\partial M_y}{\partial x} + T_o\frac{\partial M_y}{\partial t} + (T_g + \gamma^2 B_1^2)M_y = F_o\gamma B_1. \tag{3.27}$$

If the applied RF field is then designed such that

$$\gamma B_1 = \sqrt{\frac{v(x)}{M_y}\frac{dv(x)}{dx}\frac{\partial M_y}{\partial x} + T_o\frac{dv(x)}{dx} - T_g}. \tag{3.28}$$

Equation (3.27) becomes:

$$\begin{aligned}
&v^2(x)\frac{\partial^2 M_y}{\partial x^2} + v(x)\left(T_0 + \frac{dv(x)}{dx}\right)\frac{\partial M_y}{\partial x} \\
&\quad + T_0\frac{\partial M_y}{\partial t} + v(x)\frac{dv(x)}{dx}\frac{\partial M_y}{\partial x} + T_0\frac{dv(x)}{dx}M_y = F_o\gamma B_1, \\
&v^2(x)\frac{\partial^2 M_y}{\partial x^2} + 2v(x)\frac{dv(x)}{dx}\frac{\partial M_y}{\partial x} + T_0\frac{\partial M_y}{\partial t} \\
&\quad + v(x)T_0\frac{\partial M_y}{\partial x} + T_0\frac{dv(x)}{dx}M_y = F_o\gamma B_1, \\
&\frac{\partial}{\partial x}\left(v^2(x)\frac{\partial M_y}{\partial x}\right) + T_0\frac{\partial M_y}{\partial t} + \frac{\partial}{\partial x}(v(x)T_0 M_y) = F_o\gamma B_1.
\end{aligned} \tag{3.29}$$

Multiplying all through by 1/T_0 and re-arranging the termsleads to:

$$\frac{\partial}{\partial x}(vM_y) + \frac{\partial M_y}{\partial t} = \frac{\partial}{\partial x}\left(D(x)\frac{\partial M_y}{\partial x}\right) + \frac{F_o}{T_o}\gamma B_1(x,t), \tag{3.30}$$

where

$$D(x) = -\frac{v^2(x)}{T_o}.$$

3.3.2.2 Multidimensional Diffusion and Diffusion–Advection with Variable Diffusion Coefficient

The multidimensional diffusion and diffusion–advection (diffusion with transport) differential equation have been presented in (Dada and Awojoyogbe 2021) respectively. Specifying the space vector $\vec{r}$, these equations are written as follows:

$$\frac{\partial M_y}{\partial t} = \nabla \cdot \left(D(\vec{r})\nabla M_y\right) + \frac{F_o}{T_o}\gamma B_1(\vec{r}, t), \tag{3.31}$$

$$\nabla \cdot \left(v(\vec{r})M_y\right) + \frac{\partial M_y}{\partial t} = \nabla \cdot \left(D(\vec{r})\nabla M_y\right) + \frac{F_o}{T_o}\gamma B_1(\vec{r}, t). \tag{3.32}$$

Solving this problem is very tasking but possible with minimum assumptions. These assumptions are made by either setting the fluid velocity and diffusion coefficient equal to functions of only radial distance or eliminating the effects of the RF B_1 pulse.

3.3.3 *Optical Bloch Equation (OBE) and Molecular Imaging*

As pointed outin earlier sections, the Bloch equations have wide application to a number of physical effects that do not necessarily involve gyromagnetic spins (Hahn 1980). They apply directly to any two quantum level or equally spaced quantum level system (Cohen-Tannoudji et al. 1998; Eberly and Stroud 2006), and are particularly useful in predicting effects involving electric dipole laser resonance phenomena. Before now, optical and NMR applications of the Bloch equations have been established as successful separate fields. However, in this study, we shall explore the connections that may allow NMR research scientists to be able to derive optical parameters from NMR.

This could prove to be very crucial in Molecular imaging. For example, there has been considerable interest in measurements of viscoelastic properties of red blood cells (Mohanty 2012) since an understanding of this is important to investigate their passage through narrow capillaries. Optical tweezers based studies can provide critical insights into changes in the elastic properties of cell membrane during progress of disease as well as its altered adhesion properties (Mohanty et al. 2004, 2005, 2008). This optical method has been highly successful in understanding of malaria, sickle cell diseases, leukaemia, etc. However, the magnetic resonant effects in these processes have not been explored in detail. In this section, a theoretical background to this problemis presented.

Optics is the branch of physics which involves the behaviour and properties of light, including its interactions with matter and the construction of instruments that use or detect it (Parker 1997). Optics usually describes the behaviour of visible, ultraviolet, and infrared light. Because light is an electromagnetic (EM) wave, other forms of EM radiation such as X-rays, microwaves, and radio waves (this is where NMR comes in) exhibit similar properties. Since NMR deals with radio waves and radio waves have optical behaviour, it follows that NMR research is possibly yet to explore new information which are likely contained in optical behaviour of NMR radio waves.

To address this problem, atomic system in which Larmor condition (leading to optical nutations or Rabi oscillations) holdshas been considered. The optical Bloch equation in compact form is given as follows (Meystre 2006; Eberly and Stroud 2006):

$$\frac{d\vec{P}_r}{dt} = \vec{P}_r \times [\hat{x}\Xi\xi_p + \hat{z}(\Delta\omega_0 + d\phi/dt)] + \frac{(\hat{x}u + \hat{y}v)Np_0}{T_2} - \frac{\hat{z}(w - w_0)Np_0}{T_1}, \tag{3.33}$$

where ϕ is an arbitrary propagation phase assigned to a plane wave propagating in $\vec{r}$ (general spatial coordinate) direction, $\vec{P}_r$ is the radiative polarisation of the undamped system and w_0 is the equilibrium atomic inversion (Meystre 2006; Eberly and Stroud 2006). It must be noted that p_0 is the electric dipole moment, $\Xi = \frac{2|p_0|}{\hbar}$ (synonymous with the NMR gyromagnetic ratio), N is the number of atoms per unit volume and ξ_p is the applied electric field envelope (Eberly and Stroud 2006).

A set of polarisation components Np_0u, Np_0v, Np_0w appear as slowly varying function of t and $\vec{r}$ in a frame of reference rotating at frequency $\omega + \frac{d\phi}{dt}$. The radiative polarisation is given as follows (Hahn 1980):

$$\vec{P}_r = (\hat{x}u + \hat{y}v - \hat{z}w)Np_0 = \hat{x}P_x + \hat{y}P_y + \hat{z}P_z, \tag{3.34}$$

where P_x, P_y and P_z are the induced polarisation components. Hence, we shall rewrite Eq. (3.33) as follows:

$$\frac{d\vec{P}_r}{dt} = \vec{P}_r \times [\hat{x}\aleph\xi + \hat{z}(\Delta\omega_0 + d\phi/dt)] - \frac{(\hat{x}P_x + \hat{y}P_y)}{T_2} - \frac{\hat{z}(P_z - P_0)}{T_1}. \tag{3.35}$$

From Eq. (3.35), the following coupled differential equations are obtained:

$$\left.\begin{aligned}\frac{dP_x}{dt} &= \left|\vec{P}_r \times [\hat{x}\Xi\xi_p + \hat{z}(\Delta\omega_0 + d\phi/dt)]\right|_x - \frac{P_x}{T_2},\\ \frac{dP_y}{dt} &= \left|\vec{P}_r \times [\hat{x}\Xi\xi_p + \hat{z}(\Delta\omega_0 + d\phi/dt)]\right|_y - \frac{P_y}{T_2},\\ \frac{dP_z}{dt} &= \left|\vec{P}_r \times [\hat{x}\Xi\xi_p + \hat{z}(\Delta\omega_0 + d\phi/dt)]\right|_z - \frac{(P_z - P_0)}{T_1}.\end{aligned}\right\} \tag{3.36}$$

However, the cross product

$$\begin{aligned}&\vec{P}_r \times \left[\hat{x}\Xi\xi_p + \hat{z}\left(\Delta\omega_0 + \frac{d\phi}{dt}\right)\right]\\ &= \begin{vmatrix}\hat{x}\ \hat{y}\ \hat{z}\\ P_x\ P_y\ P_z\\ \Xi\xi_p\ 0\ \Delta\omega_0 + d\phi/dt\end{vmatrix} = \hat{x}\left(\Delta\omega_0 + \frac{d\phi}{dt}\right)P_y - \hat{y}\left[\left(\Delta\omega_0 + \frac{d\phi}{dt}\right)P_x - \Xi\xi P_z\right]\\ &- \hat{z}\Xi\xi P_y.\end{aligned} \tag{3.37}$$

From Eqs. (3.37), (3.36) becomes (Hahn 1980):

$$\left.\begin{aligned}\frac{dP_x}{dt} &= \left(\Delta\omega_0 + \frac{d\phi}{dt}\right)P_y - \frac{P_x}{T_2},\\ \frac{dP_y}{dt} &= -\left(\Delta\omega_0 + \frac{d\phi}{dt}\right)P_x + \Xi\xi_p P_z - \frac{P_y}{T_2},\\ \frac{dP_z}{dt} &= -\Xi\xi_p P_y - \frac{(P_z - P_0)}{T_1}.\end{aligned}\right\} \tag{3.38}$$

Since we have added the term ϕ to account for phase variations in the resonance process, we note that:

$$\left.\begin{aligned}\frac{dP_x}{dt} &= \left(\Delta\omega_0 + \frac{d\phi}{dt}\right)P_y - \frac{P_x}{T_2},\\ \frac{dP_y}{dt} &= -\left(\Delta\omega_0 + \frac{d\phi}{dt}\right)P_x + \Xi\xi_p P_z - \frac{P_y}{T_2},\\ \frac{dP_z}{dt} &= -\Xi\xi_p P_y - \frac{(P_z - P_0)}{T_1}.\end{aligned}\right\} \tag{3.39}$$

However, the deviation from resonance condition is (Cohen-Tannoudji et al. 1998):

$$\Delta\omega_0 = \omega_0 - \omega, \tag{3.40}$$

where ω_0 is the atomic frequency, ω is the frequency of the spin in the rotating frame of reference. If phase effects are not significant, Eqs. (3.38) and (3.39) becomes:

$$\left.\begin{aligned}\frac{dP_x}{dt} &= (\omega_0 - \omega)P_y - \frac{P_x}{T_2},\\ \frac{dP_y}{dt} &= -(\omega_0 - \omega)P_x + \Xi\xi_p P_z - \frac{P_y}{T_2},\\ \frac{dP_z}{dt} &= -\Xi\xi_p P_y - \frac{(P_z - P_0)}{T_1}.\end{aligned}\right\} \tag{3.41}$$

Equation (3.41) is the optical analog of the NMR Bloch equation in the rotating frame as presented in earlier studies (Awojoyogbe et al. 2011; Gupta et al. 2014).

3.3.3.1 Generalized Optical Bloch Flow Equation

Considering atoms in a system rotating at the frequency ω, spin resonance occurs such that

$$\omega_0 = \omega. \tag{3.42}$$

Therefore, Eq. (3.41) becomes:

$$\frac{dP_x}{dt} = -\frac{P_x}{T_2}, \tag{3.43}$$

$$\frac{dP_y}{dt} = \Xi\xi P_z - \frac{P_y}{T_2}, \tag{3.44}$$

$$\frac{dP_z}{dt} = -\Xi\xi P_y - \frac{(P_z - P_0)}{T_1}. \tag{3.45}$$

Therefore, from the kinematic theory of moving fluids and using the methods employed in the derivation of Bloch NMR flow equations (Awojoyogbe et al. 2011; Gupta et al. 2014), the following partial differential equation is obtained

$$\frac{\partial^2 P_y}{\partial t^2} + 2v\frac{\partial^2 P_y}{\partial x \partial t} + v^2\frac{\partial^2 P_y}{\partial x^2} + T_0\frac{\partial P_y}{\partial t} + vT_0\frac{\partial P_y}{\partial x} + \left(\Xi^2\xi_p^2 + T_g\right)P_y = \frac{P_0}{T_1}\Xi\xi. \tag{3.46}$$

3.3.3.2 Time-dependent Optical Bloch Flow Equation

As formulated in earlier sections, the case in which the transverse polarisation function $P_y(x, t)$ is slowly varying with the spatial coordinate x becomes:

$$P_z = \frac{-\Xi\xi_p P_y + \frac{P_0}{T_1}}{\frac{d}{dt} + \frac{1}{T_1}}. \tag{3.47}$$

Substituting for Eq. (3.47) in Eq. (3.44) gives:

$$\frac{d^2P_y}{dt^2} + \frac{1}{T_1}\frac{dP_y}{dt} = \frac{P_0}{T_1}\Xi\xi_p - \Xi^2\xi_p^2 P_y - \left(\frac{1}{T_2}\frac{dP_y}{dt} + \frac{P_y}{T_1T_2}\right), \tag{3.48}$$

$$\frac{d^2P_y}{dt^2} + \left(\frac{1}{T_1} + \frac{1}{T_2}\right)\frac{dP_y}{dt} + \left(\Xi^2\xi_p^2 + \frac{1}{T_1T_2}\right)P_y = \frac{P_0}{T_1}\Xi\xi_p, \tag{3.49}$$

$$\frac{d^2P_y}{dt^2} + T_0\frac{dP_y}{dt} + (\Xi^2\xi_p^2 + T_g)P_y = \frac{P_0}{T_1}\Xi\xi_p. \tag{3.50}$$

Equation (3.50) may be called time-dependent Optical Bloch flow equation. It describes the behaviour of the polarisation functionof spins in motion as a function of the electric field and relaxation times.

3.3.3.3 Time-independent Optical Bloch Flow Equation

In situations where the transverse polarisation function exhibits transient behaviour; for mobile molecular magnets (spin), the following equations hold:

$$\left.\begin{aligned} v\nabla P_x &= -\frac{P_x}{T_2}, \\ v\nabla P_y &= \Xi\xi_p P_z - \frac{P_y}{T_2}, \\ v\nabla P_z &= -\Xi\xi_p P_y - \frac{(P_z - P_0)}{T_1}. \end{aligned}\right\} \tag{3.51}$$

If the spatial gradient of the transverse polarisation is in one dimension only,

$$v\frac{dP_x}{dx} = -\frac{P_x}{T_2}, \tag{3.52}$$

$$v\frac{dP_y}{dx} = \Xi\xi_p P_z - \frac{P_y}{T_2}, \tag{3.53}$$

$$v\frac{dP_z}{dx} = -\Xi\xi_p P_y - \frac{(P_z - P_0)}{T_1}. \tag{3.54}$$

From Eq. (3.296), we may write:

$$P_z = \frac{-\Xi\xi_p P_y + \frac{P_0}{T_1}}{v\frac{d}{dx} + \frac{1}{T_1}}. \tag{3.55}$$

Substituting for Eq. (3.55) in Eq. (3.53) gives:

$$v\frac{dP_y}{dx} = \Xi\xi_p\left(\frac{-\Xi\xi_p P_y + \frac{P_0}{T_1}}{v\frac{d}{dx} + \frac{1}{T_1}}\right) - \frac{P_y}{T_2}. \tag{3.56}$$

$$v\left(v\frac{d}{dx} + \frac{1}{T_1}\right)\frac{dP_y}{dx} = \Xi\xi_p\left(-\Xi\xi_p P_y + \frac{P_0}{T_1}\right) - \left(v\frac{d}{dx} + \frac{1}{T_1}\right)\frac{P_y}{T_2}, \tag{3.57}$$

$$v^2\frac{d^2P_y}{dx^2} + \frac{v}{T_1}\frac{dP_y}{dx} = -\Xi^2\xi_p^2 P_y + \frac{P_0}{T_1}\Xi\xi_p - \left(\frac{v}{T_2}\frac{dP_y}{dx} + \frac{P_y}{T_1T_2}\right),$$

$$v^2\frac{d^2P_y}{dx^2} + v\left(\frac{1}{T_1} + \frac{1}{T_2}\right)\frac{dP_y}{dx} + \left(\Xi^2\xi_p^2 + \frac{1}{T_1T_2}\right)P_y = \frac{P_0}{T_1}\Xi\xi_p, \tag{3.58}$$

$$v^2\frac{d^2P_y}{dx^2} + vT_0\frac{dP_y}{dx} + \left(\Xi^2\xi_p^2 + T_g\right)P_y = \frac{P_0}{T_1}\Xi\xi_p. \tag{3.59}$$

3.3.3.4 Optical Bloch Flow Equation with Variable Fluid Velocity

In most of the formulation of the Bloch picture for Optical phenomenon in magnetic resonance, it is interesting to note that both nuclear and optical Bloch phenomena have similar equation. Therefore, for the case of variable fluid velocity, the Optical Bloch equations follow similar derivation as presented in Sect. 3.3.1.1; hence, Eqs. (3.46) and (3.59) may be modified according to the expressions of Eqs. (3.11) and (3.16) such that the general optical Bloch flow equation and time-independent optical Bloch flow equation become:

$$\frac{\partial^2P_y}{\partial t^2} + 2v(x)\frac{\partial^2P_y}{\partial x\partial t} + v^2(x)\frac{\partial^2P_y}{\partial x^2} + T_0\frac{\partial P_y}{\partial t} + v(x)\left(T_0 + \frac{dv(x)}{dx}\right)\frac{\partial P_y}{\partial x} + \left(\Xi^2\xi_p^2 + T_g\right)P_y = \frac{P_0}{T_1}\Xi\xi_p, \tag{3.60}$$

$$v^2(x)\frac{d^2P_y}{dx^2} + v(x)\left(T_0 + \frac{dv(x)}{dx}\right)\frac{dP_y}{dx} + \left(\Xi^2\xi_p^2 + T_g\right)P_y = \frac{P_0}{T_1}\Xi\xi_p. \tag{3.61}$$

3.3.3.5 Atomic Diffusion, Rabi Oscillations in Optical Bloch Flow Equation

Rabi cycle is the cyclic behaviour of a two-state quantum system in the presence of an applied oscillatory field. Physical processes involving quantum computing, condensed matter, atomic and molecular physics, and nuclear and particle physics can be studied quite easily in terms of two-level quantum systems. When coupled to an oscillatory driving field, they show Rabi cycle. This phenomenon named after Isidor Isaac Rabi, has proved to be very crucial in the description of processes in MRI, quantum optics and quantum computing (Cohen-Tannoudji et al. 1998; Paschotta 2008).

It is noteworthy that a two-state system has two possible states and in the scenario where they are not degenerate, the optical system absorbs a quantum of energy and become excited. Hence, when an atom (or some other two-level system) is treated with radio waves, it will cyclically absorb photons and re-emit them by stimulated emission applied from the EM field (Cohen-Tannoudji et al. 1998). This cycle is often known as the Rabi cycle and the inverse of its duration is the Rabi frequency of the applied field.

Meanwhile, the quantum fluctuations often lead to rrelaxation whenever the molecular surroundings of a biological system fluctuates randomly (Eberly and Stroud 2006). These random fluctuations comprise of all interactions that are too complex to be treated fundamentally but usually demonstrated in terms of the degradation of the degree of coherence of a specific interaction. Whenever the quantum fluctuations are significantly rapid, all dynamical systems in a macroscopic sample-experience the elaborate fluctuations within a short period of time in comparison to the time an experiment and homogeneous relaxation results. If the existence of the molecular fluctuations is random over a macroscopic sample such that there is slow relative change in time, then there is inhomogeneous relaxation.

The quantum fluctuations which lead to spin relaxation also cause a phenomenon known as momentum diffusion (Ungar et al. 1989; Berg-Sorenson et al. 1992). Momentum diffusion is the spread of momentum between particles (atoms or molecules) of matter. This is exhibited generally in the fluid state. Treatment of momentum diffusion often requires the use of difficult density-matrix elements and the computation of the diffusion rate as a function of velocity is not an easy task as well (Ungar et al. 1989). Therefore, there is a need to provide an easier means of handling atomic diffusion in optical resonance. The next section presents diffusion equations in terms of the optical resonance parameters.

The derivation of the optical resonance diffusion equation is quite similar to the ones presented earlier for NMR spins. If in Eq. (3.46), the following assumption holds

$$\frac{\partial^2 P_y}{\partial t^2} + 2v(x)\frac{\partial^2 P_y}{\partial x dt} = 0. \tag{3.62}$$

Provided that the form of the transverse polarisation function permits the existence of Eqs. (3.62), (3.60) becomes:

$$v^2(x)\frac{\partial^2 P_y}{\partial x^2} + T_0\frac{\partial P_y}{\partial t} + v(x)\left(T_0 + \frac{dv(x)}{dx}\right)\frac{\partial P_y}{\partial x} + \left(\Xi^2\xi_p^2 + T_g\right)P_y = \frac{P_0}{T_1}\Xi\xi_p. \tag{3.63}$$

If the applied electric field is then designed such that

$$\Xi\xi_p(x,t) = \sqrt{\frac{v(x)}{P_y}\left(T_0 + \frac{dv(x)}{dx}\right)\frac{\partial P_y}{\partial x} - T_g}. \tag{3.64}$$

Equation (3.63) then becomes:

$$\begin{aligned} v^2(x)\frac{\partial^2 P_y}{\partial x^2} &+ T_0\frac{\partial P_y}{\partial t} + v(x)\left(T_0 + \frac{dv(x)}{dx}\right)\frac{\partial P_y}{\partial x} \\ &+ v(x)\left(T_0 + \frac{dv(x)}{dx}\right)\frac{\partial P_y}{\partial x} = \frac{P_0}{T_1}\Xi\xi_p, \end{aligned} \tag{3.65}$$

$$v^2(x)\frac{\partial^2 P_y}{\partial x^2} + T_0\frac{\partial P_y}{\partial t} + 2v(x)\left(T_0 + \frac{dv(x)}{dx}\right)\frac{\partial P_y}{\partial x} = \frac{P_0}{T_1}\Xi\xi_p. \tag{3.66}$$

Just like in the case of NMR diffusion, if the velocity gradient changes very slowly compared to the relaxation rate of molecular dynamics, it may be assumed:

$$T_o \ll \frac{dv(x)}{dx}. \tag{3.67}$$

Then Eq. (3.66) becomes:

$$v^2(x)\frac{\partial^2 P_y}{\partial x^2} + 2v(x)\frac{dv(x)}{dx}\frac{\partial P_y}{\partial x} + T_0\frac{\partial P_y}{\partial t} = \frac{P_0}{T_1}\Xi\xi_p. \tag{3.68}$$

This equation may be re-written as follows:

$$\frac{\partial}{\partial x}\left(v^2(x)\frac{\partial P_y}{\partial x}\right) + T_0\frac{\partial P_y}{\partial t} = \frac{P_0}{T_1}\Xi\xi_p. \tag{3.69}$$

Multiplying all through by $\frac{1}{T_0}$ and re-arranging the above equations:

$$\frac{\partial P_y}{\partial t} = \frac{\partial}{\partial x}\left(-\frac{v^2(x)}{T_0}\frac{\partial P_y}{\partial x}\right) + \frac{P_0}{T_0 T_1}\Xi\xi_p. \tag{3.70}$$

If the diffusion coefficient is defined as (just like in the case of NMR diffusion)

$$D(x) = -\frac{v^2(x)}{T_0}. \tag{3.71}$$

Therefore, the equation of spin diffusion during optical resonance is given as follows:

$$\frac{\partial P_y}{\partial t} = \frac{\partial}{\partial x}\left(D(x)\frac{\partial P_y}{\partial x}\right) + \frac{P_0}{T_0 T_1}\Xi\xi_p. \tag{3.72}$$

Optical Diffusion in with Constant Diffusion Coefficient

The Optical diffusion equation has a constant diffusion coefficient whenever the fluid velocity is independent of spatial measurements. Hence, Eq. (3.72) becomes:

$$\frac{\partial P_y}{\partial t} = D\frac{\partial^2 P_y}{\partial x^2} + \frac{P_0}{T_0 T_1}\Xi\xi_p. \tag{3.73}$$

Optical Diffusion in Multidimensions with Constant Diffusion Coefficient

If the transverse polarisation function becomes dependent on more than one spatial variable, Eq. (3.73) may be generalized as follows:

$$\frac{\partial P_y}{\partial t} = D\nabla^2 P_y + \frac{P_0}{T_0 T_1}\Xi\xi_p, \tag{3.74}$$

where ∇^2 is the Laplacian in the coordinate (Wylie and Barrett 1982; Kreyszig 1996) which describes the molecular dynamic system that is being considered. In common geometries, this equation is (Awojoyogbe et al. 2011; Gupta et al. 2014)

$$\text{Cartesian Geometry:} \frac{\partial P_y}{\partial t} = D\left(\frac{\partial^2 P_y}{\partial x^2} + \frac{\partial^2 P_y}{\partial y^2} + \frac{\partial^2 P_y}{\partial z^2}\right) + \frac{P_0}{T_0 T_1}\Xi\xi_p(\vec{r}, t), \tag{3.75}$$

where $\vec{r} = f(x, y, z)$.

$$\text{Cylindrical Geometry:} \frac{\partial P_y}{\partial t} = D\left(\frac{1}{r}\frac{\partial}{\partial r}\left(r\frac{\partial P_y}{\partial r}\right) + \frac{1}{r^2}\frac{\partial^2 P_y}{\partial \phi^2} + \frac{\partial^2 P_y}{\partial z^2}\right) + \frac{P_0}{T_0 T_1}\Xi\xi_p(\vec{r}, t), \tag{3.76}$$

where $\vec{r} = f(r, \phi, z)$.

$$\text{Spherical Geometry:} \frac{\partial P_y}{\partial t} = \frac{D}{r^2}\left(\frac{\partial}{\partial r}\left(r^2\frac{\partial P_y}{\partial r}\right) + \frac{1}{\sin^2\theta}\frac{\partial^2 P_y}{\partial \phi^2} + \frac{1}{\sin\theta}\frac{\partial}{\partial\theta}\left(\sin\theta\frac{\partial P_y}{\partial\theta}\right)\right) + \frac{P_0}{T_0 T_1}\Xi\xi_p(\vec{r}, t), \tag{3.77}$$

where $\vec{r} = f(r, \phi, \theta)$.

Optical Diffusion in Multidimensions with Variable Diffusion Coefficient

If the momentum diffusion is taking place in a non-homogeneous medium such that the transverse polarisation function varies with multiple spatial variables, then Eq. (3.72) may be generalised as follows:

$$\frac{\partial P_y}{\partial t} = \nabla \cdot \left(D \nabla P_y\right) + \frac{P_0}{T_0 T_1} \Xi \xi_p(\vec{r}, t), \tag{3.78}$$

where ∇ is the Del operator (Wylie and Barrett 1982; Kreyszig 1996). In the usual geometries, Eq. (3.78) may be written as

Rectangular Cartesian Geometry:

$$\frac{\partial P_y}{\partial t} = \frac{\partial}{\partial x}\left(D \frac{\partial P_y}{\partial x}\right) + \frac{\partial}{\partial y}\left(D \frac{\partial P_y}{\partial y}\right) + \frac{\partial}{\partial z}\left(D \frac{\partial P_y}{\partial z}\right) + \frac{P_0}{T_0 T_1} \Xi \xi_p(\vec{r}, t), \tag{3.79}$$

or

$$\frac{\partial P_y}{\partial t} = \frac{\partial}{\partial x}\left(D_x \frac{\partial P_y}{\partial x}\right) + \frac{\partial}{\partial y}\left(D_y \frac{\partial P_y}{\partial y}\right) + \frac{\partial}{\partial z}\left(D_z \frac{\partial P_y}{\partial z}\right) + \frac{P_0}{T_0 T_1} \Xi \xi_p(\vec{r}, t). \tag{3.80}$$

Cylindrical Geometry:

$$\frac{\partial P_y}{\partial t} = \frac{1}{r}\frac{\partial}{\partial r}\left(D r \frac{\partial P_y}{\partial r}\right) + \frac{1}{r^2}\frac{\partial}{\partial \phi}\left(D \frac{\partial P_y}{\partial \phi}\right) + \frac{\partial}{\partial z}\left(D \frac{\partial P_y}{\partial z}\right) + \frac{P_0}{T_0 T_1} \Xi \xi_p(\vec{r}, t), \tag{3.81}$$

or

$$\frac{\partial P_y}{\partial t} = \frac{1}{r}\frac{\partial}{\partial r}\left(D_r r \frac{\partial P_y}{\partial r}\right) + \frac{1}{r^2}\frac{\partial}{\partial \phi}\left(D_\phi \frac{\partial P_y}{\partial \phi}\right) + \frac{\partial}{\partial z}\left(D_z \frac{\partial P_y}{\partial z}\right) + \frac{P_0}{T_0 T_1} \Xi \xi_p(\vec{r}, t). \tag{3.82}$$

Spherical Geometry:

$$\begin{aligned}\frac{\partial P_y}{\partial t} &= \frac{1}{r^2}\frac{\partial}{\partial r}\left(D r^2 \frac{\partial P_y}{\partial r}\right) + \frac{1}{r^2 \sin\theta}\frac{\partial}{\partial \theta}\left(D \sin\theta \frac{\partial P_y}{\partial \theta}\right) \\ &\quad + \frac{1}{r^2 \sin^2\theta}\frac{\partial}{\partial \phi}\left(D \frac{\partial P_y}{\partial \phi}\right) + \frac{P_0}{T_0 T_1} \Xi \xi_p(\vec{r}, t),\end{aligned} \tag{3.83}$$

or

$$\frac{\partial P_y}{\partial t} = \frac{1}{r^2}\frac{\partial}{\partial r}\left(D_r r^2 \frac{\partial P_y}{\partial r}\right) + \frac{1}{r^2 \sin\theta}\frac{\partial}{\partial \theta}\left(D_\theta \sin\theta \frac{\partial P_y}{\partial \theta}\right) + \frac{1}{r^2 \sin^2\theta}\frac{\partial}{\partial \phi}\left(D_\phi \frac{\partial P_y}{\partial \phi}\right) + \frac{P_0}{T_0 T_1}\Xi\xi_p(\vec{r}, t). \tag{3.84}$$

Rotational Diffusion in Optical Bloch Flow Formulation

Rotational diffusion is phenomenon in which the equilibrium statistical distribution of the overall orientation of molecules is restored. The diffusion processes that have been discussed thus far, are that of translational diffusion in which the position in space of equilibrium statistical distribution of molecules is restored or maintained. In molecular systems, there is the random re-orientation of molecules and is very important in the design of molecular probes and general molecular imaging (Goose and Sansom 2013). Due to the equipartition theorem, larger molecules re-orient more slowly than smaller molecules. Consequently, measurements of the rotational diffusion constants can give insight into the overall mass and distribution within a given molecule.

This rotation has been known to be quite crucial in optical trapping (Mohanty 2012). Optical traps now have impressive applications in both physics and biology (Mohanty et al. 2004, 2005, 2008; Williams 2002; Neuman and Block 2004); the ability to apply picoNewton-level forces to micron-sized molecules while simultaneously measuring displacement with nanometer-level precision is now applied to the study of molecular motors at the single-molecule level (Neuman and Block 2004), the physics of colloids and mesoscopic systems and the mechanical properties of polymers and biopolymers. The widespread use of optical trapping has led to the active pursuit of theoretical and experimental work on fundamental aspects of optical trapping. Therefore, a description of this molecular rotation as a function of optical resonance diffusion and nuclear relaxation may prove to be important in the formulation of theoretical fundamentals for optical trapping.

If the size (including the effective hydrodynamic radius, R) of the molecule which is to be imaged does not change significantly, then the translational diffusion coefficient as defined by the Stokes–Einstein relation (Price 1997; Loman et al. 2010):

$$D = D_{trans} = \frac{k_B T}{6\pi \eta R}. \tag{3.85}$$

As long as the hydrodynamic radius does not change significantly, the translational diffusion coefficient is constant and hence, Eq. (3.77) could be applied to solve this problem and Eq. (3.77) becomes:

$$\frac{\partial P_y}{\partial t} = \frac{D_{trans}}{r^2}\left(\frac{\partial}{\partial r}\left(r^2 \frac{\partial P_y}{\partial r}\right) + \frac{1}{\sin^2\theta}\frac{\partial^2 P_y}{\partial \phi^2} + \frac{1}{\sin\theta}\frac{\partial}{\partial \theta}\left(\sin\theta \frac{\partial P_y}{\partial \theta}\right)\right)$$

$$+\frac{P_0}{T_0T_1}\Xi\xi_p(\vec{r},t). \quad (3.86)$$

Since r = R, the term $\frac{\partial P_y}{\partial r}$ disappears and Eq. (3.86) may be written as follows:

$$\frac{\partial P_y}{\partial t}=\frac{D_{trans}}{R^2}\left(\frac{1}{\sin^2\theta}\frac{\partial^2 P_y}{\partial\phi^2}+\frac{1}{\sin\theta}\frac{\partial}{\partial\theta}\left(\sin\theta\frac{\partial P_y}{\partial\theta}\right)\right)+\frac{P_0}{T_0T_1}\Xi\xi_p(\vec{r},t). \quad (3.87)$$

The rotational diffusion coefficient is related to the translational diffusion coefficient according to the following expression (Martínez 1987; Loman et al. 2010):

$$D_r=\frac{D_{trans}}{R^2}. \quad (3.88)$$

Equation (3.86) then becomes:

$$\frac{\partial P_y}{\partial t}=D_r\left(\frac{1}{\sin^2\theta}\frac{\partial^2 P_y}{\partial\phi^2}+\frac{1}{\sin\theta}\frac{\partial}{\partial\theta}\left(\sin\theta\frac{\partial P_y}{\partial\theta}\right)\right)+\frac{P_0}{T_0T_1}\Xi\xi_p(\vec{r},t). \quad (3.89)$$

A common method of solution of this differential equation is the use of method of separation of variables and setting the electric field $\xi(\vec{r},t)$ to zero (provided the assumption is experimentally feasible). This method has the form (Cowan 1997):

$$P_y(\phi,\theta,t)=\sum_{l=1}^{\infty}\sum_{m=-1}^{l}\alpha_m^l(t)Y_l^m(\phi,\theta), \quad (3.90)$$

where $\alpha_m^l(t)$ represent the time-dependent coefficient and $Y_l^m(\phi,\theta)$ represents the Spherical harmonics. For a dimensionless parameter l, the effect of the rotational Laplacian is such that

$$\nabla^2 Y_l^m(\phi,\theta)=-l(l+1)Y_l^m(\phi,\theta). \quad (3.91)$$

Hence, the following equations ensued

$$\frac{d\alpha_m^l(t)}{dt}=D_r l(l+1)\alpha_m^l(t)=\frac{D}{R^2}l(l+1)\alpha_m^l(t), \quad (3.92)$$

$$\alpha_m^l(t)=\alpha_m^l(0)\exp\left(-\frac{t}{\tau_r}\right), \quad (3.93)$$

where τ_r is the rotational correlation time and it has the following definition:

$$\frac{1}{\tau_r}=D_r l(l+1)=\frac{D}{R^2}l(l+1). \quad (3.94)$$

3.3.4 General Mathematical Solutions to Model Equations

The process of problem solving using mathematical models comprises the setting up models in terms of differential equations, seeking mathematical solutions to the equations and then interpreting the result in physical terms (Velten et al. 2024). The immediate section has taken care of the task of model setup. It is very important to now state general mathematical solutions to some of these model equations. Such solutions could then be applied directly to specific medical problems or physiological processes or modified slightly in order to fit into the problem at hand.

3.3.4.1 Solutions to the Time-Independent NMR Bloch Equation in Terms of Bessel Functions

For flow problems in which the fluid velocity is spatially dependent, the spin dynamics of the fluid based on the time-independent Bloch NMR flow equation given in Eq. (3.16). Then, within a rotating frame of reference, Larmor condition ($f_0 = \gamma B - \omega = 0$) must hold (Awojoyogbe 2002, 2003, 2004). Meanwhile, provided that the bulk protons are moving at a variable velocity v(x) and if the RF field $B_1(x)$ is applied such that M_y is sampled at maximum magnitude, $M_0 \approx 0$. Therefore, Eq. (3.16) reduces to:

$$v^2(x)\frac{d^2M_y}{dx^2} + v(x)\left(T_0 + \frac{dv}{dx}\right)\frac{dM_y}{dx} + (\gamma^2 B_1^2(x) + T_g)M_y = 0, \tag{3.95}$$

where T_0 and T_g retains their usual meaning. Assuming that the fluid velocity and the applied field have the following forms:

$$v(x) = \frac{x}{\delta} \text{ and } B_1(x) = \zeta G x, \tag{3.96}$$

where ζ is a dimensionless parameter. Equation (3.95) becomes

$$\frac{x^2}{\delta^2}\frac{d^2M_y}{dx^2} + \frac{x}{\delta}\left(T_0 + \frac{1}{\delta}\right)\frac{dM_y}{dx} + \left(\zeta^2\gamma^2G^2x^2 + T_g\right)M_y = 0, \tag{3.97}$$

$$x^2\frac{d^2M_y}{dx^2} + x(1+\delta T_0)\frac{dM_y}{dx} + \left(\zeta^2\gamma^2G^2\delta^2x^2 + \delta^2T_g\right)M_y = 0, \tag{3.98}$$

$$x^2\frac{d^2M_y}{dx^2} + x(1+\delta T_0)\frac{dM_y}{dx} + (\zeta^2\gamma^2G^2\delta^2x^2 + \delta^2T_g)M_y = 0. \tag{3.99}$$

Equation (3.99) is an equation transformable to Bessel differential equation (DE) and the solution is (Wylie and Barrett 1982):

$$M_y(x) = x^{\alpha}(C_1 J_n(\zeta\gamma G\delta x) + C_2 Y_n(\zeta\gamma G\delta x)), \tag{3.100}$$

where $\alpha = -\frac{\delta T_0}{2}$, $n = \frac{\delta}{2}\sqrt{T_0^2 - 4T_g}$; C_1 and C_2 are constants. J_n is the Bessel functions of the first kind, while Y_n are the Bessel functions of the second kind. Since we require that the transverse magnetisation be finite at all points x, it follows that $C_2 = 0$. Hence,

$$M_y(x) = C_1 x^{\alpha} J_n(\zeta\gamma G\delta x). \tag{3.101}$$

3.3.4.2 Solutions to the Time-independent NMR Bloch Equation in Terms of Legendre Functions

If the dynamics of hydrogen atom within the fluid being investigated under the influence of radio frequency (RF) magnetic field at Lamour frequency is employed, Eq. (3.16) holds as well and as in earlier sections, under the conditions that when the RF $B_1(x)$ field is applied and the transverse magnetisation M_y, has a highest possible magnitude, the equilibrium magnetisation $M_0 \approx 0$. Hence, for variable velocity, and given that the following condition holds

$$\gamma^2 B_1^2(x) \ll \frac{1}{T_1 T_2}, \tag{3.102}$$

Equation (3.16) becomes:

$$v^2(x)\frac{d^2 M_y}{dx^2} + v(x)\left(T_0 + \frac{dv}{dx}\right)\frac{dM_y}{dx} + T_g M_y = 0. \tag{3.103}$$

Meanwhile, a different assumption for the nature of the fluid velocity is assumed and it is given as follows:

$$\frac{1}{v(x)}\left(T_0 + \frac{dv}{dx}\right) = \frac{1}{l}\cot\left(\frac{x}{l}\right), \tag{3.104}$$

where $l = l(x)$ is an experimental parameter and $\cot\left(\frac{x}{l}\right)$ is a cotangent function of $\left(\frac{x}{l}\right)$. Since the expression of Eq. (3.104) implies that the fluid velocity is a function of relaxation times, we can assume that v^2 is directly proportional to T_g such that

$$\frac{v^2(x)}{T_g} = \frac{1}{n(n+1)} l^2(x), \tag{3.103}$$

where $\frac{1}{n(n+1)}$ is the constant of proportion (n is a positive integer parameter). Hence, from Eqs. (3.105), (3.103) becomes:

$$\frac{d^2M_y}{dx^2} + \frac{1}{l}\cot\left(\frac{x}{l}\right)\frac{dM_y}{dx} + \frac{1}{l^2}n(n+1)_{M_y=0}, \quad (3.106)$$

$$\sin\left(\frac{x}{l}\right)\frac{d^2M_y}{dx^2} + \frac{1}{l}\cos\left(\frac{x}{l}\right)\frac{dM_y}{dx} + \frac{1}{l^2}\sin\left(\frac{x}{l}\right)n(n+1)M_y = 0, \quad (3.107)$$

$$\frac{d}{dx}\left(\sin\left(\frac{x}{l}\right)\frac{dM_y}{dx}\right) + \frac{1}{l^2}\sin\left(\frac{x}{l}\right)n(n\ +\ 1)_{M_y=0}. \quad (3.108)$$

Defining $\left(\frac{x}{l}\right)$ as the phase of the NMR signal and setting $\sigma = \cos\left(\frac{x}{l}\right)$, a Legendre differential equation is thus obtained from Eq. (3.108) as follows:

$$\frac{d}{d\sigma}\left[\left(1-\sigma^2\right)\frac{dM_y}{d\sigma}\right] + n(n+1)M_y = 0, \quad (3.109)$$

$$\left(1-\sigma^2\right)\frac{d^2M_y}{d\sigma^2} - 2\sigma\frac{dM_y}{d\sigma} + n(n+1)M_y = 0. \quad (3.110)$$

The solution of Eq. (3.110) is of the form:

$$M_{yn}(\sigma) = c_1P_n(\sigma) + c_2Q_n(\sigma), \quad (3.111)$$

where $P_n(\sigma)$ and $Q_n(\sigma)$ are the Legendre polynomials of the first kind and second kind respectively; C_1 and C_2 are constants. It is worthy of note that $P_n(\sigma)$ and $Q_n(\sigma)$ are two linearly independent solutions to Eq. (3.110). For the transverse magnetisation to be finite when σ very large C_2 becomes must be equal to zero. Therefore, the final solution is given as:

$$M_{yn}(\sigma) = C_1P_n(\sigma). \quad (3.112)$$

Equation (3.112) represents NMR transverse magnetisation as functions of Legendre polynomials. Traditionally, n is a positive integer and may be defined as follows:

$$n = \frac{T_1}{T_2}. \quad (3.113)$$

The experimental values for the gradient pulse parameters may be obtained by setting:

$$\gamma G\tau = \frac{1}{l}, \quad (3.114)$$

where γ is the gyromagnetic ratio, G is the gradient pulse magnitude while τ is the RF gradient pulse duration.

3.3.4.3 Solutions to the Time-dependent NMR Bloch Equation in Terms of Legendre Functions

If the dynamics of hydrogen atom within the fluid is not spatially dependent, then under the influence of a time-dependent radio frequency (RF) magnetic field at Lamour frequency has been described (Dada and Awojoyogbe 2021). Just like in earlier sections, if the RF $B_1(t)$ field is applied and the transverse magnetisation M_y has a highest possible magnitude, the equilibrium magnetisation $M_0 \approx 0$, and we can write:

$$\frac{d^2M_y}{dt^2} + T_0\frac{dM_y}{dt} + (T_g + \gamma^2 B_1^2(t))M_y = 0. \tag{3.115}$$

If the non-adiabatic condition (Awojoyogbe 2002, 2004) holds

$$\gamma^2 B_1^2(t) \ll \frac{1}{T_1 T_2}, \tag{3.116}$$

Equation (3.115) becomes:

$$\frac{d^2M_y}{dt^2} + T_0\frac{dM_y}{dt} + T_g M_y = 0. \tag{3.117}$$

If the following assumptions are true

$$\begin{aligned} T_0 &= k_0 \cot(k_0 t), \\ T_g &= k_0^2 n(n+1), \end{aligned} \tag{3.118}$$

where k_0 and k_1 are experimental time constants. Therefore, with the use of these expressions, Eq. (3.117) becomes:

$$\frac{d^2M_y}{dt^2} + k_0 \cot(k_0 t)\frac{dM_y}{dx} + k_0^2 n(n+1)M_y = 0, \tag{3.119}$$

$$\sin(k_0 t)\frac{d^2M_y}{dt^2} + k_0 \cos(k_0 t)\frac{dM_y}{dt} + k_0^2 \sin(k_0 t)n(n+1)M_y = 0, \tag{3.120}$$

$$\frac{d}{dt}\left(\sin(k_0 t)\frac{dM_y}{dt}\right) + k_0^2 \sin(k_0 t)n(n+1)M_y = 0. \tag{3.121}$$

Setting

$$\upsilon = \cos(k_0 t). \tag{3.122}$$

Equation (3.121) then becomes:

$$
\begin{aligned}
&- k_0 \sin(k_0 t)\frac{d}{d\upsilon}\left(\sin(k_0 t)(-k_0 \cos(k_0 t))\frac{dM_y}{d\upsilon}\right) \\
&\quad + k_0^2 \sin(k_0 t) n(n+1) M_y = 0, \\
&\frac{d}{d\upsilon}\left(\sin^2 k_0 t \frac{dM_y}{d\upsilon}\right) k_0^2 \sin k_0 t \\
&\quad + k_0^2 \sin k_0 t n(n+1) M_y = 0, \\
&\frac{d}{d\upsilon}\left[(1-\upsilon^2)\frac{dM_y}{d\upsilon}\right] + n(n+1)M_y = 0, \qquad (3.123)
\end{aligned}
$$

$$
(1-\upsilon^2)\frac{d^2 M_y}{d\upsilon^2} - 2\upsilon \frac{dM_y}{d\upsilon} + n(n+1)M_y = 0. \qquad (3.124)
$$

The solution of Eq. (3.124) is of the form:

$$
M_{yn}(\upsilon) = A_1 P_n(\upsilon) + A_2 Q_n(\upsilon), \qquad (3.125)
$$

where $P_n(\upsilon)$ and $Q_n(\upsilon)$ are the Legendre polynomials of the first kind and second kind respectively; A_1 and A_2 are constants. For the transverse magnetisation to be finite when υ very large, $A_2 = 0$. Therefore, the final solution is given as:

$$
M_{yn}(\upsilon) = C_1 P_n(\upsilon). \qquad (3.126)
$$

3.3.5 Mathematical Solutions to Model Equations of Specific Imaging Problem

Thus far, solutions to general Bloch NMR flow equations have been presented. These solutions are not specifically solved for any particular molecular imaging problem in medical challenges. In the following sections, model equations are solved specifically for various magnetic resonance imaging problems. The first part would deal with several computational imaging problems (disease imaging and overcoming challenges in the use of contrast agents) while the second part deals with development of theranostics methods for thermal treatment in the presence of MR imaging. Meanwhile, before the solutions are presented, the background to each problem would be given and justifications for choice of model equation are appropriately presented.

3.3.5.1 Development of Field Cycling Susceptibility Method for Molecular Imaging

Magnetic susceptibility is the magnetic response of a tissue when in a magnetic field. Susceptibility weighted magnetic resonance imaging (SWI) is a relatively new MR sequence described by a team of researchers in recent studies (Haacke et al. 2004). SWI shows susceptibility differences in tissues such as deoxygenated blood in veins, acute hemorrhage, acute thrombus, calcium, hemosiderin, and iron exhibit susceptibility differences from their surroundings (Naik et al. 2014).Most tissues in the human body possess unique magnetic relative susceptibility differences in comparison to the surrounding tissues. Biological materials such as partially deoxygenated venous blood, clot, calcium and air/tissue interfaces have magnetic susceptibilities which cannot be differentiated using chemical shifteffects.

In the case where information concerning various species occupying the same voxel is desired, imaging scientists usually obtain the spectral information by collecting a time series of data and then performing Fourier transformation of the data (Haacke et al. 2004); a process known as chemical shift imaging. However, in the scenario where only a single element or tissue component is present in a voxel (homogeneous medium), or if there is a dominant element in each voxel, it is possible in some circumstances to extract spectral information from only the phase. The phase image extracted from this information can then be used to separate the dominantspectral information on a pixel-by-pixel basis. Thisconcept has been quite successful in single point water/fat separation and velocity imaging using phase inMR angiography (Bryant et al. 1984; Haacke et al. 2004).

The challenge with phase imageshas generally been the presence of background localfields that confound the effects of local phase changes intissue. In order to address this problem, there is a need to develop a computational method which could extract spectral information from variations in magnetic susceptibility with consideration of molecular flow and the effect of local fields on NMR relaxation.

Field cycling methods are NMR techniques which are based on obtaining the frequency (or magnetic field) dependence of relaxation times (Kimmich and Anoardo 2004). This has been widely used in relaxometry (measurement of relaxation times) and in fact, field cycling relaxometry is a NMR technique which allows imaging scientists to cover several decades of the frequencywith the same instrument. In contemporary measurements, emphasis is on the spin–lattice relaxation times (T_1). However, transverse relaxation and effects due to residual dipolar couplings can also prove to be source of useful supplementary information. Hence, this section intends to bring in susceptibility mapping into NMR imaging (similar to the methods employed in field cycling techniques).

As a starting point, Eq. (3.11) is considered and if the applied radio frequency is not dependent on the sampling time t, the following expression may be written

$$\omega_1(x) = \gamma B_1(x). \tag{3.127}$$

Consequently, Eq. (3.11) becomes:

$$v^2(x)\frac{\partial^2 M_y}{\partial x^2} + 2v(x)\frac{\partial^2 M_y}{\partial x \partial t} + v(x)\left(T_o + \frac{dv(x)}{dx}\right)\frac{\partial M_y}{\partial x} + T_o\frac{\partial M_y}{\partial t} + \frac{\partial^2 M_y}{\partial t^2} + (T_g + \omega_1^2(x))M_y = \frac{M_o}{T_1}\omega_1(x), \quad (3.128)$$

where $T_o = \frac{1}{T_1} + \frac{1}{T_2}, T_g = \frac{1}{T_1 T_2}$.

The total magnetic field magnetic field in a typical NMR system is given as (Cowan 1997; Haacke et al. 1999)

$$B(x) = B_0 + B_1(x), \quad (3.129)$$

where B_0 is the static magnetic field. In the presence of RF magnetic gradient given as

$$B_1(x) = Gx, \quad (3.130)$$

The Larmor frequency becomes a function of x as given by (Gupta et al. 2014):

$$\omega(x) = \gamma B_0 + \gamma Gx = \omega_0 + \omega_1. \quad (3.131)$$

From Eq. (3.131), the following expression may be written

$$x = \frac{\omega - \omega_0}{\gamma G}. \quad (3.132)$$

Making transformations from the x-domain to the ω-domain, the following standard equations (Wylie and Barrett 1982) may be written:

$$\frac{\partial M_y}{\partial x} = \frac{\partial M_y}{\partial \omega}\frac{d\omega}{dx} = \gamma G\frac{\partial M_y}{\partial \omega}, \quad (3.133)$$

$$\frac{\partial^2 M_y}{\partial x^2} = \frac{\partial^2 M_y}{\partial \omega^2}\left(\frac{d\omega}{dx}\right)^2 + \frac{\partial M_y}{\partial \omega}\frac{d^2\omega}{dx^2} = (\gamma G)^2\frac{\partial^2 M_y}{\partial \omega^2} + \frac{\partial M_y}{\partial \omega}(0) = (\gamma G)^2\frac{\partial^2 M_y}{\partial \omega^2}, \quad (3.134)$$

$$\frac{\partial^2 M_y}{\partial x \partial t} = \frac{\partial}{\partial t}\left(\frac{\partial M_y}{\partial x}\right) = \frac{\partial}{\partial t}\left(\frac{\partial M_y}{\partial \omega}\frac{d\omega}{dx}\right) = \frac{\partial}{\partial t}\left(\gamma G\frac{\partial M_y}{\partial \omega}\right) = \gamma G\frac{\partial^2 M_y}{\partial \omega \partial t}. \quad (3.135)$$

Substituting for Eqs. (3.133)–(3.135) in (3.128) gives

$$v^2(\gamma G)^2\frac{\partial^2 M_y}{\partial \omega^2} + 2v\gamma G\frac{\partial^2 M_y}{\partial \omega \partial t} + v\left(T_o + \frac{dv}{dx}\right)\left(\gamma G\frac{\partial M_y}{\partial \omega}\right) + T_o\frac{\partial M_y}{\partial t} + \frac{\partial^2 M_y}{\partial t^2} + (T_g + (\omega - \omega_0)^2)M_y = \frac{M_o}{T_1}(\omega - \omega_0). \quad (3.136)$$

If the fluid velocity has the following form (Laub 2000; Haacke et al. 2000; Awojoyogbe and Dada 2013):

$$v = \alpha_0 \frac{x}{\delta}, \tag{3.137}$$

where δ is the pulse duration and is a dimensionless weighting parameter. According to Eq. (3.132), the fluid velocity becomes

$$v = \frac{\alpha_0(\omega - \omega_0)}{\gamma G \delta}, \tag{3.138}$$

$$v\left(T_o + \frac{dv}{dx}\right) = \alpha_0 \frac{\omega - \omega_0}{\gamma G \delta}\left(T_o + \frac{\alpha_0}{\delta}\right). \tag{3.139}$$

Hence, Eq. (3.136) may be written as:

$$\begin{aligned}&\left(\alpha_0 \frac{\omega - \omega_0}{\gamma G \delta}\right)^2 (\gamma G)^2 \frac{\partial^2 M_y}{\partial \omega^2} + 2\alpha_0 \left(\frac{\omega - \omega_0}{\gamma G \delta}\right)\gamma G \frac{\partial^2 M_y}{\partial \omega \partial t}\\ &+ \alpha_0 \left(\frac{\omega - \omega_0}{\gamma G \delta}\right)\left(T_o + \frac{\alpha_0}{\delta}\right)\left(\gamma G \frac{\partial M_y}{\partial \omega}\right)\\ &+ T_o \frac{\partial M_y}{\partial t} + \frac{\partial^2 M_y}{\partial t^2} + (T_g + (\omega - \omega_0)^2)M_y = \frac{M_o}{T_1}(\omega - \omega_0),\end{aligned} \tag{3.140}$$

$$\begin{aligned}&\left(\alpha_0 \frac{\omega - \omega_0}{\delta}\right)^2 \frac{\partial^2 M_y}{\partial \omega^2} + 2\alpha_0 \left(\frac{\omega - \omega_0}{\delta}\right)\frac{\partial^2 M_y}{\partial \omega \partial t}\\ &+ \alpha_0 \left(\frac{T_o(\omega - \omega_0)}{\delta} + \frac{\alpha_0(\omega - \omega_0)}{\delta^2}\right)\frac{\partial M_y}{\partial \omega}\\ &+ T_o \frac{\partial M_y}{\partial t} + \frac{\partial^2 M_y}{\partial t^2} + (T_g + (\omega - \omega_0)^2)M_y = \frac{M_o}{T_1}(\omega - \omega_0).\end{aligned} \tag{3.141}$$

If Eq. (3.141) is multiplied all through by $\frac{\delta^2}{\alpha_0^2}$, the following expression is obtained

$$\begin{aligned}&(\omega - \omega_0)^2 \frac{\partial^2 M_y}{\partial \omega^2} + \frac{2\delta}{\alpha_0}(\omega - \omega_0)\frac{\partial^2 M_y}{\partial \omega \partial t} + \left(T_o \frac{\delta}{\alpha_0}(\omega - \omega_0) + (\omega - \omega_0)\right)\frac{\partial M_y}{\partial \omega}\\ &+ T_o \frac{\delta^2}{\alpha_0^2}\frac{\partial M_y}{\partial t} + \frac{\delta^2}{\alpha_0^2}\frac{\partial^2 M_y}{\partial t^2} + \frac{\delta^2}{\alpha_0^2}(T_g + (\omega - \omega_0)^2)M_y = \frac{M_o}{T_1 \alpha_0^2}(\omega - \omega_0)\delta^2,\end{aligned} \tag{3.142}$$

$$\begin{aligned}&(\omega - \omega_0)^2 \frac{\partial^2 M_y}{\partial \omega^2} + \frac{2\delta}{\alpha_0}(\omega - \omega_0)\frac{\partial^2 M_y}{\partial \omega \partial t} + (\omega - \omega_0)\left(1 + \frac{\delta T_o}{\alpha_0}\right)\frac{\partial M_y}{\partial \omega}\\ &+ T_o \frac{\delta^2}{\alpha_0^2}\frac{\partial M_y}{\partial t} + \frac{\delta^2}{\alpha_0^2}\frac{\partial^2 M_y}{\partial t^2} + \frac{\delta^2}{\alpha_0^2}(T_g + (\omega - \omega_0)^2)M_y = \frac{M_o}{T_1 \alpha_0^2}(\omega - \omega_0)\delta^2.\end{aligned} \tag{3.143}$$

The solution to this expression is represented as

$$M_y(\omega, t). \tag{3.144}$$

The relationship between Eq. (3.144) and the dynamic susceptibility is given as (Abragam 1961, 1996; Cowan 1997; Slichter 2013):

$$\chi'(\omega) = \frac{M_x(\omega, t)}{2H_1} = \frac{\mu_0 M_x(\omega, t)}{2B_1} = \frac{\mu_0 \gamma M_x(\omega, t)}{2\omega_1}, \tag{3.145}$$

$$\chi''(\omega) = \frac{M_y(\omega, t)}{2H_1} = \frac{\mu_0 M_y(\omega, t)}{2B_1} = \frac{\mu_0 \gamma M_y(\omega, t)}{2\omega_1}, \tag{3.146}$$

where $\chi'(\omega)$ and $\chi''(\omega)$ are the real (in-phase response) and imaginary (quadrature response) respectively; and μ_0 is the permeability in free space. In a rotating frame, Eqs. (3.145) and (3.146) are effectively the same according to Dada and Awojoyogbe (2021). Since the susceptibility must be measurable at every point, the complex contribution to the dynamic susceptibility may be neglected such that:

$$\chi'(\omega) = \frac{\mu_0 \gamma}{2\omega_1} M_y(\omega, t). \tag{3.147}$$

Steady State Approximation

Finding solution to Eq. (3.143) is quite challenging because of the variable coefficients. Hence, if a steady state condition holds, all time dependence in Eq. (3.143) becomes insignificant and the field-dependent approximation becomes an ordinary differential equation:

$$\begin{aligned}(\omega - \omega_0)^2 \frac{\partial^2 M_y}{\partial \omega^2} &+ (\omega - \omega_0)\left(1 + \frac{\delta T_o}{\alpha_0}\right)\frac{\partial M_y}{\partial \omega} \\ &+ \frac{\delta^2}{\alpha_0^2}(T_g + (\omega - \omega_0)^2) M_y = \frac{M_o}{T_1 \alpha_0^2}(\omega - \omega_0)\delta^2.\end{aligned} \tag{3.148}$$

From Eqs. (3.127) and (3.132), the following expression may be recalled

$$\omega_1 = \omega - \omega_0, \tag{3.149}$$

$$\frac{dM_y}{d\omega} = \frac{\partial M_y}{\partial \omega_1}\frac{d\omega_1}{d\omega} = \frac{\partial M_y}{\partial \omega_1}, \tag{3.150}$$

$$\frac{d^2 M_y}{d\omega^2} = \frac{d^2 M_y}{d\omega_1^2}\left(\frac{d\omega_1}{d\omega}\right)^2 + \frac{dM_y}{d\omega_1}\frac{d^2\omega_1}{d\omega^2} = \frac{d^2 M_y}{d\omega_1^2}. \tag{3.151}$$

Equation (3.148) becomes:

$$\omega_1^2 \frac{\partial^2 M_y}{\partial \omega_1^2} + \omega_1 \left(1 + \frac{\delta T_o}{\alpha_0}\right) \frac{\partial M_y}{\partial \omega_1} + \frac{\delta^2}{\alpha_0^2}(T_g + (\omega_1)^2) M_y = \frac{M_o}{T_1 \alpha_0^2} \omega_1 \delta^2. \tag{3.152}$$

The equilibrium magnetisation has the following expression (Cowan 1997):

$$M_o = \frac{\chi_0 B_0}{\mu_0}, \tag{3.153}$$

where χ_0 represents the static magnetic susceptibility of the biological system under consideration. Using this expression, Eq. (3.152) becomes:

$$\omega_1^2 \frac{\partial^2 M_y}{\partial \omega_1^2} + \omega_1 \left(1 + \frac{\delta T_o}{\alpha_0}\right) \frac{\partial M_y}{\partial \omega_1} + \frac{\delta^2}{\alpha_0^2}(T_g + (\omega_1)^2) M_y = \frac{\chi_0 B_0}{\alpha_0^2 \mu_0 T_1} \omega_1 \delta^2. \tag{3.154}$$

Once a solution to Eq. (3.154) is obtained, the magnetic susceptibility distribution then follows directly from Eq. (3.147) and may be given as:

$$\chi'(\omega) = \frac{\mu_0 \gamma}{2\omega_1} M_y(\omega), \tag{3.155}$$

where the solution to this differential equation is given as

$$M_y(\omega_1) \text{ or } M_y(\omega). \tag{3.156}$$

Solution to the Steady State Equation

The solution to Eq. (3.154) is not easy to obtain because of the inhomogeneous nature of the equation. If the magnetic susceptibility is to be sampled at varying values of the applied radio frequency ω_1, it may not be helpful to set the right hand side of the equation to zero. Therefore, the inhomogeneous equation must be solved without any sort of reduction. Using a Wolfram Mathematica computer script, the solution to Eq. (3.154) is given as follows:

$$\begin{aligned}
M_y(\omega_1) = {} & (\omega_1)^{-\frac{\delta T_0}{2\alpha_0}} \left\{ C_1 J_{\frac{\varepsilon\delta}{2\alpha_0}}\left(\frac{\delta\omega_1}{\alpha_0}\right) + C_2 Y_{\frac{\varepsilon\delta}{2\alpha_0}}\left(\frac{\delta\omega_1}{\alpha_0}\right) \right\} \\
& + \frac{\pi \chi_0 B_o}{\alpha_0^2 \mu_0 T_1} \omega_1 \delta^2 (2)^{-\frac{\varepsilon\delta}{2\alpha_0} - 2} \left(\omega_1 \frac{\delta}{\alpha_0}\right)^{-\frac{\varepsilon\delta}{2\alpha_0}} \\
& \left\{ (2)^{\varepsilon\delta/\alpha_0} \csc\left(\frac{\pi \varepsilon\delta}{2\alpha_0}\right) \Gamma\left(\frac{\delta T_0 - \varepsilon\delta}{4\alpha_0} + \frac{1}{2}\right) J_{\frac{\varepsilon\delta}{2\alpha_0}}\left(\frac{\delta\omega_1}{\alpha_0}\right)_1 \right. \\
& \left. \tilde{F}_2\left(\left(\frac{\delta T_0 - \varepsilon\delta}{4\alpha_0} + \frac{1}{2}\right); 1 - \frac{\varepsilon\delta}{2\alpha_0}, \left(\frac{\delta T_0 - \varepsilon\delta}{4\alpha_0} + \frac{3}{2}\right); -\frac{\delta^2 \omega_1^2}{4\alpha_0^2}\right) \right\} \\
& + \frac{\pi \chi_0 B_o}{\alpha_0^2 \mu_0 T_1} \omega_1 \delta^2 (2)^{-\frac{\varepsilon\delta}{2\alpha_0} - 2} \left(\frac{\delta\omega_1}{\alpha_0}\right)^{\frac{\varepsilon\delta}{2\alpha_0}}
\end{aligned}$$

$$\left\{\Gamma\left(\frac{\delta T_0+\varepsilon\delta+2\alpha_0}{4\alpha_0}\right)Y_{\frac{\varepsilon\delta}{2\alpha_0}}\left(\frac{\delta\omega_1}{\alpha_0}\right)_1\right.$$
$$\left.\widetilde{F}_2\left(\left(\frac{\delta T_0+\varepsilon\delta}{4\alpha_0}+\frac{1}{2}\right);\left(\frac{\varepsilon\delta}{2\alpha_0}+1\right),\left(\frac{\delta T_0+\varepsilon\delta}{4\alpha_0}+\frac{3}{2}\right);-\frac{\delta^2\omega_1^2}{4\alpha_0^2}\right)\right\}$$
$$-\frac{\pi\chi_0 B_o}{\alpha_0^2\mu_0 T_1}\omega_1\delta^2(2)^{-\frac{\varepsilon\delta}{2\alpha_0}-2}\left(\frac{\delta\omega_1}{\alpha_0}\right)^{\frac{\varepsilon\delta}{2\alpha_0}}$$
$$\left\{\cot\left(\frac{\pi\varepsilon\delta}{2\alpha_0}\right)\Gamma\left(\frac{\delta T_0+\varepsilon\delta}{4\alpha_0}+\frac{1}{2}\right)J_{\frac{\varepsilon\delta}{2\alpha_0}}\left(\frac{\delta\omega_1}{\alpha_0}\right)_1\right.$$
$$\left.\widetilde{F}_2\left(\left(\frac{\delta T_0+\varepsilon\delta}{4\alpha_0}+\frac{1}{2}\right);\left(\frac{\varepsilon\delta}{2\alpha_0}+1\right),\left(\frac{\delta T_0+\varepsilon\delta}{4\alpha_0}+\frac{3}{2}\right);-\frac{\delta^2\omega_1^2}{4\alpha_0^2}\right)\right\},\tag{3.157}$$

where $\varepsilon=\delta\sqrt{T_0^2-4T_g}$, B_0 is the static magnetic field, χ_0 is the static magnetic susceptibility, μ_0 is the magnetic permeability in free space, csc is the Cosecant function, Γ is the Gamma function, $J_{\frac{\varepsilon\delta}{2\alpha_0}}(\omega_1\delta)$ is the Bessel function of first kind, $Y_{\frac{\varepsilon\delta}{2\alpha_0}}(\omega_1\delta)$ is the Bessel function of the second kind, ${}_1\widetilde{F}_2$ is the regularized generalized hypergeometric function while C_1 and C_2 are both constants. Therefore, from Eq. (3.155), the following expression is obtained

$$\chi'(\omega_1)=\frac{\gamma\mu_0}{2}(\omega_1)^{-\frac{\delta T_0}{2\alpha_0}}\left\{C_1 J_{\frac{\varepsilon\delta}{2\alpha_0}}\left(\frac{\delta\omega_1}{\alpha_0}\right)+C_2 Y_{\frac{\varepsilon\delta}{2\alpha_0}}\left(\frac{\delta\omega_1}{\alpha_0}\right)\right\}$$
$$+\frac{\pi\gamma\chi_0 B_o}{2\alpha_0^2 T_1}\delta^2(2)^{-\frac{\varepsilon\delta}{2\alpha_0}-2}\left(\omega_1\frac{\delta}{\alpha_0}\right)^{-\frac{\varepsilon\delta}{2\alpha_0}}$$
$$\left\{(2)^{\varepsilon\delta/\alpha_0}\csc\left(\frac{\pi\varepsilon\delta}{2\alpha_0}\right)\Gamma\left(\frac{\delta T_0-\varepsilon\delta}{4\alpha_0}+\frac{1}{2}\right)J_{\frac{\varepsilon\delta}{2\alpha_0}}\left(\frac{\delta\omega_1}{\alpha_0}\right)_1\right.$$
$$\left.\widetilde{F}_2\left(\left(\frac{\delta T_0-\varepsilon\delta}{4\alpha_0}+\frac{1}{2}\right);1-\frac{\varepsilon\delta}{2\alpha_0},\left(\frac{\delta T_0-\varepsilon\delta}{4\alpha_0}+\frac{3}{2}\right);-\frac{\delta^2\omega_1^2}{4\alpha_0^2}\right)\right\}$$
$$+\frac{\pi\gamma\chi_0 B_o}{\alpha_0^2 T_1}\delta^2(2)^{-\frac{\varepsilon\delta}{2\alpha_0}-2}\left(\frac{\delta\omega_1}{\alpha_0}\right)^{\frac{\varepsilon\delta}{2\alpha_0}}$$
$$\left\{\Gamma\left(\frac{\delta T_0+\varepsilon\delta+2\alpha_0}{4\alpha_0}\right)Y_{\frac{\varepsilon\delta}{2\alpha_0}}\left(\frac{\delta\omega_1}{\alpha_0}\right)_1\right.$$
$$\left.\widetilde{F}_2\left(\left(\frac{\delta T_0+\varepsilon\delta}{4\alpha_0}+\frac{1}{2}\right);\left(\frac{\varepsilon\delta}{2\alpha_0}+1\right),\left(\frac{\delta T_0+\varepsilon\delta}{4\alpha_0}+\frac{3}{2}\right);-\frac{\delta^2\omega_1^2}{4\alpha_0^2}\right)\right\}$$
$$-\frac{\pi\gamma\chi_0 B_o}{\alpha_0^2 T_1}\delta^2(2)^{-\frac{\varepsilon\delta}{2\alpha_0}-2}\left(\frac{\delta\omega_1}{\alpha_0}\right)^{\frac{\varepsilon\delta}{2\alpha_0}}$$
$$\left\{\cot\left(\frac{\pi\varepsilon\delta}{2\alpha_0}\right)\Gamma\left(\frac{\delta T_0+\varepsilon\delta}{4\alpha_0}+\frac{1}{2}\right)J_{\frac{\varepsilon\delta}{2\alpha_0}}\left(\frac{\delta\omega_1}{\alpha_0}\right)_1\right.$$

$$\widetilde{F}_2\left(\left(\frac{\delta T_0+\varepsilon\delta}{4\alpha_0}+\frac{1}{2}\right);\left(\frac{\varepsilon\delta}{2\alpha_0}+1\right),\left(\frac{\delta T_0+\varepsilon\delta}{4\alpha_0}+\frac{3}{2}\right);-\frac{\delta^2\omega_1^2}{4\alpha_0^2}\right)\Bigg\}. \tag{3.158}$$

3.3.5.2 Analytical Method for Monitoring the Dynamics of Responsive Contrast Agents

Responsive contrast agents are MR diagnostic agents whose contrasting features are sensitive to a given physic-chemical variable that characterizes the micro-environment in which the probe is found after administration (Aime et al. 2005). The crucial diagnostic parameters are pH, temperature, enzyme activity, redox potential and concentration of specific ions, and low weight metabolites.

However, their clinical use is still far-off because accurate measurement of one of the parameters above with a given probe requires a precise knowledge of the local concentration of the contrast medium in the region of interest. If this can be achieved, we can then safely attribute observed changes in relaxation rate of water protons to changes in the parameters to be measured (Aime et al. 2005, 2007).

Although novel classes of MRI paramagnetic contrast media which address this problem have been successfully developed (Aime et al. 2005), a possibility exists that the challenges above may actually be a blessing in disguise. That is, changes in the local concentration of the contrast medium in the region of interest may serve as additional source tissue contrast but provided we can infer the local concentration changes directly from measures MRI signal.

Bloch NMR Flow Model for Contrast Agents Moving in 1D Tissue Spaces

In order to address this problem, a computational model which relates the measured MRI signal in terms of the transverse magnetisation to the tissue temperature and concentration of the contrast agents is considered. Such model may be tested in-vitro for setting reference points for parameter relationship and then used to predict contrast media dynamics in-vivo. Since the contrast agents are meant to improve MR signal at different points within a physiological system, it is important we measure a spatial transverse magnetisation in relation to the expected variations in the relaxation times. In such case, the constant velocity time-independent Bloch NMR flow equation given in (Dada and Awojoyogbe 2021) shall be employed. Within the small RF limit, $\gamma^2 B_1^2 T_1 T_2 \ll 1$, we can write:

$$v^2\frac{d^2M_y}{dx^2}+vT_o\frac{dM_y}{dx}+T_gM_y=\frac{M_o}{T_1}\gamma B_1(x), \tag{3.159}$$

where T_0, T_g and all other terms retain their usual meanings. For this formulation, it is assumed that the resonance condition exists at Larmor frequency as well. For paramagnetic complexes, the Curie law becomes (Dada and Awojoyogbe 2021):

$$M_o = \rho \frac{I(I+1)\gamma^2 \hbar^2 B_0}{3k_B T}, \tag{3.160}$$

where ρ is the spin density [which cannot be altered in living beings (Brix et al. 2008)], k_B is the Boltzmann's constant, T is the measured body temperature, $h = 2\pi\hbar$ is the Planck constant, B_0 is the polarizing field and I is the spin quantum number of the paramagnetic complex to be administered.

If the applied RF field is related to the magnetic gradient field G_x and given as follows

$$\gamma B_1(x) = \gamma G_x(t)x, \tag{3.161}$$

$$v^2 \frac{d^2 M_y}{dx^2} + vT_o \frac{dM_y}{dx} + T_g M_y = \frac{\rho}{T_1} \frac{I(I+1)\gamma^3 \hbar^2 B_0}{3k_B T} G_x(t)x. \tag{3.162}$$

The complimentary function, which is a solution to the homogeneous part of Eq. (3.162), is given as:

$$M_{yc} = C_1 e^{-\frac{T_O}{2v}(1-\alpha)x} + C_2 e^{-\frac{T_O}{2v}(1+\alpha)x}, \tag{3.163}$$

where $\alpha = \sqrt{1 - \frac{4T_g}{T_0^2}}$ is a dimensionless relaxation weighting parameter. Due to the nature of the right hand side of Eq. (3.162), the following particular solution may be assumed:

$$M_{yp} = K_1 x + K_0. \tag{3.164}$$

Substituting for Eq. (3.164) in Eq. (3.162):

$$vT_o K_1 + T_g K_1 x + T_g K_0 = \frac{\rho}{T_1} \frac{I(I+1)\gamma^3 \hbar^2 B_0}{3k_B T} G_x(t)x. \tag{3.165}$$

Equating the coefficients of like powers of x, the following expressions are obtained

$$K_0 = 0,\ K_1 = \frac{\rho}{T_1 T_g} \frac{I(I+1)\gamma^3 \hbar^2 B_0}{3k_B T} G_x(t) = \rho T_2 \frac{I(I+1)\gamma^3 \hbar^2 B_0}{3k_B T} G_x(t). \tag{3.166}$$

Therefore, the general solution ($M_{yc} + M_{yp}$) is given as:

$$M_y(x) = C_1 e^{-\frac{T_O}{2v}(1-\alpha)x} + C_2 e^{-\frac{T_O}{2v}(1+\alpha)x} + \rho T_2 \frac{I(I+1)\gamma^3 \hbar^2 B_0}{3k_B T} G_x(t)x, \tag{3.167}$$

$$M_y(x) - \rho T_2 \frac{I(I+1)\gamma^3 \hbar^2 B_0}{3k_B T} G_x(t)x = e^{-\frac{T_O}{2v}x}\left(C_1 e^{\alpha x} + C_2 e^{-\alpha x}\right), \tag{3.168}$$

$$T_o = -\frac{2v}{x}\ln\left[\frac{M_y(x) - \rho T_2 \frac{I(I+1)\gamma^3\hbar^2 B_0}{3k_B T} G_x(t)x}{(C_1 e^{\alpha x} + C_2 e^{-\alpha x})}\right]. \tag{3.169}$$

Application of Model to Cardiovascular Disease

In contrast-enhanced magnetic resonance angiography (MRA), contrast agents are administered to monitor blood flow in human blood vessels. Cardiovascular diseases occur when blood vessels are blocked by fatty deposits. When a contrast agent is injected, the T_1 of blood is shortened from $T_{1\text{blood}} = 1.2$ s (at $B_0 = 1.5$ T) to less than 100 ms during the first bolus passage (Brix et al. 2008). The relaxation rate as a function of the local contrast agent concentration (C) is given by (Dada and Awojoyogbe 2021):

$$R_1 = \frac{1}{T_1} = \frac{1}{T_{1blood}} + r_1 C. \tag{3.170}$$

If the contrast agent does not have any significant influence on the transverse relaxation time T_2, Eq. (3.169) becomes:

$$r_1 C = -\frac{2v}{x}\ln\left[\frac{M_y(x) - \rho T_2 \frac{I(I+1)\gamma^3\hbar^2 B_0}{3k_B T} G_x(t)x}{(C_1 e^{\alpha x} + C_2 e^{-\alpha x})}\right] - \left(\frac{1}{T_{1blood}} + \frac{1}{T_2}\right). \tag{3.171}$$

3.3.5.3 Computational Diffusion Magnetic Resonance Imaging Based on Time-Dependent Bloch NMR Flow Equation and Bessel Functions for Tissue Imaging

MRI uses a powerful magnetic field along with radio waves and a computer to produce highly detailed "slice-by-slice" images of virtually all internal structures of the body. The results enable physicians to examine parts of the body in minute detail and identify diseases in ways that are not possible with other techniques. For example, MRI is one of the few imaging tools that can see through bone, making it an excellent tool for examining the brain and other soft tissue (Cline et al. 1987).

Pulsed-field gradient experiments provide a straightforward means of obtaining information on the translational motion of nuclear spins. However, the interpretation of the data is complicated by the effects of restricting geometries as in the case of most cancerous tissue and the mathematical concept required to account for this becomes very difficult. Most diffusion magnetic resonance techniques are based on the Stejskal-Tanner formulation usually derived from the Bloch-Torrey partial differential equations by including additional terms to accommodate the diffusion effect (Torrey 1956; Tanner 1978; Price and Kuchel 1991). Despite the early success of this technique, it has been shown that it has important limitations, the most important of which occurs when there is orientation *heterogeneity* of the fibers in the voxel of interest (VOI) (Wiegell et al. 2000).

Traditionally diffusion weighted MRI involves two gradient pulses (Price 1997, 1998). The first gradient pulse will introduce a spatially-dependent phase to each excited spin while for stationary spins, the second gradient phase will cause a reversal of the phase introduced by the first gradient pulse and hence the bipolar gradient will have no net effect. Consequently, there is a phase offset proportional to the distance travelled for spins that are displaced between the applications of the two gradient pulses. This phase dispersion, in turn, attenuates the signal exponentially according to the b-factor and the diffusion coefficient. For unrestricted flow, the generalized magnetic resonance mathematical framework (Chang et al. 1973; Rorschach et al. 1973; Cooper et al. 1974; Taylor and Bushell 1985; Hazlewood et al. 1974, 1991) for relating the signal attenuation, $\frac{S}{S_0}$; the conditional displacement distribution, the pulse duration, δ; the pulse strength, g; results in a Gaussian distribution of the form (Basser et al. 1994; Mori and Barker 1999; Le Bihan et al. 2001)

$$A = \frac{S}{S_0} = e^{-bD}, \tag{3.172}$$

where S_0 is the signal intensity without diffusion weighting while S is the diffusion weighted signal. The b-value identifies the measurement's sensitivity to diffusion and determines the strength and duration of the diffusion gradients and is measured in s/mm^2 (Basser et al. 1994; Mori and Barker 1999)

$$b = \gamma^2 g^2 \delta^2 \left(\Delta - {}^{\delta}/_{3}\right), \tag{3.173}$$

where g is the amplitude of the two diffusion gradient pulses, and Δ is the time between two pulses. Diffusion weighted Imaging (DWI) is useful in the diagnosis and monitoring of various neurological diseases such as cancer, multiple sclerosis and Alzheimer. The most profound clinical impact of diffusion contrast has been in the diagnosis of acute stroke. Diffusion imaging is the extreme sensitivity to motion artifacts and thus, motion correction is an important component of any in vivo multi-shot DWI pulse sequence (Anderson and Gore 1994).

The expression in Eq. (3.173) shows how the variations in the spatial scale parameter (b) and diffusion time (Δ) independently affect the signal attenuation. In a medium in which water does not move freely such that it is confined within internal compartments, the molecules are reflected by obstacles but still free to meander relatively freely throughout the medium (hindered) or experiences higher diffusivity in some directions and lower in others (anisotropic) the simple relationship in Eq. (3.173) no longer holds (Awojoyogbe et al. 2011).

To overcome this difficulty, there is a need to specify the diffusion coefficients as functions of spatial coordinates consistent with the prevalent compartmental conditions which can be analyzed accurately by solving the time-dependent Bloch NMR flow equations analytically. Consequently, this study intends to develop a mathematical formulation of MRI flow sequence in restricted geometry. This method is based on a general second order partial differential equation derived directly from the fundamental Bloch NMR flow equations. The NMR signal is obtained completely in terms of NMR experimental parameters. The process is described based on Bessel functions/properties for distinguishing various tissues. A typical example of brain tumor distinguished from Gray matter, white matter and cerebrospinal fluid is demonstrated. Bessel functions and properties are specifically needed to show the direct effect of the instantaneous velocity on the NMR signal originating from normal and abnormal tissues.

In this investigation, the general Bloch Flow partial differential Eqs. (3.1) and (3.2) is applied to this problem. If resonance condition exists at Larmor frequency (Awojoyogbe 2002, 2003, 2004), the following assumption could be made:

$$M_y(x,t) = C\psi(x)e^{\zeta t}. \tag{3.174}$$

With the use of this expression, Eq. (3.173) yields:

$$v^2\frac{d^2\psi(x)}{dx^2} + v(2\zeta + T_o)\frac{d\psi(x)}{dx} + \left(\zeta^2 + T_o\zeta + \gamma^2 B_1(x) + \frac{1}{T_1T_2}\right)\psi(x) = \frac{M_o\gamma B_1(x)}{T_1}e^{-\zeta t}, \tag{3.175}$$

$$e^{\zeta t} \neq \frac{M_o\gamma B_1(x)}{V^2 T_1}, \tag{3.176}$$

where ς is any constant. RF $B_1(x)$ is the applied spatially varying magnetic field and v is the fluid flow velocity and $t = \left(\Delta - {}^{\delta}/_{3}\right)$. Assuming the flow velocity and RF are given as:

$$v = \frac{x}{\delta}, \tag{3.177}$$

$$\gamma B_1(x) = \gamma G x. \tag{3.178}$$

In addition to this, the pulse time δ is assumed to be infinitesimally short and negligible displacements occur during the pulse period as compared with the diffusion time (that is, $\delta \ll t$). In a typical MRI procedure, G is the pulsed gradient applied for the length of time t. Substituting for Eqs. (3.177) and (3.178) in Eq. (3.175), the following expression is obtained

$$x^2\frac{d^2\psi(x)}{dx^2} + \delta(2\zeta + T_o)x\frac{d\psi(x)}{dx} + \left[\gamma^2 g^2 t^2 x^2 + \delta^2\left(\zeta^2 + T_o\zeta + \frac{1}{T_1T_2}\right)\right]\psi(x)$$

$$= \frac{M_o \delta^2 \gamma B_1(x)}{T_1} e^{-\zeta t}. \tag{3.179}$$

When the NMR signal is sampled at maximum magnitude, $M_o \approx 0$. Hence, Eq. (3.179) becomes:

$$x^2 \frac{d^2 \psi(x)}{dx^2} + \delta(2\zeta + T_o)x \frac{d\psi(x)}{dx} + \left[\gamma^2 g^2 t^2 x^2 + \delta^2 \left(\zeta^2 + T_o \zeta + \frac{1}{T_1 T_2}\right)\right] \psi(x) = 0. \tag{3.180}$$

Equation (3.180) is transformable into Bessel's equation of order β and parameter k with general solution in the form (Wylie and Barrett 1982):

$$\psi(x) = x^{-p}[AJ_n(u_0 x) + A_1 Y_n(u_0 x)], \tag{3.181}$$

where u_o, η, β are all constants defined as:

$$n = \sqrt{p^2 - \beta^2}, \tag{3.182}$$

$$u_o = \gamma \delta g, \tag{3.183}$$

$$\beta^2 = \delta^2 \left(\zeta^2 + T_o \zeta + \frac{1}{T_1 T_2}\right), \tag{3.184}$$

$$p = \frac{\delta(2\zeta + T_o) - 1}{2}, \tag{3.185}$$

$$T_0 = \left(\frac{1}{T_1} + \frac{1}{T_2}\right) = \frac{(T_1 + T_2)}{T_1 T_2}. \tag{3.186}$$

General Behaviour of the Transverse Magnetisation

Since transverse magnetisation is required to be measurable at all values of x, $A_1 = 0$. Therefore, the Stejskal and tanner formulation in Eq. (3.170) for diffusion NMR/MRI systems is now in terms of Bessel function of the first kind where $\zeta = -u_0^2 D_0$. The transverse magnetisation is then given as follows:

$$M_y(x, t) = Ax^{-p} J_n(u_0 x) e^{-u_0^2 D_0 t}, \tag{3.187}$$

$$\frac{M_y(x, t)}{A} = x^{-p} J_n(u_0 x) e^{-u_0^2 D_0 t}, \tag{3.188}$$

where D_0 is the apparent diffusion coefficient (ADC) of the tissue.

Direct NMR Signal Analysis

$$\frac{M_y(x,t)}{A} = J_n(kx)e^{-u_0^2 D_0 t}, \tag{3.189}$$

where

$$p = \frac{t(2\zeta + T_o) - 1}{2} = 0, \tag{3.190}$$

and

$$\frac{M_y(x,t)}{A} = xJ_n(kx)e^{-u_0^2 D_0 t}, \tag{3.191}$$

where

$$p = \frac{t(2\zeta + T_o) - 1}{2} = -1. \tag{3.192}$$

The NMR signal based of Eqs. (3.189–3.192) can be written as

$$\frac{S(x,t)}{A} = v\int_0^\infty tJ_0(u_0 vt)e^{-u_0^2 D_0 t}dt = \frac{u_0 v^2}{\left(u_o^4 D_o^2 + u_o^2 v^4\right)^{3/2}}, \tag{3.193}$$

for n = 0 and p = − 1, and

$$\frac{S(x,t)}{A} = v\int_0^\infty tJ_1(u_0 vt)e^{-u_0^2 D_0 t}dt = \frac{u_0^3 D_0}{\left(u_o^4 D_o^2 + u_o^2 v^4\right)^{3/2}}, \tag{3.194}$$

for n = 1 and p = − 1.

The physical interpretation of parameter n is apparent from Eqs. (3.182), (3.184) and (3.185), where n is in term of NMR relaxation in the presence of diffusion. Parameter n is related to the pulse angle (Haacke et al. 1999); in the presence of flow and diffusion, a parameter may be defined as follows

$$\cos(\alpha_{dNMR}) = \exp(-n) = \exp\left(-\sqrt{\frac{(2\zeta\delta + \delta T_o - 1)^2}{4} - \delta^2\left(\zeta^2 + T_o\zeta + \frac{1}{T_1 T_2}\right)}\right), \tag{3.195}$$

where α_{dNMR} is the relaxation angle in the presence of flow and diffusion. Image analysis based Eq. (3.195) and the choice of pulse angle is critical for determining both signal intensity as well as image contrast. It also depends on the particular pulse sequence utilized.

3.3.5.4 Computational Model of Diffusion Imaging and Diffusion Tensor Imaging for Studying Biological Systems

Diffusion magnetic resonance imaging has the potential to provide information on the organization of tissues at cellular level (Le Bihan and Iima 2015; Basser 2008; Le Bihan et al. 2001). In fact, all biological processes such as DNA replication, RNA transcription, protein translation, protein/enzyme activity, cross membrane transport require molecules to interact and this usually occurs where they are close proximity. It is interesting to note that diffusion appears to be the universal process through which these biological processes occur. Hence, diffusion MRI is a great tool for exploring the process in vitro and in vivo (Le Bihan and Iima 2015). Despite the numerous interesting applications of diffusion imaging, the mechanisms governing diffusion of water in tissues, especially in the brain is still not well understood. Since tissues are complex, organized structures in which diffusion largely differs from the Brownian motion, a more robust understanding of such mechanisms is required to exploit the full potential of the technique and to obtain information directly on tissue microstructure.

In order to address this problem, the NMR diffusion equation with variable diffusion coefficientsis considered. With such equation, some of the complexities of the tissue may be characterized from point to point. In the Cartesian coordinate, this equation is given as (Awojoyogbe et al. 2011):

$$\frac{\partial M_y}{\partial t} = \frac{\partial}{\partial x}\left(D_x \frac{\partial M_y}{\partial x}\right) + \frac{\partial}{\partial y}\left(D_y \frac{\partial M_y}{\partial y}\right) + \frac{\partial}{\partial z}\left(D_z \frac{\partial M_y}{\partial z}\right) + \frac{F_o}{T_o}\gamma B_1(\vec{r}, t). \quad (3.196)$$

If the following assumption holds (Gupta et al. 2014):

$$M_y(x, y, z, t) = M_{y1}(x, y, z, t) + \Lambda(t). \quad (3.197)$$

Then, using Eq. (3.197) in (3.196)

$$\frac{\partial M_{y1}}{\partial t} + \frac{d\Lambda(t)}{dt} - \frac{F_o}{T_o}\gamma B_1(t) = \frac{\partial}{\partial x}\left(D_x \frac{\partial M_{y1}}{\partial x}\right) + \frac{\partial}{\partial y}\left(D_y \frac{\partial M_{y1}}{\partial \phi}\right) + \frac{\partial}{\partial z}\left(D_z \frac{\partial M_{y1}}{\partial z}\right), \quad (3.198)$$

and provided that

$$\Lambda'(t) = \frac{F_o}{T_o}\gamma B_1(t). \quad (3.199)$$

Hence, Eq. (3.198) becomes:

$$\frac{\partial M_{y1}}{\partial t} = \frac{\partial}{\partial x}\left(D_x \frac{\partial M_{y1}}{\partial x}\right) + \frac{\partial}{\partial y}\left(D_y \frac{\partial M_{y1}}{\partial \phi}\right) + \frac{\partial}{\partial z}\left(D_z \frac{\partial M_{y1}}{\partial z}\right). \quad (3.200)$$

If we make the following assumptions (Zoppou and Knight 1999):

$$\begin{aligned} x &= \exp(K_x X), \\ y &= \exp(K_y Y), \\ z &= \exp(K_z Z), \end{aligned} \tag{3.201}$$

where

$$K_x = \frac{T_2}{T_1}\gamma G_x \delta,\ K_y = \frac{T_2}{T_1}\gamma G_y \delta,\ K_z = \frac{T_2}{T_1}\gamma G_z \delta. \tag{3.202}$$

These expressions then lead to

$$\left.\begin{aligned} \partial x &= K_x \exp(K_x X)\partial X = K_x x \partial X \\ \partial y &= K_y \exp(K_y Y)\partial Y = K_y y \partial Y \\ \partial z &= K_z \exp(K_z Z)\partial Z = K_z z \partial Z \end{aligned}\right\}, \tag{3.203}$$

$$\begin{aligned} \partial x^2 &= K_x^2 x^2 \partial X^2, \\ \partial y^2 &= K_y^2 y^2 \partial Y^2, \\ \partial z^2 &= K_z^2 z^2 \partial Z^2. \end{aligned} \tag{3.204}$$

Equation (3.200) then becomes:

$$\begin{aligned} \frac{\partial M_{y1}}{\partial t} &= \frac{\partial D_x}{\partial x}\frac{\partial M_{y1}}{\partial x} + D_x \frac{\partial^2 M_{y1}}{\partial x^2} + \frac{\partial D_y}{\partial y}\frac{\partial M_{y1}}{\partial y} \\ &\quad + D_y \frac{\partial^2 M_{y1}}{\partial y^2} + \frac{\partial D_z}{\partial z}\frac{\partial M_{y1}}{\partial z} + D_z \frac{\partial^2 M_{y1}}{\partial z^2}, \\ \frac{\partial M_{y1}}{\partial t} &= \frac{\partial D_x}{\partial x}\frac{\partial M_{y1}}{K_x x \partial X} + D_x \frac{\partial^2 M_{y1}}{K_x^2 x^2 \partial X^2} + \frac{\partial D_y}{\partial y}\frac{\partial M_{y1}}{K_y y \partial Y} \\ &\quad + D_y \frac{\partial^2 M_{y1}}{K_y^2 y^2 \partial Y^2} + \frac{\partial D_z}{\partial z}\frac{\partial M_{y1}}{K_z z \partial Z} + D_z \frac{\partial^2 M_{y1}}{K_z^2 z^2 \partial Z^2}. \end{aligned} \tag{3.205}$$

Another assumption could also be made as follows (Zoppou and Knight 1999; De Marsily 1986)

$$\left.\begin{aligned} D_x &= D_0 K_x^2 x^2 \\ D_y &= D_0 K_y^2 y^2 \\ D_z &= D_0 K_z^2 z^2 \end{aligned}\right\}, \tag{3.206}$$

where D_0 is taken to be the free diffusion coefficient of the diffusing molecules.

$$\left.\begin{array}{l}\frac{\partial D_x}{\partial x} = \frac{dD_x}{dx} = 2D_0K_x^2x \\ \frac{\partial D_y}{\partial y} = \frac{dD_y}{dy} = 2D_0K_y^2y \\ \frac{\partial D_z}{\partial z} = \frac{dD_z}{dz} = 2D_0K_z^2z\end{array}\right\}. \tag{3.207}$$

Equation (3.205) now becomes:

$$\frac{\partial M_{y1}}{\partial t} = 2D_0K_x\frac{\partial M_{y1}}{\partial X} + D_0\frac{\partial^2 M_{y1}}{\partial X^2} + 2D_0K_y\frac{\partial M_{y1}}{\partial Y} + D_0\frac{\partial^2 M_{y1}}{\partial Y^2} + 2D_0K_z\frac{\partial M_{y1}}{\partial Z} + D_0\frac{\partial^2 M_{y1}}{\partial Z^2}, \tag{3.208}$$

$$\frac{\partial M_{y1}}{\partial t} = D_0\left(\frac{\partial^2 M_{y1}}{\partial X^2} + \frac{\partial^2 M_{y1}}{\partial Y^2} + \frac{\partial^2 M_{y1}}{\partial Z^2}\right) + 2D_0K_x\frac{\partial M_{y1}}{\partial X} + 2D_0K_y\frac{\partial M_{y1}}{\partial Y} + 2D_0K_z\frac{\partial M_{y1}}{\partial Z}. \tag{3.209}$$

For an instantaneous release of water (which holds also for any diffusing molecule in biological cavities), we seek a solution to Eq. (3.209) in the following form:

$$M_{y1}(X, Y, Z, t) = \Omega_1(X, t; X_0)\Omega_2(Y, t; Y_0)\Omega_3(Z, t; Z_0), \tag{3.210}$$

where Ω_1, Ω_2, Ω_3 (which are not tensors) are the solution to the 1D constant coefficient modified diffusion equation in the transformed space X, Y, Z. These solutions are assumed to have the following from:

$$\begin{aligned}\Omega_1(X, t; X_0) &= \frac{A_{01}}{2K_x\sqrt{\pi D_0 t}}\exp\left(-\frac{[(X - X_0) - 2D_0K_xt]^2}{4D_0t}\right), \\ \Omega_2(Y, t; Y_0) &= \frac{A_{02}}{2K_y\sqrt{\pi D_0 t}}\exp\left(-\frac{[(Y - Y_0) - 2D_0K_yt]^2}{4D_0t}\right), \\ \Omega_3(Z, t; Z_0) &= \frac{A_{03}}{2K_z\sqrt{\pi D_0 t}}\exp\left(-\frac{[(Z - Z_0) - 2D_0K_zt]^2}{4D_0t}\right).\end{aligned} \tag{3.211}$$

In order to determine the constant coefficients A_{01}, A_{02} and A_{03}, we shall consider a carefully selected reduced transverse magnetisation and volume such that the associated magnetic moment is unitary:

$$\int_0^\infty\int_0^\infty\int_0^\infty M_{y1}(x, y, z, t)dxdydz = 1\forall t. \tag{3.212}$$

In the transformed spatial coordinates, Eq. (3.212) becomes (Zoppou and Knight 1999; De Marsily 1986):

$$\int_0^\infty\int_0^\infty\int_0^\infty M_{y1}(x,y,z,t)dxdydz$$
$$=\int_{-\infty}^{\infty}\int_{-\infty}^{\infty}\int_{-\infty}^{\infty} K_xK_yK_z e^{(K_xX+K_yY+K_zZ)}M_{y1}$$
$$(X,Y,Z,t)dXdYdZ=1. \tag{3.213}$$

or

$$\int_{-\infty}^{\infty} K_x e^{K_xX}\Omega_1(X,t;X_0)dX\int_{-\infty}^{\infty} K_y e^{K_yY}\Omega_2(Y,t;Y_0)dY\int_{-\infty}^{\infty} K_z e^{K_zZ}\Omega_3(Z,t;Z_0)dZ=1. \tag{3.214}$$

Computing the integral in Eq. (3.214), we have:

$$\begin{aligned}
&A_{01}\exp(K_x(X_0+3D_0K_xt))A_{02}\exp(K_y(X_0+3D_0K_yt))\\
&A_{03}\exp(K_z(X_0+3D_0K_zt))=1,\\
&A_{01}A_{02}A_{03}\exp(K_xX_0)\exp(3D_0K_x^2t)\exp(K_yY_0)\\
&\exp(3D_0K_y^2t)\exp(K_zZ_0)\exp(3D_0K_z^2t)=1.
\end{aligned} \tag{3.215}$$

According to Eq. (3.201), this expression becomes:

$$A_{01}A_{02}A_{03}x_0y_0z_0\exp(3D_0K_x^2t)\exp(3D_0K_y^2t)\exp(3D_0K_z^2t)=1. \tag{3.216}$$

Obviously, the constant coefficients may be given as follows:

$$A_{01}=\frac{\exp(3D_0K_x^2t)}{x_0},\ A_{02}=\frac{\exp(3D_0K_y^2t)}{y_0},\ A_{03}=\frac{\exp(3D_0K_z^2t)}{z_0}. \tag{3.217}$$

Then, the reduced transverse magnetisation in the transformed spatial coordinates is

$$\begin{aligned}
M_{y1}(X,Y,Z,t)=&\frac{1}{8K_xK_yK_z\sqrt{\pi D_0t}}\frac{\exp(3D_0K_x^2t)}{x_0}\frac{\exp(3D_0K_y^2t)}{y_0}\frac{\exp(3D_0K_z^2t)}{z_0}\\
&\exp\left(-\frac{[(X-X_0)-2D_0K_xt]^2}{4D_0t}\right)\exp\left(-\frac{[(Y-Y_0)-2D_0K_yt]^2}{4D_0t}\right)
\end{aligned}$$

$$\exp\left(-\frac{[(Z-Z_0)-2D_0K_zt]^2}{4D_0t}\right), \tag{3.218}$$

$$M_{y1}(X,Y,Z,t)=\frac{1}{8x_0y_0z_0K_xK_yK_z\sqrt{\pi D_0t}}\exp\left(3D_0K_x^2t-\frac{[(X-X_0)-2D_0K_xt]^2}{4D_0t}\right)$$
$$\times\exp\left(3D_0K_y^2t-\frac{[(Y-Y_0)-2D_0K_yt]^2}{4D_0t}\right)$$
$$\exp\left(3D_0K_z^2t-\frac{[(Z-Z_0)-2D_0K_zt]^2}{4D_0t}\right), \tag{3.219}$$

$$M_{y1}(X,Y,Z,t)=\frac{1}{8x_0y_0z_0K_xK_yK_z\sqrt{\pi D_0t}}$$
$$\exp\left(3D_0K_x^2t-\frac{[(X-X_0)^2-4D_0K_xt(X-X_0)+4D_0^2K_x^2t^2]}{4D_0t}\right)$$
$$\times\exp\left(3D_0K_y^2t-\frac{[(Y-Y_0)^2-4D_0K_yt(Y-Y_0)+4D_0^2K_x^2t^2]}{4D_0t}\right)$$
$$\times\exp\left(3D_0K_z^2t-\frac{[(Z-Z_0)-4D_0K_zt(Z-Z_0)+4D_0^2K_z^2t^2]}{4D_0t}\right),$$

$$M_{y1}(X,Y,Z,t)=\frac{1}{8x_0y_0z_0K_xK_yK_z\sqrt{\pi D_0t}}$$
$$\exp\left(3D_0K_x^2t-\frac{[(X-X_0)^2-4D_0K_xt(X-X_0)+4D_0^2K_x^2t^2]}{4D_0t}\right)$$
$$\times\exp\left(3D_0K_y^2t-\frac{[(Y-Y_0)^2-4D_0K_yt(Y-Y_0)+4D_0^2K_y^2t^2]}{4D_0t}\right)$$
$$\times\exp\left(3D_0K_z^2t-\frac{[(Z-Z_0)^2-4D_0K_zt(Z-Z_0)+4D_0^2K_z^2t^2]}{4D_0t}\right),$$

$$M_{y1}(X,Y,Z,t)=\frac{1}{8x_0y_0z_0K_xK_yK_z\sqrt{\pi D_0t}}$$
$$\exp\left(3D_0K_x^2t-D_0K_x^2t-\frac{[(X-X_0)^2-4D_0K_xt(X-X_0)]}{4D_0t}\right)$$
$$\times\exp\left(3D_0K_y^2t-D_0K_y^2t-\frac{[(Y-Y_0)^2-4D_0K_yt(Y-Y_0)]}{4D_0t}\right)$$
$$\times\exp\left(3D_0K_z^2t-D_0K_z^2t-\frac{[(Z-Z_0)^2-4D_0K_zt(Z-Z_0)]}{4D_0t}\right),$$

$$M_{y1}(X,Y,Z,t)=\frac{1}{8x_0y_0z_0K_xK_yK_z\sqrt{\pi D_0t}}$$
$$\exp\left(2D_0K_x^2t+K_x(X-X_0)-\frac{(X-X_0)^2}{4D_0t}\right)$$

$$\times \exp\left(2D_0K_y^2t + K_y(Y - Y_0) - \frac{(Y - Y_0)^2}{4D_0t}\right)$$
$$\exp\left(2D_0K_z^2t + K_z(Z - Z_0) - \frac{(Z - Z_0)^2}{4D_0t}\right). \tag{3.220}$$

From Eq. (3.201), we have the following:

$$X = \frac{1}{K_x}\ln x, Y = \frac{1}{K_y}\ln y, Z = \frac{1}{K_z}\ln z,$$
$$X_0 = \frac{1}{K_x}\ln x_0, Y_0 = \frac{1}{K_y}\ln y_0, Z_0 = \frac{1}{K_z}\ln z_0. \tag{3.221}$$

In actual spatial coordinates, Eq. (3.220) becomes:

$$M_{y1}(x, y, z, t) = \frac{1}{8x_0y_0z_0K_xK_yK_z\sqrt{\pi D_0t}}$$
$$\exp\left(2D_0K_x^2t + K_x\left(\frac{1}{K_x}\ln x - \frac{1}{K_x}\ln x_0\right) - \frac{\left(\frac{1}{K_x}\ln x - \frac{1}{K_x}\ln x_0\right)^2}{4D_0t}\right)$$
$$\times \exp\left(2D_0K_y^2t + K_y\left(\frac{\ln y}{K_y} - \frac{\ln y_0}{K_y}\right) - \frac{\left(\frac{\ln y}{K_y} - \frac{\ln y_0}{K_y}\right)^2}{4D_0t}\right)$$
$$\times \exp\left(2D_0K_z^2t + K_z\left(\frac{\ln z}{K_z} - \frac{\ln z_0}{K_z}\right) - \frac{\left(\frac{\ln z}{K_z} - \frac{\ln z_0}{K_z}\right)^2}{4D_0t}\right),$$

$$M_{y1}(x, y, z, t) = \frac{1}{8x_0y_0z_0K_xK_yK_z\sqrt{\pi D_0t}}$$
$$\exp\left(2D_0K_x^2t + \ln\left(\frac{x}{x_0}\right) - \frac{\left(\frac{1}{K_x}\ln\left(\frac{x}{x_0}\right)\right)^2}{4D_0t}\right)$$
$$\times \exp\left(2D_0K_y^2t + \ln\left(\frac{y}{y_0}\right) - \frac{\left(\frac{1}{K_y}\ln\left(\frac{y}{y_0}\right)\right)^2}{4D_0t}\right)$$
$$\exp\left(2D_0K_z^2t + \ln\left(\frac{z}{z_0}\right) - \frac{\left(\frac{1}{K_z}\ln\left(\frac{z}{z_0}\right)\right)^2}{4D_0t}\right), \tag{3.222}$$

$$M_{y1}(x, y, z, t) = \frac{1}{8x_0y_0z_0K_xK_yK_z\sqrt{\pi D_0t}}\left(\frac{xyz}{x_0y_0z_0}\right)$$

$$\exp\left(2D_0K_x^2t + 2D_0K_y^2t + 2D_0K_z^2t - \frac{\left(\frac{1}{K_x}\ln\left(\frac{x}{x_0}\right)\right)^2}{4D_0t} - \frac{\left(\frac{1}{K_y}\ln\left(\frac{y}{y_0}\right)\right)^2}{4D_0t} - \frac{\left(\frac{1}{K_z}\ln\left(\frac{z}{z_0}\right)\right)^2}{4D_0t}\right), \tag{3.223}$$

$$M_{y1}(x, y, z, t) = \frac{xyz}{8(x_0y_0z_0)^2K_xK_yK_z\sqrt{\pi D_0t}}\exp\left(2(K_x^2 + K_y^2 + K_z^2)D_0t - \frac{P_{xyz}^2}{4D_0t}\right), \tag{3.224}$$

where

$$P_{xyz}^2 = \left(\frac{1}{K_x}\ln\left(\frac{x}{x_0}\right)\right)^2 + \left(\frac{1}{K_y}\ln\left(\frac{y}{y_0}\right)\right)^2 + \left(\frac{1}{K_z}\ln\left(\frac{z}{z_0}\right)\right)^2. \tag{3.225}$$

Finally from Eqs. (3.197), (3.199) and (3.224), the final solution becomes:

$$M_y(x, y, z, t) = \frac{xyz}{8(x_0y_0z_0)^2K_xK_yK_z\sqrt{\pi D_0t}} \exp\left(2(K_x^2 + K_y^2 + K_z^2)D_0t - \frac{P_{xyz}^2}{4D_0t}\right) + \frac{F_o}{T_o}\int_0^{TE}\gamma B_1(t)dt. \tag{3.226}$$

Evaluation of MR Diffusion Tensor

Since the transverse magnetisation is actually a tensor, a more general form of Eq. (3.194) is given as follows (Awojoyogbe et al. 2011):

$$\frac{\partial \vec{M}_y}{\partial t} = \nabla(\vec{D}\nabla\vec{M}_y(\vec{r}, t)) + \frac{F_o}{T_o}\gamma B_1(\vec{r}, t). \tag{3.227}$$

Equation (3.227) may alternatively be written as follows (Price 1997):

$$\begin{bmatrix} \frac{\partial M_y(x,t)}{\partial t} \\ \frac{\partial M_y(y,t)}{\partial t} \\ \frac{\partial M_y(z,t)}{\partial t} \end{bmatrix} = \begin{bmatrix} \frac{\partial}{\partial x} \\ \frac{\partial}{\partial y} \\ \frac{\partial}{\partial z} \end{bmatrix} \begin{bmatrix} D_{xx} & D_{xy} & D_{xz} \\ D_{yx} & D_{yy} & D_{yz} \\ D_{zx} & D_{zy} & D_{zz} \end{bmatrix} \begin{bmatrix} \frac{\partial M_y(x,t)}{\partial x} \\ \frac{\partial M_y(y,t)}{\partial x} \\ \frac{\partial M_y(z,t)}{\partial z} \end{bmatrix} + \begin{bmatrix} \frac{F_o}{T_o}\gamma B_1(x, t) \\ \frac{F_o}{T_o}\gamma B_1(y, t) \\ \frac{F_o}{T_o}\gamma B_1(z, t) \end{bmatrix}. \tag{3.228}$$

On performing diagonalization on the diffusion tensor and provided that D_x, D_y and D_z represent the Eigen value of the tensor, it follows that (Jellison et al. 2004)

$$\begin{bmatrix} D_{xx} & D_{xy} & D_{xz} \\ D_{yx} & D_{yy} & D_{yz} \\ D_{zx} & D_{zy} & D_{zz} \end{bmatrix} \xrightarrow{diagonalization} \begin{bmatrix} D_x & 0 & 0 \\ 0 & D_y & 0 \\ 0 & 0 & D_z \end{bmatrix} \to D_0 \begin{bmatrix} K_x^2x^2 & 0 & 0 \\ 0 & K_y^2y^2 & 0 \\ 0 & 0 & K_z^2z^2 \end{bmatrix}. \tag{3.229}$$

From the matrix in Eq. (3.227), for quantitative DTI, the following expressions are feasible (Le Bihan et al. 2001).

Trace of the Diffusion Tensor:

$$Tr(\vec{D}) = D_x + D_y + D_z = D_0K_x^2x^2 + D_0K_y^2y^2 + D_0K_z^2z^2. \tag{3.230}$$

Mean Diffusivity (MD)

$$MD = \frac{Tr(\vec{D})}{3} = \frac{D_0}{3}(K_x^2x^2 + K_y^2y^2 + K_z^2z^2). \tag{3.231}$$

Volume Ratio (VR)

$$VR = \frac{D_xD_yD_z}{MD^3} = \frac{D_0^3(K_x^2x^2)(K_y^2y^2)(D_0K_z^2z^2)}{MD^3}. \tag{3.232a}$$

Relative Anisotropy (RA)

$$\begin{aligned} RA &= \sqrt{\frac{(D_x - MD)^2 + (D_y - MD)^2 + (D_z - MD)^2}{D_x + D_y + D_z}} \\ &= \sqrt{\frac{(D_0K_x^2x^2 - MD)^2 + (D_0K_y^2y^2 - MD)^2 + (D_0K_z^2z^2 - MD)^2}{D_0K_x^2x^2 + D_0K_y^2y^2 + D_0K_z^2z^2}}. \end{aligned} \tag{3.232b}$$

Fractional Anisotropy (FA)

$$\begin{aligned} FA &= \sqrt{\frac{3}{2}\frac{(D_x - MD)^2 + (D_y - MD)^2 + (D_z - MD)^2}{D_x^2 + D_y^2 + D_z^2}} \\ &= \sqrt{\frac{3}{2}\frac{(D_0K_x^2x^2 - MD)^2 + (D_0K_y^2y^2 - MD)^2 + (D_0K_z^2z^2 - MD)^2}{(D_0K_x^2x^2)^2 + (D_0K_y^2y^2)^2 + (D_0K_z^2z^2)^2}} \end{aligned} \tag{3.233}$$

Diffusion Ellipsoid

For anisotropic tissues, we can always rotate the laboratory (x, y, z) coordinate axes so that they are aligned with the local principal (x′, y′, z′) axes of the diffusion ellipsoid in each voxel (Basser 2008). Within this frame, all off-diagonal elements of the tensor $\vec{D}$ vanish as in the expression of Eq. (3.226) and the equation describing the diffusion ellipsoid is given as follows (Basser 2008; Le Bihan et al. 2001):

$$\left(\frac{x'}{\sqrt{2D'_x t}}\right)^2 + \left(\frac{y'}{\sqrt{2D'_y t}}\right)^2 + \left(\frac{z'}{\sqrt{2D'_z t}}\right)^2 = 1, \tag{3.234}$$

where D'_x, D'_y and D'_z are the principal diffusivities and since they represent the eigen values of the diffusion tensor $\vec{D}$. The terms $\sqrt{2D'_x t}$, $\sqrt{2D'_y t}$ and $\sqrt{2D'_z t}$ give the RMS displacements along the three principal directions at time t, respectively. It is quite important to note that the directions of the principal axes (x′, y′, z′) of the diffusion ellipsoid are usually not known before hand and do not generally coincide with the laboratory coordinate (x, y, z). Hence, the orientation of the ellipsoid must be assumed to change in each voxel and must be determined experimentally (Basser 2008).

3.3.5.5 Computational Model of Diffusion Magnetic Resonance Imaging for Studying Biological Porous Systems

Transport phenomena through porous media have had increased research interests in recent decades because of the importance of having a detailed understanding of the accompanying associated transport processes. Such knowledge has been shown to be immensely useful in various practical applications that can be considered as transport through porous media (Khanafer and Vafai 2006).

The application of porous media theory tomodelling biomedical applications has been successful over the years. Such applications comprises of computational biology, tissue replacement production, drug delivery, advancedmedical imaging, porous scaffolds for tissue engineering, effective tissue replacement and transport in biological tissues (Yang and Vafai 2006; Ai and Vafai 2006; Yang et al. 2003).

Another application of porous media to biological system includesdiffusion process in the extracellular space (ECS) which is very important in the investigation of central nervoussystem physiology and disease conditions (Khanafer and Vafai 2006). By physical observation, theECS of the brain looks like a porous medium (Kuffler and Potter 1964) and it has been considered to mimickthe water phase of a foam. Themicroenvironment of brain cells is made up of ECS and it occupies about 20% of nervous tissue volume. The ECS serves as a pathway forcellular metabolities, a channel for chemical signalling mediated by volume transmission and a route for drugdelivery (Prokopová-Kubinová et al. 2001; Nicholson and Rice 1991; Sykova 1991).

Magnetic resonance imaging is fast becoming an invaluable imaging tool in clinical diagnostic radiology, porous materialcharacterization, phase change and dynamics ofcompounds confined within porous media (Barrie 1995; Khanafer and Vafai 2006). Although most MRI research studies have concentrated on clinical applications, MRI is currently becoming a powerful technique for other applications that are not clinical in nature. MRI has proved to be useful in thein vivo measurement of the diffusion of water andintracellular metabolites; consequently paving way for its extensive use in the studies of biological porous media.

An equation suitable for the evaluation of NMR signal in heterogeneous biological systems in which the pores have cylindrical nature is Eq. (2.297). Since the architecture of biological systems slows down the movement of fluids (Nicholson and Syková 1998; Syková and Nicholson 2008), it could be said that the spatial diffusion coefficient vary very slowly with spatial coordinates such that $D \approx D_r \approx D\phi \approx D_z$ (that is the constant diffusion coefficient).

Hence, for a porous medium which has a constant diffusion coefficient, D along its channels, the equation diffusion may be written as (Dada and Awojoyogbe 2021):

$$\frac{\partial M_y}{\partial t} - \frac{F_0}{T_0}\gamma B_1(t) = D\left(\frac{\partial^2 M_y}{\partial r^2} + \frac{1}{r}\frac{\partial M_y}{\partial r} + \frac{1}{r^2}\frac{\partial^2 M_y}{\partial \phi^2} + \frac{\partial^2 M_y}{\partial z^2}\right), \tag{3.235}$$

where $F_o = \frac{M_0}{T_1}$, $T_g = \frac{1}{T_1 T_2}$, $T_0 = \frac{1}{T_1} + \frac{1}{T_2}$, D is the diffusion coefficient, γ is the gyromagnetic ratio, T_1 is the spin lattice relaxation time, T_2 is the spin–spin relaxation time, M_o is the equilibrium magnetisation, RF $B_1(x, t)$ is the applied radiofrequency magnetic field and M_y is the transverse magnetisation, T_1 is the spin lattice relaxation rate and T_2 is the spin–spin relaxation time.

However, if a very small voxel is considered, the same amount of RF power is always delivered to the voxel such that the position of the magnetic particle is not a factor significant enough to be considered, the following assumption may be considered:

$$M_y(r, \phi, z, t) = M_{y0}(r, \phi, z, t) + A_{y0}(t). \tag{3.236}$$

Equation (3.235) becomes

$$\frac{1}{D}\frac{\partial M_{y0}}{\partial t} = \frac{\partial^2 M_{y0}}{\partial r^2} + \frac{1}{r}\frac{\partial M_{y0}}{\partial r} + \frac{1}{r^2}\frac{\partial^2 M_{y0}}{\partial \phi^2} + \frac{\partial^2 M_{y0}}{\partial z^2}. \tag{3.237}$$

Assuming that

$$\frac{dA_{y0}(t)}{dt} = \frac{F_0}{T_0}\gamma B_1(t). \tag{3.238}$$

If the term M_{y0} does not vary significantly with ϕ and z (Spiegel, 1974), Eq. (3.235) becomes:

$$\frac{1}{D}\frac{\partial M_{y0}}{\partial t} = \frac{\partial^2 M_{y0}}{\partial r^2} + \frac{1}{r}\frac{\partial M_{y0}}{\partial r}. \tag{3.239}$$

A fundamental expression for the permeability (k) of a porous medium is given by the modified Kozeny-Carman equation (Pape et al. 1998):

$$k = \frac{r^2}{8F}, \tag{3.240}$$

where r is the effective radii of the tissue section approximated to a hydraulic capillary, F is the formation factor, which is defined as the ratio of tortuosity (τ) and porosity (p):

$$F = \frac{\tau}{p}, \tag{3.241}$$

$$k = \frac{r^2 p}{8\tau}. \tag{3.242}$$

For general consideration of the tortuosity of porous materials, we shall assume that (Dada et al. 2015):

$$\tau = \frac{\alpha}{8p}. \tag{3.243}$$

Equation (3.242) may then be written as:

$$k = \frac{r^2 p^2}{\alpha}, \tag{3.244}$$

$$p = \frac{\sqrt{\alpha k}}{r}. \tag{3.245}$$

The transverse magnetisation is proportional to the number of protons in the sensitive region and may be scaled to give porosity in porous rocks (Morriss et al. 1994). If this behaviour is assumed to hold for biological porous media, thenusing Eq. (3.245), the effective radii-dependent transverse magnetisation can be transformed (Kreyszig 1996) into a porosity dependent transverse magnetisation in a biological system. Therefore, the following equations may be obtained

$$\frac{\partial M_{y0}}{\partial r} = \frac{\partial M_{y0}}{\partial p}\frac{dp}{dr}, \tag{3.246}$$

$$\frac{\partial^2 M_{y0}}{\partial r^2} = \frac{\partial^2 M_{y0}}{\partial p^2}\left(\frac{dp}{dr}\right)^2 + \frac{\partial M_{y0}}{\partial p}\frac{d^2 p}{dr^2}, \tag{3.247}$$

$$\frac{dp}{dr} = -\frac{\sqrt{\alpha k}}{r^2},\ \frac{d^2 p}{dr^2} = \frac{2\sqrt{\alpha k}}{r^3},\ \left(\frac{dp}{dr}\right)^2 = \frac{\alpha k}{r^4}.$$

Hence, Eq. (3.239) becomes:

$$\frac{1}{D}\frac{\partial M_{y0}}{\partial t} = \frac{\partial^2 M_{y0}}{\partial p^2}\left(\frac{dp}{dr}\right)^2 + \frac{\partial M_{y0}}{\partial p}\frac{d^2 p}{dr^2} + \frac{1}{r}\frac{\partial M_{y0}}{\partial p}\frac{dp}{dr}, \tag{3.248}$$

$$\frac{1}{D}\frac{\partial M_{y0}}{\partial t} = \frac{\alpha k}{r^4}\frac{\partial^2 M_{y0}}{\partial p^2} + \frac{2\sqrt{\alpha k}}{r^3}\frac{\partial M_{y0}}{\partial p} - \frac{1}{r}\frac{\sqrt{\alpha k}}{r^2}\frac{\partial M_{y0}}{\partial p}, \tag{3.249}$$

$$\frac{1}{D}\frac{\partial M_{y0}}{\partial t} = \frac{p^4}{\alpha k}\frac{\partial^2 M_{y0}}{\partial p^2} + \frac{p^3}{\alpha k}\frac{\partial M_{y0}}{\partial p}, \tag{3.250}$$

$$\frac{\alpha k}{D}\frac{\partial M_{y0}}{\partial t} = p^4\frac{\partial^2 M_{y0}}{\partial p^2} + p^3\frac{\partial M_{y0}}{\partial p}. \tag{3.251}$$

By the method of separation of variables, the following product may be assumed (Kreyszig 1996):

$$M_{y1} = G(t)W(p), \tag{3.252}$$

$$\frac{\alpha k}{DG}\frac{dG}{dt} = \frac{p^4}{W}\frac{d^2 W}{dp^2} + \frac{p^3}{W}\frac{dW}{dp}. \tag{3.253}$$

Equation (3.253) must be equal to a constant $-\beta_0^2$; consequently, the following equations are obtained:

$$\frac{dG}{dt} = -\frac{\beta_0^2 D}{\alpha k}G, \tag{3.254}$$

$$p^4\frac{d^2 W}{dp^2} + p^3\frac{dW}{dp} + \beta_0^2 W = 0. \tag{3.255}$$

The solution to Eq. (3.254) is given as:

$$G(t) = A_{00}e^{-\frac{\beta_0^2 D}{\alpha k}t}, \tag{3.256}$$

where A_{00} is a constant. In solving Eq. (3.255), assume:

$$p = \frac{1}{\nu}, \tag{3.257}$$

Equation (2.253) becomes:

$$p^4\left(\frac{1}{p^4}\frac{d^2 W}{d\nu^2} + \frac{2}{p^3}\frac{dW}{d\nu}\right) + p^3\left(-\frac{1}{p^2}\frac{dW}{d\nu}\right) + \beta_0^2 W = 0, \tag{3.258}$$

$$\frac{d^2 W}{d\nu^2} + 2p\frac{dW}{d\nu} - p\frac{dW}{d\nu} + \beta_0^2 W = 0, \tag{3.259}$$

$$\frac{d^2 W}{d\nu^2} + p\frac{dW}{d\nu} + \beta_0^2 W = 0, \tag{3.260}$$

$$\frac{d^2W}{dv^2} + \frac{1}{v}\frac{dW}{dv} + \beta_0^2 W = 0. \tag{3.261}$$

Equation (3.261) is the Bessel differential equation where W is the Bessel function of the zeroth order (Kreyszig 1996):

$$W(v) = C_{00}J_0(\beta_0 v), \tag{3.262}$$

where C_{00} is a constant. From Eqs. (3.258), (3.262) becomes:

$$W(p) = C_0 J_0\left(\frac{\beta_0}{p}\right). \tag{3.263}$$

Based on Eqs. (3.252), (3.256) and (3.263), M_y becomes

$$M_y(p,t) = A_p e^{-\frac{\beta_0^2 D}{\alpha k}t} J_0\left(\frac{\beta_0}{p}\right) + \frac{F_0}{T_0}\int_0^t \gamma B_1(t)dt;\ A_p = A_{00}C_{00}. \tag{3.264}$$

Equation (3.264) gives a single analytical expression relating the porosity, permeability, diffusion coefficient, relaxation times (T_1, T_2) and the transverse magnetisation for a biological porous medium. According to the prevailing physiological situation, Eqs. (3.240)–(3.243) may be used to express the transverse magnetisation as a function of the effective radii of a biological channel.

Whenever the pore volume is the same as the total volume, porosity equals 1 and since it is now nearly impossible for fluids to flow the channel, the NMR signal may be sampled such that transverse magnetisation from the fluids spins tends to zero. Hence,

$$M_{y0}(1,t) = 0. \tag{3.265}$$

Secondly, before the introduction of the RF field, the fluid spins rotate about the B_0 field and generate extremely small transverse wave which is usually very difficult to detect (Haacke et al. 1999). If this transverse magnetisation component is defined as $-pM_{p0}$, then the following initial condition may hold:

$$M_{y0}(p,0) = -pM_{p0}(p). \tag{3.266}$$

Using these conditions in Eq. (3.264) gives:

$$A_p e^{-\frac{\beta_0^2 D}{\alpha k}t} J_0\left(\frac{\beta_0}{p}\right) = 0. \tag{3.267}$$

Equation (3.37) implies:

$J_0(\beta) = 0$ and $\beta = \beta_{01}, \beta_{02}, \beta_{03}, \ldots$, are the positive roots (Wylie and Barrett 1982). Thus a solution is given as:

$M_{y0}(p,t) = A_p e^{-\frac{\beta_0^2 D}{\alpha k}t} J_0\left(\frac{\beta_{0m}}{p}\right)$; $m = 1, 2, 3, \ldots$.

By superposition of all solutions,

$$M_{y0}(p,t) = \sum_{m=1}^{\infty} A_{pm} e^{-\frac{\beta_0^2 D}{\alpha k}t} J_0\left(\frac{\beta_{0m}}{p}\right). \tag{3.268}$$

From Eq. (3.266),

$$M_{y0}(p,0) = -pM_{p0}(p) = \sum_{m=1}^{\infty} A_{pm} J_0\left(\frac{\beta_{0m}}{p}\right). \tag{3.269}$$

Using the expressions of Eq. (3.257) and integration of Bessel functions (Spiegel 1974), the constant becomes:

$$A_{pm} = -\frac{2}{J_1^2(\beta_{0m})} \int_{\infty}^{1} \frac{1}{p^2} M_{p0}(p) J_0\left(\frac{\beta_{0m}}{p}\right) dp, \tag{3.270}$$

$$M_{y0}(p,t) = \sum_{m=1}^{\infty} \left[\frac{2M_{p0}}{J_1^2(\beta_{0m})} \int_{\infty}^{1} \frac{1}{p^2} J_0\left(\frac{\beta_{0m}}{p}\right) dp \right] e^{-\frac{\beta_0^2 D}{\alpha k}t} J_0\left(\frac{\beta_{0m}}{p}\right); \; m = 1, 2, 3, \ldots \tag{3.271}$$

However,

$$\frac{2M_{p0}}{J_1^2(\beta_{0m})} \int_{\infty}^{1} \frac{1}{p^2} J_0\left(\frac{\beta_{0m}}{p}\right) dp = \frac{2M_{\rho 0}[\pi J_1(\beta_{0m}) H_0(\beta_{0m}) + J_0(\beta_{0m})\{2 - \pi J_1 H_0(\beta_{0m})\}]}{J_1^2(\beta_{0m})}, \tag{3.272}$$

where M_{p0} is a constant and $H_n(\beta_m)$ are Struve functions of order n. Therefore, the final solution becomes:

$$M_y(p,t) = \sum_{m=1}^{\infty} \left\{ \frac{2M_{\rho 0}[\pi J_1(\beta_{0m}) H_0(\beta_{0m}) + 2J_0(\beta_{0m}) - \pi J_0(\beta_m) J_1 H_0(\beta_{0m})]}{J_1^2(\beta_{0m})} \right\} e^{-\frac{\beta_0^2 D}{\alpha k}t} J_0\left(\frac{\beta_{0m}}{p}\right) + \frac{F_0}{T_0} \int_0^t \gamma B_1(t) dt. \tag{3.273}$$

The NMR signal as presented in Eq. (3.273) is dependent on the nature of the RF field. If we assume the RF field is given as follows (Price 1997, 1998):

$$\gamma B_1(t) = \gamma G(t)r. \tag{3.274}$$

Hence, the integral in Eq. (3.273) becomes:

$$\frac{F_o}{T_o}\int_0^t \gamma B_1(t)dt = \gamma M_0 r\frac{1}{T_1 T_0}\int_0^t G(t)dt. \tag{3.275}$$

If the gradient pulse G(t) is designed such that G(t) undergoes exponential rise and fall, then the following equation is possible (Price 1997, 1998):

$$G(t) = ge^{-kt}, \tag{3.276}$$

where k is the exponential rate constant (Price 1998). Setting $k = \frac{T_g}{T_0}$ and the integral in Eq. (3.275) has a new lower limit as the gradient pulse time δ and this is given as:

$$\frac{F_0}{T_0}\int_\delta^t \gamma B_1(t)dt = -\frac{M_0\gamma gr}{T_1 T_g}e^{-\frac{T_g}{T_0}(t-\delta)} = -(M_0 T_2\gamma gr)e^{-\frac{T_g}{T_0}(t-\delta)}. \tag{3.277}$$

Therefore, Eq. (3.273) becomes:

$$M_y(p,t) = \sum_{m=1}^{\infty}\left\{\frac{2M_{p0}[\pi J_1(\beta_{0m})H_0(\beta_{0m}) + 2J_0(\beta_{0m}) - \pi J_0(\beta_{0m})J_1 H_0(\beta_{0m})]}{J_1^2(\beta_{0m})}\right\}$$
$$e^{-\frac{\beta_0^2 D}{\alpha k}t}J_0\left(\frac{\beta_{0m}}{p}\right) - \{M_0 T_2\gamma gr\}e^{-\frac{T_g}{T_0}(t-\delta)}. \tag{3.278}$$

Equation (3.278) is the analytical expression for porosity-dependent transverse magnetisation describing NMR signal in biological porous materials.

3.3.6 Mathematical Models for Theranostics in Molecular Imaging

Theranostics has been regarded as a key part of personalized medicine and requires considerable advances in predictive medicine; novel theranostic agents are developed and carefully designed for in vivo quantitative assessment of the amount of drug reaching a pathological region and the visualization of molecular changes due to the therapeutic effects of the delivered drugs (Dada et al. 2013a, b).

The basic idea of Theranostics consists on creating a nanocapsule containing a drug and an imaging contrast agent as a single formulation, supporting the follow-up in vivo of the action of the drug (Cabal et al. 2013). Theranostic ideas and methods are producing a revolution, creating original technology platforms and challenging the established concepts. Three fields of drug development are covered by imaging as Theranostics: delivery, release and efficacy evaluation (Cabal et al. 2013).

3.3.6.1 Computational Modelling of Theranostics MR Energy Absorption

The mechanism of energy absorption in magnetic resonance could be used to mathematically model a closely knitted theranostic method in which a specially selected radiofrequency field is used to heat up a tissue and at the same time cause the spins of the tissue to emit MR signals.

The key to achieving this is the specific absorption rate (SAR) which drives both RF power heating within a tissue and is related directly to the B_1 field which is needed to cause spin resonance (Brix et al. 2002). Considering bioheat flow in one direction (Shih et al. 2007) with uniform tissue properties, the following equation holds

$$\rho c \frac{\partial T}{\partial t} = k \frac{\partial^2 T}{\partial x^2} + SAR - w_b c_b (T - T_b), \tag{3.279}$$

where ρ is tissue density, c is the specific heat of tissue, T is the tissue temperature, t is the time, w_b is the blood perfusion rate, c_b is the specific heat of blood, T_b is the supplying arterial blood temperature, k is the thermal conductivity of tissue, and x is the distance from the skin surface. SAR is the applied RF power per unit volume. If the tissue temperature changes very slowly with x, Eq. (3.279) becomes:

$$\rho c \frac{\partial T}{\partial t} = SAR - w_b c_b (T - T_b). \tag{3.280}$$

The solution to Eq. (3.281) is given as follows:

$$T(t) = T_b + \frac{SAR}{w_b c_b} + A \exp\left(-\frac{w_b c_b}{\rho c} t\right). \tag{3.281}$$

If the tissue temperature before the application of the RF field does not defer significantly from the arterial temperature, the initial the condition for this problem is given as:

$$T(t = 0) = T_b, \tag{3.282}$$

The final solution then becomes:

$$T(t) = T_b + \frac{SAR}{w_b c_b}\left\{1 - \exp\left(-\frac{w_b c_b}{\rho c}t\right)\right\}. \tag{3.283}$$

SAR and Oscillating Magnetic Field: The RF power for the voxel volume V_{vox} is $P_{rf} = (SAR)\ V_{vox}$. The energy of the oscillating radio wave is given as $E_{rf} = \hbar\gamma B_I$, while its rate of change is given as

$$P_{rf} = \frac{dE_{rf}}{dt} \text{and}\, E_{rf} = \int_{t_0}^{t} (SAR) V_{vox} dt, \tag{3.284}$$

$$\gamma B_1(t) = \frac{V_{vox}}{\hbar}(SAR)(t - t_0). \tag{3.285}$$

SAR and Time-dependent NMR Flow Equation: The time dependent MRI signal is related to the SAR using the time independent NMR equation as given by (Dada and Awojoyogbe 2021). If the signal is sampled when the transverse component of the magnetisation has the *highest possible amplitude*, it follows that $M_0 \approx 0$ (Awojoyogbe et al. 2011). Provided that the condition $T_g \ll \gamma^2 B_1^2(t)$ holds, we can write:

$$\frac{d^2 M_y}{dt^2} + T_0 \frac{dM_y}{dt} + \gamma^2 B_1^2(t) M_y = 0. \tag{3.286}$$

From Eqs. (3.285), (3.286) becomes:

$$\frac{d^2 M_y}{dt^2} + T_0 \frac{dM_y}{dt} + \left(\frac{V_{vox}}{\hbar}\right)^2 (SAR)^2 (t - t_0)^2 M_y = 0. \tag{3.287}$$

Setting

$$\alpha = \beta(t - t_0), \tag{3.288}$$

where $\beta = SAR$ is assumed to be fairly constant in time. Therefore

$$\begin{aligned} &\frac{d\alpha}{dt} = \beta,\ \frac{d^2\alpha}{dt^2} = 0,\ \frac{dM_y}{dt} = \frac{dM_y}{d\alpha}\frac{d\alpha}{dt} = \beta\frac{dM_y}{d\alpha}, \\ &\frac{d^2 M_y}{dt^2} = \frac{d^2 M_y}{d\alpha^2}\left(\frac{d\alpha}{dt}\right)^2 + \frac{dM_y}{d\alpha}\frac{d^2\alpha}{dt^2} = \beta^2\frac{d^2 M_y}{d\alpha^2}. \end{aligned} \tag{3.289}$$

With these expressions, Eq. (3.287) becomes

$$\beta^2 \frac{d^2 M_y}{d\alpha^2} + T_o \beta \frac{dM_y}{d\alpha} + \left(\frac{V_{vox}}{\hbar}\right)^2 \alpha^2 M_y = 0. \tag{3.290}$$

However, from Eq. (3.288), $\beta = \frac{\alpha}{(t-t_0)}$

$$\frac{\alpha^2}{(t-t_0)^2}\frac{d^2M_y}{d\alpha^2} + \frac{\alpha T_o}{(t-t_0)}\frac{dM_y}{d\alpha} + \left(\frac{V_{vox}}{\beta\hbar}\right)^2\alpha^2 M_y = 0. \tag{3.291}$$

Multiplying all through by $(t-t_0)^2$, we have

$$\alpha^2\frac{d^2M_y}{d\alpha^2} + T_o(t-t_0)\alpha\frac{dM_y}{d\alpha} + \left(\frac{V_{vox}(t-t_0)}{\beta\hbar}\right)^2\alpha^2 M_y = 0, \tag{3.292}$$

$$\alpha^2\frac{d^2M_y}{d\alpha^2} + T_o(t-t_0)\alpha\frac{dM_y}{d\alpha} + \left(\frac{V_{vox}}{\beta^2\hbar}\right)^2\alpha^4 M_y = 0. \tag{3.293}$$

Equation (3.293) is a generalized Emden–Fowler differential equation of the form (Demir and Süngü 2009; Panayotounakos and Kravvaritis 2006)

$$\alpha^2\frac{d^2M_y}{d\alpha^2} + a\alpha\frac{dM_y}{d\alpha} + b\alpha^{m+1}M_y^n = 0. \tag{3.294}$$

That is, Eq. (3.294) is the Lane-Fowler differential equation for which

$$n = 1, m = 3, a = T_o(t-t_0), b = \left(\frac{V_{vox}}{\beta^2\hbar}\right)^2.$$

Equation (3.250) is transformable to Bessel differential equation with solution given as (Demir and Süngü 2009; Panayotounakos and Kravvaritis 2006; Wylie and Barrett 1982):

$$\begin{aligned} M_y(t) &= (SAR)^{\frac{1-T_0(t-t_0)}{2}}(t-t_0)^{\frac{1-T_0(t-t_0)}{2}} \\ &\left[C_1 J_n\left(\frac{V_{vox}}{2\hbar}(t-t_0)^2\right) + C_2 Y_n\left(\frac{V_{vox}}{2\hbar}(t-t_0)^2\right)\right]; \\ n &= \frac{1-T_0 t}{2}. \end{aligned} \tag{3.295}$$

This solution is valid for $T_0(t-t_0) \le 1$.
If the RF B_1 field is applied at time $t_0 = 0$, Eq. (3.295) becomes:

$$M_y(t) = (SAR)^{\frac{1-T_0 t}{2}}(t)^{\frac{1-T_0 t}{2}}\left[C_1 J_n\left(\frac{V_{vox}}{2\hbar}t^2\right) + C_2 Y_n\left(\frac{V_{vox}}{2\hbar}t^2\right)\right]. \tag{3.296}$$

It is always required that the transverse magnetisation be finite (NMR signal must always be measurable) as time tends to infinity, therefore $C_2 = 0$ and

$$M_y(t) = C_1(SAR)^{\frac{1-T_0 t}{2}}(t)^{\frac{1-T_0 t}{2}} J_{\frac{1-T_0 t}{2}}\left(\frac{V_{vox}}{2\hbar}t^2\right). \tag{3.297}$$

Therefore, for this theranostics model, the system behaves according to the expression:

$$\left.\begin{aligned} T(t) &= T_b + \frac{SAR}{w_b c_b}\left\{1 - \exp\left(-\frac{w_b c_b}{\rho c}t\right)\right\} \\ M_y(t) &= C_1(SAR)^{\frac{1-T_0 t}{2}}(t)^{\frac{1-T_0 t}{2}} J_{\frac{1-T_0 t}{2}}\left(\frac{V_{vox}}{2\hbar}t^2\right) \end{aligned}\right\}. \tag{3.298}$$

3.3.6.2 Computational Design of RF Controlled Theranostic Model for Evaluation of Tissue Bio Thermal Response

Human disease management strategies are gradually shifting from the traditional approach towards personalized medicine, in which specific drugs are specifically administered to a patient at the right time. Personalized medicine is now been explored because of the discovery that heterogeneous diseases may be attacked effectively by developing specific treatment plans which are developed specifically for a given disease (Rai et al. 2010a, b; Blanchet 2010). Theranostics has proved to be very crucial in these treatment plans because it could increase the efficiency of drug development processes by elucidating mapping of patients to benefits of new drugs (Falk 2008); thus improving health economic planning since physicians now have more control over the choice of optimal and cost effective therapy to be employed. Thermal therapy comprising of clinical hyperthermia and therapeutic hypothermia has been subject of recent research (Hanusch et al. 2013). However, both have suffered in clinical trials because medical monitoring is not impressive and delivering the appropriate amount of heat to the appropriate part of the patient's tissues is challenging (Wust et al. 2002; Falk and Issels 2001).

Hence, this model is aimed at addressing these challenges by technically merging the principles of MR relaxation to the Bioheat transfer phenomenon. Analytic solutions to the Penne's Bioheat equation were obtained and the expressions are connected to RF power absorption in tissue MRI which is dependent on the tissue relaxation times. A carefully selected RF field is used to control the the thermal process for hyperthermia and therapeutic hypothermia. The solutions to the Bioheat equation are then applied to hyperthermia treatment of tumor and therapeutic hypothermia of the brain.

Changes in temperature within different tissues of the body during MR scans are stilldifficult to quantify in clinical practice (Brix et al. 2002). Hence, to estimate exposure of patients to RF field, the specific absorption rate (SAR) is normally used. SAR is the average energy per unit mass and time absorbed delivered to a region of the body per unit of mass and time. However, the temperature rise in localized tissue regions is not only dependent on the localized SAR and the duration of the MR sequence applied, but also on thermal conductivity and microvascular blood flow leading to thermal equilibration within the body. For the purpose of assessing microvascular heat transfer in biological tissues, different continuum models have been proposed in the form of a modified heat diffusion equation in which the effects of microvascular blood flow properly considered (Brix et al. 2002). However, continuum approaches often neglects the local contribution to the temperature field (Shih et al. 2007). Hence, heat transfer analysis often requires simultaneous consideration of transient and spatial heating in physiological systems. The Pennes bioheat transfer equation (Brix et al. 2002; Shih et al. 2007) has been used extensively to describe temperature distribution in various tissues and in thermal therapy (Chopra et al. 2006; Chen and Xu 2002; Liu et al. 2003). This distribution can be defined mathematically based on the Pennes's Bioheat equation in the presence of a source of RF heating as follows (Dada et al. 2013b):

$$\rho c\frac{\partial T}{\partial t} = \nabla(k\nabla T) + \rho SAR - W_0\rho_b c_b(T - T_b), \tag{3.299}$$

where$T$ is tissue temperature, c_b is specific heat of blood, c is specific heat, ρ is density, t is time of exposure, W_0 is the volumetric perfusion rate, k isthermal conductivity, T_b is arterial blood temperature and SAR is the local specific absorption rate. Considering the transient problem involving the Bioheat equation within a small voxel of human tissue and if x represents the reference depth of a tissue, then this is a 1D problem and Eq. (3.299) becomes:

$$\rho c\frac{\partial T}{\partial t} = k\frac{\partial^2 T}{\partial x^2} + \rho SAR - W_0\rho_b c_b(T - T_b). \tag{3.300}$$

Introducing a constant Q_0, we may set:

$$SAR = \frac{W_0\rho_b c_b Q_0}{\rho} = \frac{\beta_b Q_0}{\rho}, \tag{3.301}$$

And Eq. (3.300) becomes

$$\rho c\frac{\partial T}{\partial t} = k\frac{\partial^2 T}{\partial x^2} - \beta_b(T - T_b - Q_0). \tag{3.302}$$

Setting (Shih et al. 2007):

$$\theta = T - T_b - Q_0, \tag{3.303}$$

Equation (3.302) then becomes:

$$\rho c\frac{\partial \theta}{\partial t} = k\frac{\partial^2 \theta}{\partial x^2} - \beta_b\theta. \tag{3.304}$$

According to variable separation technique (Spiegel 1983):

$$\theta(x,t) = X(x)\Omega(t). \tag{3.305}$$

$$\rho c X\frac{d\Omega}{dt} = k\Omega\frac{d^2X}{dx^2} - \beta_b X\Omega. \tag{3.306}$$

$$\frac{\rho c}{k\Omega}\frac{d\Omega}{dt} + \frac{\beta_b}{k} = \frac{1}{X}\frac{d^2X}{dx^2}. \tag{3.307}$$

Equation (3.307) must be a constant (Spiegel 1983) $-\lambda^2$ such that:

$$\frac{d\Omega}{dt} = -\frac{k}{\rho c}\left(\frac{\beta_b}{k} + \lambda^2\right)\Omega, \tag{3.308}$$

$$\frac{d^2X}{dx^2} = -\lambda^2 X. \tag{3.309}$$

The solution to Eqs. (3.307) and (3.308) are given as follows:

$$\Omega(t) = A_0 e^{-\alpha\left(\frac{\beta_b}{k}+\lambda^2\right)t}, \tag{3.310}$$

$$X(x) = P_1\cos\lambda x + P_2\sin\lambda x. \tag{3.311}$$

According to Eqs. (3.305), (3.300) and (3.301), (3.303) becomes:

$$\theta(x,t) = e^{-\alpha\left(\frac{\beta_b}{k}+\lambda^2\right)t}(A_1\cos\lambda x + A_2\sin\lambda x), \tag{3.312}$$

where $\alpha = \frac{k}{\rho c}$ represents thermal diffusivity, $A_1 = A_0P_1$ and $A_2 = A_0P_2$.

Using initial boundary conditions that are realistic human tissues Bioheat transfer, it may be assumed that if at $x = 0$, only arterial blood is contributing to the measured temperature, then

$$T(x,t)|_{x=0} = T_b + Q_0. \tag{3.313}$$

Hence, from Eq. (3.313) it follows that:

$$\theta(x,t)|_{x=0} = 0. \tag{3.314}$$

Similarly, if at $t = 0$,

$$T(x,t)|_{t=0} = T_b + Q_0 + f(x). \tag{3.315}$$

$$f(x) = \theta(x,t)|_{t=0}. \tag{3.316}$$

According to the condition in Eq. (3.316), $A_1 = 0$ and

$$\theta(x,t) = A_2 e^{-\alpha\left(\frac{\beta_b}{k}+\lambda^2\right)t} \sin \lambda x. \tag{3.317}$$

Since there is no stringent restriction on λ, P_2may be replaced with $P_2(\lambda)$. Also, the integral could be evaluated over λ from 0 to ∞ and still give a solution (Spiegel 1983). Hence

$$\theta(x,t) = \int_0^\infty A_2(\lambda) e^{-\alpha\left(\frac{\beta_b}{k}+\lambda^2\right)t} \sin \lambda x d\lambda. \tag{3.318}$$

Then from Eqs. (3.314) and (3.318), *f(x)* becomes:

$$f(x) = \int_0^\infty A_2(\lambda) \sinh \lambda x d\lambda. \tag{3.319}$$

If f(x)is an odd function (Spiegel 1983),

$$A_2(\lambda) = \frac{2}{\pi}\int_0^\infty f(x) \sin \lambda x dx = \frac{2}{\pi}\int_0^\infty f(u) \sin \lambda u du. \tag{3.320}$$

Using Eq. (3.318),

$$\theta(x,t) = \frac{2}{\pi}\int_0^\infty \int_0^\infty f(u) e^{-\alpha\left(\frac{\beta_b}{k}+\lambda^2\right)t} \sin \lambda x \sin \lambda u d\lambda du. \tag{3.321}$$

Since $\frac{1}{2}[\cosh \lambda(u-x) - \cosh \lambda(u+x)] = \sin \lambda u \sin \lambda x$, the results in Eq. (3.321) is

$$\theta(x,t) = \frac{1}{\pi}\int_0^\infty \int_0^\infty f(u) e^{-\alpha\left(\frac{\beta_b}{k}+\lambda^2\right)t} [\cos \lambda(u-x) - \cos \lambda(u+x)] d\lambda du, \tag{3.322}$$

$$\theta(x,t) = \frac{1}{\pi}$$

$$\int_0^\infty f(v)\left\{\int_0^\infty e^{-\alpha\left(\frac{\beta_b}{k}+\lambda^2\right)t}\cos\lambda(u-x)d\lambda-\int_0^\infty e^{-\alpha\left(\frac{\beta_b}{k}+\lambda^2\right)t}\cos\lambda(u+x)d\lambda\right\}dv, \tag{3.323}$$

$$\theta(x,t)=\frac{1}{\pi}\int_0^\infty f(v)e^{-\frac{\alpha\beta_b}{k}t}\left\{\int_0^\infty e^{-\alpha\lambda^2 t}\cos\lambda(u-x)d\lambda-\int_0^\infty e^{-\alpha\lambda^2 t}\cos\lambda(u+x)d\lambda\right\}dv. \tag{3.324}$$

The integrals in Eq. (3.323) are (Spiegel 1983):

$$\int_0^\infty e^{-a\lambda^2}\cos b\lambda d\lambda=\frac{1}{2}\sqrt{\frac{\pi}{a}}e^{-\frac{b^2}{4a}}. \tag{3.325}$$

Therefore, Eq. (3.324) becomes:

$$\theta(x,t)=\frac{1}{2\sqrt{\pi\alpha t}}e^{-\frac{\alpha\beta_b}{k}t}\left[\int_0^\infty f(u)e^{-\frac{(u-x)^2}{4\alpha t}}du-\int_0^\infty f(u)e^{-\frac{(u+x)^2}{4\alpha t}}du\right]. \tag{3.326}$$

If $\frac{(u-x)}{2\sqrt{\alpha t}}=\varepsilon$ in the first integral and $\frac{(u+x)}{2\sqrt{\alpha t}}=\varepsilon$ in the second integral, $du=2\sqrt{\alpha t}d\varepsilon$:

$$\theta(x,t)=\frac{1}{\sqrt{\pi}}e^{-\frac{\alpha\beta_b}{k}t}\left[\int_{\frac{-x}{2\sqrt{\alpha t}}}^\infty e^{-\varepsilon^2}f(2\varepsilon\sqrt{\alpha t}+x)d\varepsilon-\int_{\frac{x}{2\sqrt{\alpha t}}}^\infty f(2\varepsilon\sqrt{\alpha t}-x)e^{-\varepsilon^2}d\varepsilon\right]. \tag{3.327}$$

In this study, f(x) is taken to be the tissue metabolic heat contribution to the estimated temperature. For localised hyperthermia treatment, a constant initial temperature is assumed such that $T_m=f(x)$:

$$\theta(x,t)=\frac{T_m}{\sqrt{\pi}}e^{-\frac{\alpha\beta_b}{k}t}\sqrt{\pi}\,erf\left(\frac{x}{2\sqrt{\alpha t}}\right)=T_m e^{-\frac{\alpha\beta_b}{k}t}erf\left(\frac{x}{2\sqrt{\alpha t}}\right). \tag{3.328}$$

From Eqs. (3.303), (3.305) and (3.328), we write:

$$T(x,t)=T_b+T_m e^{-\frac{\alpha\beta_b}{k}t}erf\left(\frac{x}{2\sqrt{\alpha t}}\right)+Q_0, \tag{3.329}$$

$$T(x,t)=T_b+T_m e^{-\frac{\alpha\beta_b}{k}t}erf\left(\frac{x}{2\sqrt{\alpha t}}\right)+\frac{\rho SAR}{W_0 c_b\rho_b}. \tag{3.330}$$

The thermal therapeutic part of the model has been dealt with already. Meanwhile, in order to introduce the diagnostic part of model, it is worthy of note that the absorbed RF power per unit volume of the tissue being investigated (including tumor) is given as (Abragam 1961):

$$\frac{P_{rf}}{V} = -\frac{\gamma B_1^2 \omega T_2}{(\omega - \omega_0)^2 T_2^2 + \gamma^2 B_1^2 T_1 T_2 + 1} M_0, \tag{3.331}$$

where ω is the irradiation frequency, $\omega_0 = \gamma B_0$ is the Larmor frequency corresponding to the static field, $\omega_1 = \gamma B_1$ is the frequency of RF-induced rotation, γ is the gyromagnetic ratio, T_1 is the transverse relaxation time, T_2 is the longitudinal relaxation and M_0 is the equilibrium magnetisation (Abragam 1961; Slichter 2013; Brix et al. 2008; Cowan 1997).

Specific absorption rate (SAR) is the rate at which energy is absorbed by human tissues when exposed to a RF field (Brix et al. 2008; Laurent et al. 2008). It is the power absorbed per mass of tissue or

$$SAR = \frac{P_{rf}}{m} = \frac{P_{rf}}{\rho V}, \tag{3.332}$$

where V represents volume of the tissues, m is the mass while is ρ the tissue density. Introducing Eq. (3.332) in Eq. (3.331),

$$\rho SAR = -\frac{\omega \gamma B_1^2 T_2 M_0}{1 + (\omega - \omega_0)^2 T_2^2 + \gamma^2 B_1^2 T_1 T_2}. \tag{3.333}$$

Finally, from Eq. (3.330),

$$T(x,t) = T_b - \left(\frac{\omega \gamma B_1^2 T_2}{W_0 \rho_b c_b}\right) \frac{M_0}{1 + (\omega - \omega_0)^2 T_2^2 + \gamma^2 B_1^2 T_1 T_2} + T_m e^{-\frac{\alpha \beta_b}{k} t} erf\left(\frac{x}{2\sqrt{\alpha t}}\right). \tag{3.334}$$

However, the solution which obtained in Eq. (3.334) is not rigid but rather; it is strictly dependent on the prevailing NMR condition. The following conditions are possible (Slichter 2013; Cowan 1997):

(i) For the adiabatic condition ($\gamma^2 B_1^2 T_1 T_2 >> 1$),

$$T(x,t) = T_b - \left(\frac{\gamma B_1^2 \omega T_2}{W_0 c_b \rho_b}\right) \frac{M_0}{(\omega - \omega_0)^2 T_2^2 + \gamma^2 B_1^2 T_1 T_2 + 1} + T_m e^{-\frac{\alpha \beta_b}{k} t} erf\left(\frac{x}{2\sqrt{\alpha t}}\right). \tag{3.335}$$

(ii) For non-adiabatic condition ($\gamma^2 B_1^2 T_1 T_2 \ll 1$),

$$T(x,t) = T_b - \left(\frac{\gamma B_1^2 \omega T_2}{W_0 c_b \rho_b}\right)\frac{M_0}{(\omega-\omega_0)^2 T_2^2 + 1} + T_m e^{-\frac{\alpha\beta_b}{k}t} erf\left(\frac{x}{2\sqrt{\alpha t}}\right). \tag{3.336}$$

(iii) At saturation ($\gamma^2 B_1^2 T_1 T_2 = 1$),

$$T(x,t) = T_b - \left(\frac{\gamma B_1^2 \omega T_2}{W_0 c_b \rho_b}\right)\frac{M_0}{(\omega-\omega_0)^2 T_2^2 + 2} + T_m e^{-\frac{\alpha\beta_b}{k}t} erf\left(\frac{x}{2\sqrt{\alpha t}}\right), \tag{3.337}$$

$$T(x,t) = T_b - \left(\frac{\omega_1^2 \omega T_2}{\gamma W_0 c_b \rho_b}\right)\frac{M_0}{(\omega-\omega_0)^2 T_2^2 + 2} + T_m e^{-\frac{\alpha\beta_b}{k}t} erf\left(\frac{x}{2\sqrt{\alpha t}}\right). \tag{3.338}$$

For RF pulse (Gupta et al. 2014),

$$\omega_1 = \gamma B_1 = \gamma G\xi x;\ \omega - \omega_0 = \gamma G\xi x, \tag{3.339}$$

where ξ is a dimensionless term. Hence, for a system with non-adiabatic behaviour:

$$T(x,t) = T_b - \left(\frac{\gamma^2 G^2 x^2 (B_0 + G\xi x) T_2}{W_0 \rho_b c_b}\right)\frac{\xi^2 M_0}{1 + (\gamma G\xi x)^2 T_2^2 + T_1 T_2 (\gamma G\xi x)^2} + T_m e^{-\frac{\alpha\beta_b}{k}t} erf\left(\frac{x}{2\sqrt{\alpha t}}\right). \tag{3.340}$$

3.3.7 *Quantum Mechanical Model of Transverse Magnetisation and Transverse Polarisation for Molecular Imaging*

It has been noted in chapter two that quantum mechanics is the most often used principle used to describe matter on the scale of molecules, atoms or elementary particles. Therefore, molecular imaging applications for which single particles/atoms are considered may be properly described by a quantum mechanical model. This is necessary because most of the applications which have been considered thus far dealt with spin aggregates in which molecular behaviours/dynamics are imprinted on transversemagnetisation serving as the aggregate map.

Meanwhile, an outline of quantum mechanical formulation for NMR system has been presented in chapter two while such formulation has been presented for optical NMR system (in terms of electric polarisation) as two-level atomic model (Hahn 1980; Cohen-Tannoudji et al. 1998; Eberly and Stroud 2006; Metcalf and Straten 2007; Chang and Minogin 2002). The study would address quantum mechanical phenomena in NMR and optical NMR from the perspective of the Bloch flow equations.

3.3.7.1 Quantum Mechanical Model of Transverse Magnetisation for Molecular Imaging

It is interesting to note that equation of motion for magnetic moment as presented in (Dada and Awojoyogbe 2021) can be obtained using the Heisenberg equation of motion for the magnetic moment. This section will concentrate on the Schrödinger formulation in order to access some interesting features of the quantum spin system.

The Schrödinger equation which describes the motion of a spin I (in terms of a wave function $\phi(t)$) (Slichter 2013; Callaghan 1994) is given as:

$$i\frac{\partial\phi(t)}{\partial t} = \frac{\aleph}{\hbar}\phi(t) = -\gamma\left\{B_0 I_z + B_1(I_x\cos\omega t + I_y\sin\omega t)\right\}\phi(t). \tag{3.341}$$

The transformation as given in (Dada and Awojoyogbe 2021) is modified to read

$$\psi(t) = e^{i\omega_z t I_z}\phi(t). \tag{3.342}$$

The Schrödinger equation becomes (Slichter 2013):

$$i\frac{\partial\psi(t)}{\partial t} = -\left[(\omega_z + \gamma B_0)I_z + \gamma B_1 I_x\right]\psi(t). \tag{3.343}$$

Based on the existence of resonance conditions at $\omega_z \approx -\gamma B_0$, the Hamiltonian in the rotating frame of reference is:

$$\aleph' = -\hbar\left[(\omega_z + \gamma B_0)I_z + \gamma B_1 I_x\right]. \tag{3.344}$$

However, setting (Abragam 1961)

$$\omega_z = -\omega. \tag{3.345}$$

Then, Eq. (3.244) may be written as:

$$\aleph' = -\hbar\left[(\omega_0 - \omega)I_z + \gamma B_1 I_x\right]. \tag{3.346}$$

Therefore, another form of Eq. (2.217) becomes:

$$-\frac{\hbar}{i}\frac{\partial\psi}{\partial t} = \aleph'\psi. \tag{3.347}$$

Since the Hamitonian operator is $\aleph'$ is commuting with the angular momentum operator $\vec{I}$, the equation of motion of the spin operator is given by Heisenberg equation as:

$$\frac{d\vec{I}}{dt} = -\frac{i}{\hbar}[\vec{I}, \aleph']. \tag{3.348}$$

Equation (3.348) leads to the following expressions (Merzbacher 1998; Abragam 1961; Slichter 2013):

$$\begin{aligned}\frac{dI_x}{dt} &= -\frac{i}{\hbar}[I_x, \aleph'] = -i(\omega_0 - \omega)[I_x, I_z] - i\gamma B_1[I_x, I_x]\\ &= -i(\omega_0 - \omega)(-iI_y) - i\gamma B_1(0) = -(\omega_0 - \omega)I_y,\end{aligned}$$

$$\begin{aligned}\frac{dI_y}{dt} &= -\frac{i}{\hbar}[I_y, \aleph'] = -i(\omega_0 - \omega)[I_y, I_z] - i\gamma B_1[I_y, I_x]\\ &= -i(\omega_0 - \omega)(iI_x) - i\gamma B_1(-iI_z) = (\omega_0 - \omega)I_x - \gamma B_1 I_z,\end{aligned}$$

$$\begin{aligned}\frac{dI_z}{dt} &= -\frac{i}{\hbar}[I_z, \aleph'] = -i(\omega_0 - \omega)[I_z, I_z] - i\gamma B_1[I_z, I_x]\\ &= -i(\omega_0 - \omega)(0) - i\gamma B_1(iI_y) = \gamma B_1 I_y,\end{aligned}$$

or

$$\frac{dI_x}{dt} = -(\omega_0 - \omega)I_y, \tag{3.349}$$

$$\frac{dI_y}{dt} = (\omega_0 - \omega)I_x - \gamma B_1 I_z, \tag{3.350}$$

$$\frac{dI_z}{dt} = \gamma B_1 I_y. \tag{3.351}$$

It is important to find an expression for I_z and I_x in terms of the relevant relaxation parameters. In terms of the angular momentum operator, the original Bloch equation (equivalent to the classical formulation) is (Cowan 1997; Haacke et al. 1999) is

$$\frac{d\vec{I}}{dt} = \gamma(\vec{I} \times \vec{B}_{eff}). \tag{3.352}$$

Provided a constant I_0 represents a form of the angular momentum related to the equilibrium magnetisation, relaxation process may be added to Eq. (3.353) to give

$$\frac{d\vec{I}}{dt} = \gamma(\vec{I} \times \vec{B}_{eff}) - \frac{\hat{x}I_x + \hat{y}I_y}{T_2} - \frac{\hat{z}(I_z - I_0)}{T_1}, \tag{3.353}$$

where the effective magnetic field is defined as

$$\vec{B}_{eff} = -\hat{z}(B_0 - B) - \hat{x}B_1. \tag{3.354}$$

If the spin angular momentum vector is defined as

$$\vec{I} = \hat{x}I_x + \hat{y}I_y + \hat{z}I_z. \tag{3.355}$$

Equation (3.354) becomes:

$$\frac{d\vec{I}}{dt} = \gamma(\vec{I} \times [-\hat{z}(B_0 - B) - \hat{x}B_1]) - \frac{\hat{x}I_x + \hat{y}I_y}{T_2} - \frac{\hat{z}(I_z - I_0)}{T_1}. \tag{3.356}$$

From Eq. (3.357), the following coupled differential equations are obtained:

$$\begin{aligned}
\frac{dI_x}{dt} &= \gamma \left| \vec{I} \times [-\hat{z}(B_0 - B) - \hat{x}B_1] \right|_x - \frac{I_x}{T_2}, \\
\frac{dI_y}{dt} &= \gamma \left| \vec{I} \times [-\hat{z}(B_0 - B) - \hat{x}B_1] \right|_y - \frac{I_y}{T_2}, \\
\frac{dI_z}{dt} &= \gamma \left| \vec{I} \times [-\hat{z}(B_0 - B) - \hat{x}B_1] \right|_z - \frac{(I_z - I_0)}{T_1}.
\end{aligned} \tag{3.357}$$

The cross product in the expression is given as:

$$\begin{aligned}
\vec{I} \times [-\hat{z}(B_0 - B) - \hat{x}B_1] &= \begin{vmatrix} \hat{x}\ \hat{y}\ \hat{z} \\ I_x\ I_y\ I_z \\ -B_1\ 0\ -(B + B_0) \end{vmatrix} \\
&= -\hat{x}(B_0 - B)I_y - \hat{y}(-(B_0 - B)I_x + B_1 I_z) + \hat{z}B_1 I_y \\
&= -\hat{x}(B_0 - B)I_y + \hat{y}((B_0 - B)I_x - B_1 I_z) + \hat{z}B_1 I_y.
\end{aligned} \tag{3.358}$$

From Eqs. (3.358), (3.357) becomes:

$$\begin{aligned}
\frac{dI_x}{dt} &= -\gamma(B_0 - B)I_y - \frac{I_x}{T_2}, \\
\frac{dI_y}{dt} &= \gamma((B_0 - B)I_x - B_1 I_z) - \frac{I_y}{T_2}, \\
\frac{dI_z}{dt} &= \gamma B_1 I_y - \frac{(I_z - I_0)}{T_1}.
\end{aligned}$$

or

$$\frac{dI_x}{dt} = -(\omega_0 - \omega)I_y - \frac{I_x}{T_2}, \tag{3.359}$$

$$\frac{dI_y}{dt} = (\omega_0 - \omega)I_x - \gamma B_1 I_z - \frac{I_y}{T_2}, \quad (3.360)$$

$$\frac{dI_z}{dt} = \gamma B_1 I_y - \frac{(I_z - I_0)}{T_1}. \quad (3.361)$$

It could be seen therefore that Eqs. (3.359)–(3.361) is the complete Bloch equation for the spin system. From these equations,

$$\frac{dI_x}{dt} + \frac{I_x}{T_2} = -(\omega_0 - \omega)I_y, \quad (3.362)$$

$$\frac{dI_y}{dt} + \frac{I_y}{T_2} = (\omega_0 - \omega)I_x - \gamma B_1 I_z, \quad (3.363)$$

$$\frac{dI_z}{dt} + \frac{(I_z - I_0)}{T_1} = \gamma B_1 I_y. \quad (3.364)$$

In the absence of NMR relaxation, Eqs. (3.363–3.365) reduce to Eqs. (3.350–3.352). By close observation, it would be seen that adding the relaxation terms to Eq. (3.353) did not modify the nature of $\frac{d\vec{I}}{dt}$. That is, if Eq. (3.354) is re-arranged as follows:

$$\frac{d\vec{I}}{dt} + \frac{\hat{x}I_x + \hat{y}I_y}{T_2} + \frac{\hat{z}(I_z - I_0)}{T_1} = \gamma(\vec{I} \times \vec{B}_{eff}), \quad (3.365)$$

then, the natural behaviour of Eqs. (3.353) and (3.366) must be nearly the same. Hence, if a new rate $\frac{d\vec{I}^r}{dt}$ is defined to describe the spin system in which NMR relaxation is present, Eq. (3.368) becomes:

$$\frac{d\vec{I}^r}{dt} = \frac{d\vec{I}}{dt} + \frac{\hat{x}I_x + \hat{y}I_y}{T_2} + \frac{\hat{z}(I_z - I_0)}{T_1}, \quad (3.366)$$

and according to Eq. (3.356)

$$\frac{d\vec{I}^r}{dt} = \hat{x}\left(\frac{dI_x}{dt} + \frac{I_x}{T_2}\right) + \hat{y}\left(\frac{dI_y}{dt} + \frac{I_y}{T_2}\right) + \hat{z}\left(\frac{dI_z}{dt} + \frac{(I_z - I_0)}{T_1}\right). \quad (3.367)$$

From this expression,

$$I_x^r = I_x + \int \frac{I_x}{T_2} dt, \quad (3.368)$$

$$I_y^r = I_y + \int \frac{I_y}{T_2} dt, \quad (3.369)$$

$$I_z^r = I_z + \int \frac{(I_z - I_0)}{T_1} dt. \tag{3.370}$$

Therefore, Eqs. (3.362–3.364) become:

$$\frac{dI_x^r}{dt} = -(\omega_0 - \omega) I_y, \tag{3.371}$$

$$\frac{dI_y^r}{dt} = (\omega_0 - \omega) I_x - \gamma B_1 I_z, \tag{3.372}$$

$$\frac{dI_z^r}{dt} = \gamma B_1 I_y. \tag{3.373}$$

Meanwhile, Eqs. (3.348) and (3.352) shows that

$$\gamma(\vec{I} \times \vec{B}_{eff}) = -\frac{i}{\hbar}[\vec{I}, \aleph']. \tag{3.374}$$

Hence, Eqs. (3.366–3.373) show that

$$\frac{d\vec{I}^r}{dt} = -\frac{i}{\hbar}[\vec{I}^r, \aleph']. \tag{3.375}$$

This is in agreement with studies on the density matrix formulation for spin systems (Redfield 1996). The time evolution operator converts a state vector at a given time, $|\psi(t = t_0)\rangle = |\psi\rangle$ into the state vector at the time $t, |\psi(t)\rangle$. Then the expectation value of an operator $\vec{I}^r$ is associated with a time-dependent state vectors $|\psi(t)\rangle$. This is given as follows (Burkhardt and Leventhal 2008; Bransden and Joachain 2000; Cohen-Tannoudji et al. 1977; Merzbacher 1998; Peleg et al. 1998)

$$\left\langle \vec{I}^r(t) \right\rangle = \langle \psi(t) | \vec{I}^r | \psi(t) \rangle. \tag{3.376}$$

Using the time evolution operator $\hat{U}(t, t_0)$, the state vector becomes:

$$|\psi(t)\rangle = \hat{U}(t, t_0) |\psi(t_0)\rangle. \tag{3.377}$$

Hence, Eq. (3.377) becomes (Burkhardt and Leventhal 2008; Abragam 1961):

$$\left\langle \vec{I}^r(t) \right\rangle = \langle \psi(t_0) | \hat{U}^{-1}(t, t_0) \vec{I}^r \hat{U}(t, t_0) | \psi(t_0) \rangle, \tag{3.378}$$

where $\hat{U}^{-1}(t, t_0)$ is the adjoint of $\hat{U}(t, t_0)$. In order to obtain a time-dependent operator, from Eq. (3.378),

$$\vec{I}^r(t) = \hat{U}^{-1}(t, t_0) \vec{I}^r \hat{U}(t, t_0). \tag{3.379}$$

Since the time evolution operators may be defined as follows

$$\hat{U}(t, t_0) = e^{i\frac{\aleph'}{\hbar}(t-t_0)},$$
$$\hat{U}^{-1}(t, t_0) = e^{-i\frac{\aleph'}{\hbar}(t-t_0)}. \tag{3.380}$$

Equation (3.380) becomes:

$$e^{i\frac{\aleph'}{\hbar}(t-t_0)}\vec{I}^r e^{-i\frac{\aleph'}{\hbar}(t-t_0)} = \vec{I}^r(t). \tag{3.381}$$

Hence, the two ways of representing the time-dependent expectation value of a quantum mechanical angular spin momentum has been shown. This concept of quantum mechanics represents the Schrödinger picture (Burkhardt and Leventhal 2008). Fortunately, the alternative formulation in which the state vectors remain fixed in time and the operators are dynamic, which is the Heisenberg picture has been used extensively in the generation of the Bloch equations. Therefore, it is interesting to state that the Schrödinger picture, which is themost widely used formulationof quantum mechanics, is equivalent to the Heisenberg formulation (Burkhardt and Leventhal 2008; Bransden and Joachain 2000; Cohen-Tannoudji et al. 1977; Merzbacher 1998; Peleg et al. 1998; Sakurai 1994).

Within the frameworks of the Heisenberg picture, the time dependence of the observables and operators completely describe the physical system. State vectors do not have significant influence on the system because theyare evaluated at a fixed time. It is worthy of note that theeigenvalues are the same in both pictures while the eigenvectors are time dynamic in the Heisenberg formulation (Burkhardt and Leventhal 2008; Bransden and Joachain 2000).

Now let the eigenvalue of the spin angular momentum operator $\vec{I}^r$ be represented as a_i in the Schrödinger formulation. The corresponding operator of the Heisenberg formulation is $\vec{I}^r(t)$. If the eigenkets of $\vec{I}^r$ in the Schrödinger picture is given as $|a_i\rangle$, the eigenvalue equation at $t = 0$ is:

$$\vec{I}^r|a_i\rangle = a_i|a_i\rangle. \tag{3.382}$$

Operating the time evolution function at $t_0 = 0$ on Eq. (3.383) gives:

$$\hat{U}^{-1}\vec{I}^r(\hat{U}\hat{U}^{-1})|a_i\rangle = a_i\hat{U}^{-1}|a_i\rangle, \tag{3.383}$$

$$\hat{U}^{-1}\vec{I}^r\hat{U}(\hat{U}^{-1}|a_i\rangle) = a_i(\hat{U}^{-1}|a_i\rangle). \tag{3.384}$$

According to Eqs. (3.379), (3.384) becomes:

$$\vec{I}^r(t)(\hat{U}^{-1}|a_i\rangle) = a_i(\hat{U}^{-1}|a_i\rangle). \tag{3.385}$$

This means that Eq. (3.386) is the eigenvalue equation for the spin operator $\vec{I}^r(t)$ having the eigenvectors $\hat{U}^{-1}|a_i\rangle$. This shows thatthe eigenvalues a_i in Heisenberg picture are the same as that in the Schrödinger formulation and this is in line with Postulate I of quantum mechanics; because the possible results of a measurement of the observable $\vec{I}^r$ is independent of the representation.

Then, if the result of the Schrödinger picture presented in Eq. (3.382) is differentiated with respected to time,

$$\frac{d\vec{I}^r(t)}{dt} = i\frac{\aleph'}{\hbar}\left(e^{i\frac{\aleph'}{\hbar}(t-t_0)}\vec{I}^r e^{-i\frac{\aleph'}{\hbar}(t-t_0)}\right) - i\frac{\aleph'}{\hbar}\left(e^{i\frac{\aleph'}{\hbar}(t-t_0)}\vec{I}^r e^{-i\frac{\aleph'}{\hbar}(t-t_0)}\right),$$
$$\frac{d\vec{I}^r(t)}{dt} = \frac{i}{\hbar}\left\{\aleph'\left(e^{i\frac{\aleph'}{\hbar}(t-t_0)}\vec{I}^r e^{-i\frac{\aleph'}{\hbar}(t-t_0)}\right) - \aleph'\left(e^{i\frac{\aleph'}{\hbar}(t-t_0)}\vec{I}^r e^{-i\frac{\aleph'}{\hbar}(t-t_0)}\right)\right\}. \quad (3.386)$$

since the Hamiltonian $\aleph'$ commutes with itself,

$$\frac{d\vec{I}^r(t)}{dt} = \frac{i}{\hbar}\left\{\aleph'\left(e^{i\frac{\aleph'}{\hbar}(t-t_0)}\vec{I}^r e^{-i\frac{\aleph'}{\hbar}(t-t_0)}\right) - \left(e^{i\frac{\aleph'}{\hbar}(t-t_0)}\vec{I}^r e^{-i\frac{\aleph'}{\hbar}(t-t_0)}\right)\aleph'\right\},$$
$$\frac{d\vec{I}^r(t)}{dt} = \frac{i}{\hbar}\left\{\aleph'\vec{I}^r(t) - \vec{I}^r(t)\aleph'\right\},$$
$$\frac{d\vec{I}^r(t)}{dt} = -\frac{i}{\hbar}\left\{\vec{I}^r(t)\aleph' - \aleph'\vec{I}^r(t)\right\},$$
$$\frac{d\vec{I}^r(t)}{dt} = -\frac{i}{\hbar}[\vec{I}^r(t), \aleph']. \quad (3.387)$$

It could be observed that Eqs. (3.388) and (3.376) are exactly the same. Therefore, if for the NMR relaxation system defined by Eq. (3.376), there is a quantum NMR wavefunction associated with the system defined as ψ_q; then the Schrödinger equation for such system is

$$\frac{d\psi_q}{dt} = -\frac{i}{\hbar}\aleph'\psi_q. \quad (3.388)$$

Since the system has been demonstrated to behave quantum mechanically, the energy equation may be written as follows:

$$\aleph'\psi_q = E_q\psi_q, \quad (3.389)$$

where E_q is the eigenvalue obtainable from the term (Abragam 1961). According to Eqs. (3.389), (3.388) becomes:

$$\frac{d\psi_q}{dt} = -\frac{i}{\hbar}E_q\psi_q. \quad (3.390)$$

The dynamics of molecular magnets which are usually studied in NMR systems is not only dependent on time but also on spatial coordinates (Slichter 2013; Callaghan 1991; Haacke et al. 1999). Therefore, it is may be assumed that there is a component of the wavefunction in Eq. (3.391) which is spatially dependent such that

$$\Psi_q(x, t) = \phi(x)\psi_q(t). \tag{3.391}$$

This implies that $\phi(x)$ is the component of a generalized wavefunction $\Psi_q(x, t)$ that has been reduced to $\psi_q(t)$ in the quantum formulation of NMR spin system. In fact, Eq. (3.392) effectively reduces to $\psi_q(t)$ if a particle is evaluated at any point x_0 in space such that $\phi(x_0)$ now becomes part constant of solution to the time-dependent Schrödinger equation. Hence, the transformed Hamiltonian operator for nuclear magnetic resonance now becomes (Slichter 2013; Borgonovi et al. 2006)

$$\aleph' = -\hbar\big[(\omega_0 - \omega)I_z + \gamma B_1 I_x\big] = \frac{1}{2m}\Big(\hat{p} - \frac{q}{c}\vec{A}\Big)^2 + V(x, t), \tag{3.392}$$

where q represents the particle charge, V(x,t) is sytem potential, c represents the speed of electromagnetic wave, m is mass of the particle, $\hat{p}$ is the momentum operator and $\vec{A}$ is the vector potential for both static magnetic field and the RF field. If another assumption is made such that the potential function is not dependent on time and the effects of $\hat{p}$ is dominant, Eq. (3.393) may be written as:

$$\aleph' = \frac{\hat{p}'^2}{2m} + V(x);\ \hat{p}' = \hat{p} - \frac{q}{c}\vec{A} \approx -\frac{i\hbar}{2m}\frac{\partial}{\partial x}. \tag{3.393}$$

For generalized problems,

$$\aleph' = \frac{\hat{p}'^2}{2m} + V(\vec{r});\ \hat{p}' = \hat{p} - \frac{q}{c}\vec{A} \approx -\frac{i\hbar}{2m}\nabla. \tag{3.394}$$

Hence, the time-dependent Schrödinger equation is given as

$$\frac{\partial \Psi_q(x, t)}{\partial t} = -\frac{i}{\hbar}\bigg(-\frac{\hbar^2}{2m}\frac{\partial^2 \Psi_q(x, t)}{\partial x^2} + V(x)\Psi_q(x, t)\bigg), \tag{3.395}$$

$$\frac{\partial \Psi_q(x, t)}{\partial t} = \frac{1}{i\hbar}\bigg(-\frac{\hbar^2}{2m}\frac{\partial^2 \Psi_q(x, t)}{\partial x^2} + V(x)\Psi_q(x, t)\bigg),$$

$$i\hbar\frac{\partial \Psi_q(x, t)}{\partial t} = -\frac{\hbar^2}{2m}\frac{\partial^2 \Psi_q(x, t)}{\partial x^2} + V(x)\Psi_q(x, t). \tag{3.396}$$

This is the general time-dependent Schrödinger equation in a spin system. It is worthy of note that if the condition stated in Eq. (3.394) does not hold, the form of Eq. (3.396) would not change drastically; few terms may just be required to be added to the potential function. Using Eq. (3.392) in Eq. (3.397) gives

$$i\hbar\frac{\partial}{\partial t}\phi(x)\psi_q(t) = -\frac{\hbar^2}{2m}\frac{\partial^2}{\partial x^2}\phi(x)\psi_q(t) + V(x)\phi(x)\psi_q(t), \tag{3.397}$$

$$i\hbar\frac{1}{\psi_q(t)}\frac{d\psi_q(t)}{dt} = -\frac{\hbar^2}{2m}\frac{1}{\phi(x)}\frac{d^2\phi(x)}{dx^2} + V(x).$$

According to the principles of separable solutions to differential equations, both sides of the above equation must be equal to the eigenvalue energy of $\aleph'$. That is,

$$-\frac{\hbar^2}{2m}\frac{1}{\phi(x)}\frac{d^2\phi(x)}{dx^2} + V(x) = E_q = \frac{i\hbar}{\psi_q(t)}\frac{\partial\psi_q(t)}{\partial t}, \tag{3.398}$$

where the following equations could be obtained

$$\frac{\partial\psi_q(t)}{\partial t} = \frac{1}{i\hbar}E_q\psi_q(t) = -\frac{i}{\hbar}E_q\psi_q(t), \tag{3.399}$$

$$\frac{d^2\phi(x)}{dx^2} = -\frac{2m}{\hbar^2}(E_q - V(x))\phi(x). \tag{3.400}$$

Equation (3.400) is essentially equal to Eq. (3.91). Equation (3.401) is the time-independent Schrödinger equation. In multi-dimensions, Eq. (3.401) becomes

$$\nabla^2\phi(x) = -\frac{2m}{\hbar^2}(E_q - V(x))\phi(x). \tag{3.401}$$

The Roles of Macroscopic Magnetisation in Quantum Measurements

In quantum mechanics, it is a single spin that is considered. In most MRI applications, the macroscopic magnetisation is measured except in magnetic resonance microscopy. According to the formulations in earlier sections [especially in Eq. (3.377)], the expectation value of the angular momentum of a single spin is given as:

$$\langle\vec{I}^r\rangle = \langle\Psi_q(x,t)|\vec{I}^r|\Psi_q(x,t)\rangle. \tag{3.402}$$

Since the magnetic moment of the spin is (Slichter 2013) $\vec{\mu}^r = \gamma\hbar\vec{I}^r$, the corresponding expectation value is given as follows:

$$\gamma\hbar\langle\vec{I}^r\rangle = \gamma\hbar\langle\Psi_q(x,t)|\vec{I}^r|\Psi_q(x,t)\rangle, \tag{3.403}$$

$$\langle\vec{\mu}^r\rangle = \langle\Psi_q(x,t)|\vec{\mu}^r|\Psi_q(x,t)\rangle. \tag{3.404}$$

The magnetisation of a single spin is obtained by simply dividing Eq. (3.405) all through by the volume of the voxel (Cowan 1997; Slichter 2013) to be imaged:

$$M_q(t) = \frac{1}{V}\langle \vec{\mu}^r \rangle = \langle \Psi_q(x,t) | \vec{M}^r | \Psi_q(x,t) \rangle. \tag{3.405}$$

Summing over all the N spins in a voxel gives the macroscopic magnetisation given as:

$$M(t) = NM_q(t) = \sum_{n=1}^{N} M_q(t) = \sum_{n=1}^{N} \langle \Psi_q(x,t) | \vec{M}^r | \Psi_q(x,t) \rangle. \tag{3.406}$$

Since the transverse component of this magnetisation is usually involved in the induction of energy in the signal coil, transverse magnetisation has the form

$$\sum_{n=1}^{N} \langle \Psi_q(x,t) | \vec{M}_y^r | \Psi_q(x,t) \rangle M_y(t). \tag{3.407}$$

Once the wave function is defined, it is quite easy to evaluate the contribution of a quantum spin to the macroscopic magnetisation as follows:

$$\int \Psi_q(x,t) \vec{M}_y^r \Psi_q^*(x,t) dx = \langle \Psi_q(x,t) | \vec{M}_y^r | \Psi_q(x,t) \rangle, \tag{3.408}$$

$$M_y(t) = \sum_{n=1}^{N} \int \Psi_q(x,t) \vec{M}_y^r \Psi_q^*(x,t) dx. \tag{3.409}$$

Derivation of the Nonlinear Schrödinger Equation for NMR Quantum Imaging

Nonlinear quantum mechanics is a theory for studying properties and dynamics of microscopic particles in nonlinear systems that exhibit quantum features. It was so named in relation to the quantum mechanics established by Bohr, Heisenberg, Schrödinger, and many others (Xiao-Feng and Yuan-Ping 2005). TheSchrödingerequation considered thus far deals with only properties and motion of microscopic particles in linear systems and it is usually known as linear quantum mechanics.

A system which does not satisfy the superposition principle is known as a nonlinear system. The meaning of this is that the output of a nonlinear system is not directly proportional to the input (Khalil and Grizzle 1996). Nonlinearity is usually found in physical phenomena such as gas dynamics, fluid mechanics, ecology, plasma mechanics, elasticity, chemical reactions, relativity, combustion and biomechanics. Then for molecular processes with significant level of unpredictability, the nonlinear Schrödinger equation may be necessary for describing the motion of a single spin.

In order to derive the Nonlinear Schrödinger equation, it is worthy of note that the NMR system deals with the application and analysis of electromagnetic waves

(Cowan 1997; Haacke et al. 1999). Such electromagnetic waves must obey the general wave equation (Merzbacher 1998; Metcalf and Straten 2007; Meystre 2006):

$$\nabla^2\Theta(\vec{r},t) - \frac{1}{c^2}\frac{\partial^2\Theta(\vec{r},t)}{\partial t^2} = 0, \quad (3.410)$$

where $\Theta(\vec{r},t)$ is a function of the wave motion and c is the propagation speed. A plane wave solution to Eq. (3.410) may be given as:

$$\Theta(\vec{r},t) = \Theta_0 e^{\frac{i}{\hbar}(p\vec{r}-E_q t)}, \quad (3.411a)$$

where p is the momentum of the EM wave, E_q is the total energy of the wave. Substituting for this solution in Eq. (3.411) gives:

$$\left(\nabla^2 - \frac{1}{c^2}\frac{\partial^2}{\partial t^2}\right)\Theta_0 e^{\frac{i}{\hbar}(p\vec{r}-E_q t)} = 0, \quad (3.411b)$$

$$-\frac{1}{\hbar^2}\left(p^2 - \frac{E_q^2}{c^2}\right)\Theta_0 e^{\frac{i}{\hbar}(p\vec{r}-E_q t)} = 0. \quad (3.412)$$

This shows

$$p^2 - \frac{E_q^2}{c^2} = 0,$$

$$E_q^2 = p^2c^2. \quad (3.413)$$

Since radio waves are a type of electromagnetic radiation, they travel at the speed of light. Hence, a relativistic correction is needed for the total energy given in Eq. (3.413). Therefore, this modification is given as follows (Field 2004; Nelson 1966; Fritsche and Haugk 2003):

$$E_q^2 = p^2c^2 + m^2c^4. \quad (3.414)$$

Consequently, Eq. (3.412) becomes

$$-\frac{1}{\hbar^2}\left(p^2 - \frac{E_q^2}{c^2} + m^2c^2\right)\Theta_0 e^{\frac{i}{\hbar}(p\vec{r}-E_q t)} = 0, \quad (3.415)$$

$$\left(-\frac{1}{\hbar^2}\left(p^2 - \frac{E_q^2}{c^2}\right) - \frac{m^2c^2}{\hbar^2}\right)\Theta_0 e^{\frac{i}{\hbar}(p\vec{r}-E_q t)} = 0. \quad (3.416)$$

According to Eqs. (3.411b) and (3.412):

$$\nabla^2 - \frac{1}{c^2}\frac{\partial^2}{\partial t^2} = -\frac{1}{\hbar^2}\left(p^2 - \frac{E_q^2}{c^2}\right). \tag{3.417}$$

Equation (3.417) becomes

$$\left(\left(\nabla^2 - \frac{1}{c^2}\frac{\partial^2}{\partial t^2}\right) - \frac{m^2c^2}{\hbar^2}\right)\Theta_0 e^{\frac{i}{\hbar}(p\vec{r} - E_q t)} = 0, \tag{3.418}$$

$$\nabla^2\Theta - c^2\frac{m^2}{\hbar^2}\Theta = \frac{1}{c^2}\frac{\partial^2\Theta}{\partial t^2}. \tag{3.419}$$

Equation (3.419) is known as the Klein–Gordon equation (Field 2004; Nelson 1966; Fritsche and Haugk 2003). However, Schrödinger equation is not relativistic just like most NMR systems. Consequently, Eq. (3.414) needs to be modified as follows

$$E_q^2 = m^2c^4\left(1 + \frac{p^2}{m^2c^2}\right),$$
$$E_q = mc^2\sqrt{\left(1 + \frac{p^2}{m^2c^2}\right)}. \tag{3.420}$$

Expanding the parameter $\sqrt{\left(1 + \frac{p^2}{m^2c^2}\right)}$ and neglecting higher other terms lead to

$$\sqrt{\left(1 + \frac{p^2}{m^2c^2}\right)} \approx \left(1 + \frac{p^2}{2m^2c^2}\right), \tag{3.421}$$

$$E_q \approx mc^2\left(1 + \frac{p^2}{2m^2c^2}\right) \approx mc^2 + \frac{p^2}{2m} \approx mc^2 + E_k. \tag{3.422}$$

where E_k is the macroscopic kinetic energy describing aggregates of spins. Therefore, Eq. (3.411a) becomes:

$$\Theta(\vec{r}, t) = \Theta_0 e^{\frac{i}{\hbar}(p\vec{r} - mc^2 t - E_k t)}. \tag{3.423}$$

In real NMR systems, the spin velocity is small (Cowan 1997; Haacke et al. 1999) such that

$$mv \ll mc. \tag{3.424}$$

This condition implies

$$p^2 \ll m^2c^2. \tag{3.425}$$

The implication of Eqs. (3.424) and (3.425) is that the term $e^{\frac{i}{\hbar}(-mc^2t)}$ oscillates much faster than $e^{\frac{1}{\hbar}(-E_kt)}$. Assuming a reduced form of Eq. (3.423) is related to Eq. (3.392) such that

$$\Theta(x,t) = e^{\frac{-imc^2t}{\hbar}}\Psi_q(x,t);\ \Psi_q(x,t) = \Theta_0 e^{\frac{i}{\hbar}(px-E_kt)}. \tag{3.426}$$

Differentiating Eq. (3.426) gives

$$\frac{\partial\Theta(x,t)}{\partial t} = \frac{-imc^2}{\hbar}e^{\frac{-imc^2t}{\hbar}}\Psi_q + e^{\frac{-imc^2t}{\hbar}}\frac{\partial\Psi_q}{\partial t}, \tag{3.427}$$

$$\frac{\partial^2\Theta(x,t)}{\partial t^2} = -\frac{m^2c^4}{\hbar^2}e^{\frac{-imc^2t}{\hbar}}\Psi_q - \frac{2imc^2}{\hbar}e^{\frac{-imc^2t}{\hbar}}\frac{\partial\Psi_q}{\partial t} + e^{\frac{-imc^2t}{\hbar}}\frac{\partial^2\Psi_q}{\partial t^2}. \tag{3.428}$$

However, $-\frac{m^2c^4}{\hbar^2}e^{\frac{-imc^2t}{\hbar}}\Psi_q - \frac{2imc^2}{\hbar}e^{\frac{-imc^2t}{\hbar}}\frac{\partial\Psi_q}{\partial t} >> e^{\frac{-imc^2t}{\hbar}}\frac{\partial^2\Psi_q}{\partial t^2}$ and hence,

$$\frac{\partial^2\Theta(x,t)}{\partial t^2} \approx -\frac{m^2c^4}{\hbar^2}e^{\frac{-imc^2t}{\hbar}}\Psi_q - \frac{2imc^2}{\hbar}e^{\frac{-imc^2t}{\hbar}}\frac{\partial\Psi_q}{\partial t}, \tag{3.429}$$

$$\frac{\partial\Theta(x,t)}{\partial x} = e^{\frac{-imc^2t}{\hbar}}\frac{\partial\Psi_q}{\partial x},\ \frac{\partial^2\Theta(x,t)}{\partial x^2} = e^{\frac{-imc^2t}{\hbar}}\frac{\partial^2\Psi_q}{\partial x^2}. \tag{3.430}$$

Substituting for Eqs. (3.429) and (3.430) in a 1D form of Eq. (3.419) gives

$$e^{\frac{-imc^2t}{\hbar}}\frac{\partial^2\Psi_q}{\partial x^2} - \frac{m^2c^2}{\hbar^2}e^{\frac{-imc^2t}{\hbar}}\Psi_q = \frac{1}{c^2}\left(-\frac{m^2c^4}{\hbar^2}e^{\frac{-imc^2t}{\hbar}}\Psi_q - \frac{2imc^2}{\hbar}e^{\frac{-imc^2t}{\hbar}}\frac{\partial\Psi_q}{\partial t}\right),$$

$$e^{\frac{-imc^2t}{\hbar}}\left(\frac{\partial^2\Psi_q}{\partial x^2} - \frac{m^2c^2}{\hbar^2}\Psi_q\right) = e^{\frac{-imc^2t}{\hbar}}\left(-\frac{m^2c^2}{\hbar^2}\Psi_q - \frac{2im}{\hbar}\frac{\partial\Psi_q}{\partial t}\right),$$

$$\frac{\partial^2\Psi_q}{\partial x^2} = -\frac{2im}{\hbar}\frac{\partial\Psi_q}{\partial t}.$$

Multiplying all through by $\frac{1}{i\hbar}$ and re-arranging the equation gives

$$-\frac{\hbar^2}{2m}\frac{\partial^2\Psi_q}{\partial x^2} = i\hbar\frac{\partial\Psi_q}{\partial t}. \tag{3.431}$$

The wave Eq. (3.410) has been shown to produce the linear Schrödinger equation. It is now important to shown how the nonlinear Schrödinger equation may be obtained from the same equation. Noting that in Cartesian coordinate (Bao 2007)

$$\vec{r} = (x, y, z), \tag{3.432}$$

and the propagation speed is given as

$$c = c(\vec{r}, |\Theta|). \tag{3.433}$$

Equation (3.410) has time harmonic solution and using Eq. (3.423), the solution has the form

$$\Theta(\vec{r}, t) = e^{-\frac{i}{\hbar}(mc^2+E_k)t}u(\vec{r}) = e^{i\omega_q t}u(\vec{r}); \omega = -\frac{(mc^2 + E_k)}{\hbar}. \tag{3.434}$$

Using Eq. (3.434) in Eq. (3.410) gives

$$e^{i\omega_q t}\nabla^2 u(\vec{r}) + e^{i\omega_q t}\frac{\omega_q^2}{c^2}u(\vec{r}) = 0,$$

$$\nabla^2 u(\vec{r}) + \frac{\omega_q^2}{c^2}u(\vec{r}) = 0. \tag{3.435}$$

For a uniform reference speed c_0, the index of refraction and wave number are:

$$\begin{aligned} k &= \frac{\omega_q}{c_0}, \\ n(\vec{r}, |u|) &= \frac{c_0}{c(\vec{r}, |u|)}. \end{aligned} \tag{3.436}$$

Hence, the reduced wave equation may be written as:

$$\nabla^2 u(\vec{r}) + k^2 n^2(\vec{r}, |u|)u(\vec{r}) = 0. \tag{3.437}$$

It is quite important to note the following (Bao 2007):

a. For uniform media,

$$n(|u|, \vec{r}) = 1, \tag{3.438}$$

b. For linear media,

$$n(\vec{r}) = n(|u|, \vec{r}), \tag{3.439}$$

c. In a Kerr medium (a medium in which a nonlinear polarisation generated such that propagation properties a wave is modified),

$$n(\vec{r}, |u|) = \sqrt{1 + \frac{4n_k|u|^2}{n_0}}, \tag{3.440}$$

where n_0 is the linear refractive index and n_k is the Kerr coefficient.

For plane spin waves, the following wave (in Cartesian coordinate) is assumed a solution to Eq. (3.437)

$$u(\vec{r}) = e^{ikz}\Psi_{qq}(x, y, z). \tag{3.441}$$

Hence, Eq. (3.437) becomes

$$2ik\frac{\partial \Psi_{qq}}{\partial z} + \nabla^2\Psi_{qq} + k^2(n^2(x, y, z, |\Psi_{qq}|) - 1)\Psi_{qq} + \frac{\partial^2\Psi_{qq}}{\partial z^2} = 0. \tag{3.442}$$

Now it is noteworthy that the direction of propagation has technically absorbed time dependence (Bao 2007) such that the term $-k^2(n^2(x, y, z, |\Psi_{qq}|)-1)$ represents the time-dependent potential. In order to bring out time dependence, it is assumed that there is a new frame where Eq. (3.442) has the following form

$$2ik\frac{\partial \Psi_{qq}}{\partial z'} + \nabla'^2\Psi_{qq} + k^2(n^2(x', y', z', |\Psi_{qq}|) - 1)\Psi_{qq} + \frac{\partial^2\Psi_{qq}}{\partial z'^2} = 0. \tag{3.443}$$

Defining the following non-dimensional variables

$$x = \frac{x'}{r_0}, y = \frac{y'}{r_0}, t = \frac{z'}{c_0 r_0^2}, \Psi_{qq}(x', y', z') = \Psi_0\Psi_q(x, y, t), \tag{3.444}$$

where r_0 and Ψ_0 are dimensionless scaling parameters. Using Eq. (3.444) in Eq. (3.443) and re-arranging the resulting equation leads to (Bao 2007)

$$i\frac{\partial \Psi_q}{\partial t} = -\frac{1}{2}\nabla^2\Psi_q + \wp(\vec{r}, t, |\Psi_q|)\Psi_q - \frac{\eta}{2}\frac{\partial^2\Psi_q}{\partial t^2}, \tag{3.445}$$

where $\eta = \frac{1}{r_0^2 k^2}$ and the real-valued function $\wp(\vec{r}, t, |\Psi_q|)$ is dependent on the term $k^2(n^2(x', y', z', |\Psi_{qq}|) - 1)$. Provided that

$$r_0 >> \frac{2\pi}{k}, \tag{3.446}$$

Consequently, it follows that

$$\frac{\eta}{2} = \left(\frac{1}{k}\right)^2 \frac{1}{2r_0^2} \ll 1. \tag{3.447}$$

Equation (3.447) implies that the term $\frac{\eta}{2}\frac{\partial^2\Psi_q}{\partial t^2}$ is negligible. Therefore, Eq. (3.445) becomes:

$$i\frac{\partial \Psi_q}{\partial t} = -\frac{1}{2}\nabla^2\Psi_q + \wp(\vec{r}, t, |\Psi_q|)\Psi_q. \tag{3.448}$$

Equation (3.445) is simply an approximation to the complete-reduced wave equation, which is valid for smooth variation of refractive index and wave energies not lying on the boundaries (Bao et al. 2013; Antoine et al. 2013). From this expression, very important results may be obtained as follows:

(1) When $n(x, y, z, |\Psi_{qq}|) = 1$, $\wp(\vec{r}, t, |\Psi_q|) = 0$ and then

$$i\frac{\partial \Psi_q}{\partial t} = -\frac{1}{2}\nabla^2\Psi_q. \tag{3.449}$$

(2) When $n(x, y, z, |\Psi_{qq}|) = n(x, y, z)$, $\wp(\vec{r}, t, |\Psi_q|) = V(\vec{r}, t)$ and Eq. (3.446) collapses to the linear Schrödinger equation

$$i\frac{\partial \Psi_q}{\partial t} = -\frac{1}{2}\nabla^2\Psi_q + \wp(\vec{r}, t, |\Psi_q|)\Psi_q. \tag{3.450}$$

(3) For $n(x, y, z, |u|) = \sqrt{1 + \frac{4n_k|u|^2}{n_0}}$, $\Psi_0 = \frac{1}{r_0 k}\sqrt{\frac{n_0}{2n_k}}$, $\wp(\vec{r}, t, |\Psi_q|) = -|\Psi_q|^2$, Eq. (3.446) collapses to the nonlinear Schrödinger equation with a cubic focussing nonlinearity

$$i\frac{\partial \Psi_q}{\partial t} = -\frac{1}{2}\nabla^2\Psi_q - |\Psi_q|^2\Psi_q. \tag{3.451}$$

(4) For generalized consideration, $n(r, |u|) = n(x, y, z, |u|)$, $\alpha|\Psi_q|^{p-1} + V(\vec{r}, t) = \wp(\vec{r}, t, |\Psi_q|)$, the nonlinear equation becomes:

$$i\frac{\partial \Psi_q}{\partial t} = -\frac{1}{2}\nabla^2\Psi_q + V(\vec{r}, t)\Psi_q + \alpha|\Psi_q|^{p-1}\Psi_q, \tag{3.452}$$

where α is the nonlinear coefficient and $p \geq 3$ (Yan and Konotop 2009).

3.3.7.2 Quantum Mechanical Model of Transverse Polarisation for Optical Molecular MR Imaging

The quantum mechanical formulation developed in the preceding section is quite similar to the quantum formulation for optical resonance. Therefore, with such similarities, details of derivation are avoided in this section and in most cases; the final result may simply be stated as appropriate. It is worthy of note that the total Hamiltonian operator for this system is a sum of the Hamiltonian at thermal equilibrium and the interaction Hamiltonian. This is given as follows:

$$\aleph_p = \aleph_0 + \aleph_1. \tag{3.453}$$

According to Eq. (3.33), this equation becomes (Hahn 1980):

$$\aleph_p = -\hbar\left[(\Delta\omega_0 + d\phi/dt)p_{0z} + \Xi\xi_p p_{0x}\right]. \tag{3.454}$$

Since the electric polarisation is the expectation value of the electric dipole moment per unit volume given as (Meystre 2006; Eberly and Stroud 2006)

$$P = \frac{\langle \hat{p}_0 \rangle}{V}. \tag{3.455}$$

Hence, the optical Bloch Eq. (3.38) becomes

$$\left.\begin{aligned} \frac{dp_{0x}}{dt} &= \left(\Delta\omega_0 + \frac{d\phi}{dt}\right)p_{0y} - \frac{p_{0x}}{T_2}, \\ \frac{dp_{0y}}{dt} &= -\left(\Delta\omega_0 + \frac{d\phi}{dt}\right)p_{0x} + \Xi\xi_p p_{0z} - \frac{p_{0y}}{T_2}, \\ \frac{dp_{0z}}{dt} &= -\Xi\xi_p p_{0y} - \frac{(p_{0z} - p_{00})}{T_1}, \end{aligned}\right\} \tag{3.456}$$

where

$$\hat{p}_0 = \hat{x}p_{0x} + \hat{y}p_{0y} + \hat{z}p_{0z}. \tag{3.457}$$

Then, Eq. (3.456) may be modified to read

$$\left.\begin{aligned} \frac{dp_{0x}}{dt} + \frac{p_{0x}}{T_2} &= \left(\Delta\omega_0 + \frac{d\phi}{dt}\right)p_{0y}, \\ \frac{dp_{0y}}{dt} + \frac{p_{0y}}{T_2} &= -\left(\Delta\omega_0 + \frac{d\phi}{dt}\right)p_{0x} + \Xi\xi_p p_{0z}, \\ \frac{dp_{0z}}{dt} + \frac{(p_{0z} - p_{00})}{T_1} &= -\Xi\xi_p p_{0y}. \end{aligned}\right\} \tag{3.458}$$

In the absence of relaxation, Eq. (3.458) reduces to the ordinary optical Bloch equation. Hence, when relaxation is not negligible,

$$\frac{d\hat{p}_0}{dt} + \frac{\hat{x}p_{0x} + \hat{y}p_{0y}}{T_2} + \frac{\hat{z}(p_{0z} - p_{00})}{T_1} = \Xi(\hat{p}_0 \times E_{eff}), \tag{3.459}$$

where $\mathrm{E}_{\mathrm{eff}}$ (which is dependent on the field envelope ξ_p) is the effective applied field which appears in the system. Since the general behaviour of an optical NMR system in the presence of relaxation is not expected to differ very significantly from that in the absence of relaxation, a new relaxation-dependent rate may be defined as follows:

$$\frac{d\hat{p}_0^r}{dt} = \frac{d\hat{p}_0}{dt} + \frac{\hat{x}p_{0x} + \hat{y}p_{0y}}{T_2} + \frac{\hat{z}(p_{0z} - p_{00})}{T_1}. \tag{3.460}$$

Hence, according to Eq. (3.461),

$$\frac{d\hat{p}_0^r}{dt} = \hat{x}\left(\frac{dp_{0x}}{dt} + \frac{p_{0x}}{T_2}\right) + \hat{y}\left(\frac{dp_{0y}}{dt} + \frac{p_{0y}}{T_2}\right) + \hat{z}\left(\frac{dI_z}{dt} + \frac{(I_z - I_0)}{T_1}\right). \tag{3.461}$$

From this expression, the following terms could be obtained

$$p_{0x}^r = p_{0x} + \int \frac{p_{0x}}{T_2} dt, \tag{3.462}$$

$$p_{0y}^r = p_{0y} + \int \frac{p_{0y}}{T_2} dt, \tag{3.463}$$

$$p_{0z}^r = p_{0z} + \int \frac{(p_{0z} - p_{00})}{T_1} dt. \tag{3.464}$$

Therefore, from Eqs. (3.462–3.464), (3.458) becomes:

$$\left.\begin{aligned} \frac{dp_{0x}^r}{dt} &= \left(\Delta\omega_0 + \frac{d\phi}{dt}\right)p_{0y}, \\ \frac{dp_{0y}^r}{dt} &= -\left(\Delta\omega_0 + \frac{d\phi}{dt}\right)p_{0x} + \Xi\xi_p p_{0z}, \\ \frac{dp_{0z}^r}{dt} &= -\Xi\xi_p p_{0y}. \end{aligned}\right\} \tag{3.465}$$

According to Eq. (3.454),

$$\Xi(\hat{p}_0 \times \vec{E}_{eff}) = -\frac{i}{\hbar}[\hat{p}_0, \aleph_p]. \tag{3.466}$$

Therefore, it follows that

$$\frac{dp_0^r}{dt} = -\frac{i}{\hbar}[\hat{p}_0^r, \aleph_p]. \tag{3.467}$$

Then according to the proofs in the preceding section, the electric dipole moment in Eq. (3.467) is associated with a wave function which shall be given as state $|\psi_p(t = t_0)\rangle = |\psi_p\rangle$. Hence, the expectation value of the operator $\hat{p}_0^r$ is associated with a time-dependent state vectors $|\psi_p(t)\rangle$. This is given (Straten and Metcalf 2016; Eberly and Stroud 2006)

$$\langle\hat{p}_0^r(t)\rangle = \langle\psi_p(t)|\hat{p}_0^r|\psi_p(t)\rangle. \tag{3.468}$$

Therefore, the Schrödinger equation for an optical NMR system may be expressed as

$$\frac{d\psi_p}{dt} = -\frac{i}{\hbar}\aleph_p\psi_p. \tag{3.469}$$

Since the system has been demonstrated to behave quantum mechanically, the energy equation may be written as follows:

$$\aleph_p\psi_p = E_p\psi_p, \tag{3.470}$$

where E_p is the eigenvalue obtainable from the term (Hahn 1980). If the wave function given above may be expressed as a separable function:

$$\Psi_p(x,t) = \phi_p(x)\psi_p(t). \tag{3.471}$$

The general time-dependent Schrödinger equation is given as

$$-\frac{\hbar^2}{2m}\frac{\partial^2\Psi_p(x,t)}{\partial x^2} + V(x)\Psi_p(x,t) = i\hbar\frac{\partial\Psi_p(x,t)}{\partial t}. \tag{3.472}$$

and the associated time-independent Schrödinger equation is:

$$\frac{d^2\phi_p(x)}{dx^2} = -\frac{2m}{\hbar^2}(E_p - V(x))\phi_p(x). \tag{3.473}$$

In multi-dimensions, Eq. (3.473) becomes

$$\nabla^2\phi_p(x) = -\frac{2m}{\hbar^2}(E_p - V(x))\phi_p(x). \tag{3.474}$$

The Roles of Macroscopic Polarisation in Quantum Optical Measurements

In quantum mechanical treatment of electrons, the expectation value of the electron dipole moment for a single electron is given as (Eberly and Stroud 2006; Chang and Minogin 2002):

$$\langle\hat{p}_0^r\rangle = \langle\Psi_p(x,t)|\hat{p}_0^r|\Psi_p(x,t)\rangle. \tag{3.475}$$

If Eq. (3.475) is summed over all N electrons present in the system per unit the volume of the system, the expectation value becomes

$$\frac{N}{V}\langle\hat{p}_0^r\rangle = \frac{N}{V}\langle\Psi_p(x,t)|\hat{p}_0^r|\Psi_p(x,t)\rangle. \tag{3.476}$$

This is the expression for the macroscopic polarisation. That is,

$$P(x,t) = -\frac{N}{V}\langle\Psi_p(x,t)|\hat{p}_0^r|\Psi_p(x,t)\rangle. \tag{3.477}$$

If the frequency of the applied field is ω, Eq. (3.477) becomes (Eberly and Stroud 2006):

$$P(x,t) = \Re\{p_0^r(u - iv)e^{-i\omega t}\} = -\frac{N}{V}\langle\Psi_p(x,t)|\hat{p}_0^r|\Psi_p(x,t)\rangle, \tag{3.478}$$

where u and v are the Bloch vectors (time-dependent amplitudes of the in-phase and in-quadrature dipole components).In a sample where the various atoms have different resonant frequencies, a detuning distribution may be defined as $g(\omega_d)$. The detuning frequency is given as:

$$\omega_d = \frac{E_e - E_g}{\hbar} - \omega. \tag{3.479}$$

If the impressed field is a quasi-monochromatic wave (Eberly and Stroud 2006),

$$\xi_p(t) = \frac{\xi_{po}}{2}e^{-i\omega t}, \tag{3.480}$$

where the complex terms in the applied field have been neglected. Then the Rabi frequency is given as

$$\omega_{p1} = \frac{p_0^r \xi_{po}}{\hbar}. \tag{3.481}$$

In terms of the detuning distribution,

$$P(x,t) = -\frac{N}{V}\langle\Psi_p(x,t)|\hat{p}_0^r|\Psi_p(x,t)\rangle = Np_0^r \int g(\omega_d)\Re[(u - iv)e^{-i\omega t}]d\omega_d. \tag{3.482}$$

In RFR, x-component of a vector rotates into the y-component after a time $\frac{\pi}{2\omega}$ according to Eq. (2.82). Consequently, Eq. (4.480) becomes

$$P_y(x,t) = -\frac{N}{V}\langle\Psi_p(x,t)|p_{0y}^r|\Psi_p(x,t)\rangle = Np_{0y}^r \int g(\omega_d)\Re[ve^{-i\omega t}]d\omega_d;\ u \leftrightarrow v. \tag{3.483}$$

Resonant Wave Propagation in Optical NMR

The time-dependent atomic dipole moments created by impressed fields are themselves a source of fields or coherent transient (Eberly and Stroud 2006). If RF wave is a plane wave propagation in the z-direction such that Eq. (3.480) is modified as follows:

$$\xi_p(z,t) = \frac{\xi_{po}}{2}e^{i(k_p z - \omega t)}. \tag{3.484}$$

Equation (3.483) becomes

$$P_y(z,t) = Np_{0y}^r \int g(\omega_d)\Re[ve^{i(k_p z-\omega t)}]d\omega_d;\ u \leftrightarrow v. \tag{3.485}$$

If the field envelope ξ_{po} is slowly varying, the second derivatives are negligible when ξ_{po} is substituted into the Maxwell's wave equation. From these results, it follows that.

i. For dispersive reduced wave equation,

$$\begin{aligned}(k_p^2 - k^2)\xi_{po} &= 4\pi k^2 Np_0^r \int g(\omega_d)u(z,t,\omega_d)d\omega_d \\ &= 4\pi k^2 \int g(\omega_d)P_x(z,t,\omega_d)d\omega_d;\ k = \frac{\omega}{c}.\end{aligned} \tag{3.486}$$

ii. For absorptive reduced wave equation

$$\begin{aligned}k_p\frac{\partial \xi_{po}}{\partial z} - k\frac{\partial \xi_{po}}{\partial (ct)} &= 2\pi k^2 Np_0^r \int g(\omega_d)v(z,t,\omega_d)d\omega_d \\ &= 2\pi k^2 \int g(\omega_d)P_y(z,t,\omega_d)d\omega_d;\ k = \frac{\omega}{c}.\end{aligned} \tag{3.487}$$

Substituting the Optical Bloch equation into Eq. (4.484), the dispersive gives the index of refraction

$$n^2 - 1 = \frac{4\pi N(p_0^r)^2}{\hbar}\int \frac{\omega_d g(\omega_d)}{\omega_d^2 + (1/T_2)^2}d\omega_d;\ n = \frac{k_p}{k}. \tag{3.488}$$

The absorptive equation predicts steady state field attenuation during the propagation

$$\frac{\partial \xi_{po}}{\partial z} = -\frac{1}{2}\alpha_B \xi_{po}, \tag{3.489}$$

α_B is called the extinction coefficient or reciprocal Beers length (Eberly and Stroud 2006). This parameter is given as

$$\alpha_B = \frac{4\pi N(p_0^r)^2\omega}{\hbar c T_2}\int \frac{g(\omega_d)}{\omega_d^2 + (1/T_2)^2}d\omega_d. \tag{3.490}$$

According to the Beers law, the intensity is

$$I(z,t) = I(0,t)e^{-\alpha_B z}. \tag{3.491}$$

Nonlinear Schrödinger Equation for Optical NMR Phenomenon

The Nonlinear Schrödinger which has been derived for the NMR spin system may be placed side by side with its optical NMR counterpart. Therefore, taking a cue from earlier derivations, the following may be given.

(1) When $n(x, y, z, |\Psi_{pp}|) = 1$, $\wp(\vec{r}, t, |\Psi_p|) = 0$ and then

$$i\frac{\partial \Psi_p}{\partial t} = -\frac{1}{2}\nabla^2\Psi_p. \tag{3.492}$$

(2) When $n(x, y, z, |\Psi_{pp}|) = n(x, y, z)$, $\wp(\vec{r}, t, |\Psi_p|) = V(\vec{r}, t)$ and the optical form of Eq. (3.445) collapses to the linear Schrödinger equation

$$i\frac{\partial \Psi_p}{\partial t} = -\frac{1}{2}\nabla^2\Psi_p + \wp(\vec{r}, t, |\Psi_p|)\Psi_p. \tag{3.493}$$

(3) When $n(x, y, z, |u|) = \sqrt{1 + \frac{4n_k|u|^2}{n_0}}$, $\Psi_0 = \frac{1}{r_0 k}\sqrt{\frac{n_0}{2n_k}}$, $\wp(\vec{r}, t, |\Psi_p|) = -|\Psi_p|^2$, the optical counterpart of Eq. (3.445) collapses to the nonlinear Schrödinger equation with a cubic focusing nonlinearity

$$i\frac{\partial \Psi_p}{\partial t} = -\frac{1}{2}\nabla^2\Psi_p - |\Psi_p|^2\Psi_p. \tag{3.494}$$

(4) For generalized consideration, $n(r, |u|) = n(x, y, z, |u|)$, $\wp(\vec{r}, t, |\Psi_p|) = V(\vec{r}, t) + \alpha|\Psi_p|^{\varpi-1}$, the nonlinear equation becomes:

$$i\frac{\partial \Psi_p}{\partial t} = -\frac{1}{2}\nabla^2\Psi_p + V(\vec{r}, t)\Psi_p + \alpha|\Psi_p|^{\varpi-1}\Psi_p, \tag{3.495}$$

where α is the nonlinear coefficient and $\varpi \geq 3$ $p \geq 3$ (Yan and Konotop 2009).

3.4 Conclusion

This chapter has shown the possibilities of computational molecular magnetic resonance imaging based on Eqs. (3.1) and (3.2). Several models are developed one of which is the Optical Bloch equation (OBE) for cellular imaging. This reveals the hidden applications of optical imaging in biological flow according to the NMR Bloch flow equation. The possibility of performing molecular imaging without the need for all the big and expensive imaging machines is the positive essence of this chapter. All that is required is the acquisition of few NMR relaxation data of tissues being investigated (which are much cheaper to obtain and the machines for measurement are relatively smaller and easier to maintain). The theoretical and computational analyses need to be translated to interesting clinical applications.

Biological flow at the molecular level in the presence of diffusion was studied theoretically and will be simulated with experimental data in the next chapters. The formulation in Eqs. (3.1) and (3.2) is developed from the fundamental NMR Bloch flow equation and new analytical algorithms were developed for the MRI signals in terms of Bessel functions and properties. These algorithms have been demonstrated to show sensitivity for significantly restricted or complex fluid flow in accordance with higher orders of Bessel functions.

It is interesting to note that Bessel functions and properties are useful in characterising restricted diffusion. For example, the growth of tumors is dependent on nutrients provision via direct diffusion through the circulatory system. Cancerous cells normally restrict physiological flow and the associated displacement can be quantified with the formulations of this chapter in order to obtain information on the genesis, type, age and size and of metastatic developments in tissues. In this chapter, it has been demonstrated in details the usefulness of flow velocity and resonance signal for contrasting different tissues. Definitely, this forms are entirely new methods [our book) of differentiating cancerous cells from normal cells according to Bessel properties and the velocity of blood flow in tumors.

Most of the contemporary NMR/MRI detection systems rely on the use of exponential, Gaussian function and difficult sinusoidal RF excitation wave. However, this chapter has demonstrated a new system design that is dependent on the Bessel functions if NMR/MRI parameters are chose appropriately. This is particularly easier to handle and offer more flexibility to MRI detection system programmers for image contrasts improvement and unique image voxel manipulations.

Physiological processes in living systems are controlled by molecular dynamics and organised molecular interactions. The velocities that are associated with these processes are normally in the micrometre range (as demonstrated in the next chapter) and the implication of this is that early stage detection of human diseases conditions can be achieved in accordance to the results obtained in the next chapters. One of the reasons the methods employed in this study are special is that the resonance signal in terms of the transverse magnetisation is consistent with the analytical expression of the NMR Bloch flow equation in terms of Bessel functions, Legendre function and other related mathematical functions. Their properties are of utmost importance in the translation of these results into clinical applications. This is because analytical expressions make use of minimum assumptions and are very close to situations observed in real systems as in chapter four of this book. Also, each element in the expressions could easily be controlled and flexible in terms of function manipulation. This means that MRI sequence design (for process-specific or tissue-specific molecular imaging) becomes a much easier task and new clinical applications may be developed with ease. For example, the order of the Bessel function (in terms of the MRI pulse) could be used to explicitly control the spatial resolution (or temporal resolution) of the MRI system. This is simple and easily adaptable according to the requirements of the problems to be solved in the clinic or laboratories.

Acknowledgements The authors acknowledge the motivated research efforts of Udeme Iniobong Nicholas (Department of Physics, Federal University of Technology, Minna, Nigeria) as a promising young researcher.

References

Abragam, A.: The Principles of Nuclear Magnetism. Clarendon Press, Oxford, United Kingdom (1961)
Abragam, A.: Nuclear ferromagnetism and antiferromagnetism. In: Grant, D.M., Harris, R.K. (eds.) *The Encyclopaedia of Nuclear Magnetic Resonance*. Wiley, Chichester, United Kingdom (1996)
Ai, L., Vafai, K.: A coupling model for macromolecule transport in a stenosed arterial wall. Int. J. Heat Mass Transf. **49**(9), 1568–1591 (2006)
Aime, S., Botta, M., Terreno, E.: Gd (III)—based contrast agents for MRI. Adv. Inorg. Chem. **57**, 173–227 (2005)
Aime, S., Baranyai, Z., Gianolio, E., Terreno, E.: Paramagnetic contrast agents. In: Modo, M.J.M., Bulte, J.W.M. (eds.) Molecular and Cellular MR Imaging, pp. 1–9. CRC Press, Taylor & Francis Group, New York (2007)
Anderson, A.W., Gore, J.C.: Analysis and correction of motion artifacts in diffusion weighted imaging. Magn. Reson. Med. **32**(3), 379–387 (1994)
Antoine, X., Bao, W., Besse, C.: Computational methods for the dynamics of the nonlinear Schrödinger/Gross–Pitaevskii equations. Comput. Phys. Commun. **184**(12), 2621–2633 (2013)
Awojoyogbe, O.B., Dada, O.M.: Mathematical design of a magnetic resonance imaging sequence based on bloch NMR flow equations and bessel functions. Chin. J. Magn. Resonance Imag. **4**(5) (2013)
Awojoyogbe, O.B., Dada, O.M., Faromika, O.P., Dada, O.E.: Mathematical concept of the bloch flow equations for general magnetic resonance imaging: a review. Concepts Magn. Reson. Part A **38A**(3), 85–101 (2011)
Awojoyogbe, O.B.: Analytical solution of the time-dependent bloch NMR flow equations: a translational mechanical analysis. Physica A **339**, 437–460 (2004)
Awojoyogbe, O.B.: A mathematical model of bloch NMR equations for quantitative analysis of blood flow in blood vessels of changing cross-section-Part II. Physica A **323**, 534–550 (2003)
Awojoyogbe, O.B.: A Mathematical model of bloch NMR equations for quantitative analysis of blood flow in blood vessels of changing cross-section I. Physica A **303**, 163–175 (2002)
Bao, W.: The nonlinear Schrödinger equation and applications in Bose-Einstein condensation and plasma physics. Dyn. Models Coarsening Coagulation Condens. Quantization **9**, 141–239 (2007)
Bao, W., Tang, Q., Xu, Z.: Numerical methods and comparison for computing dark and bright solitons in the nonlinear Schrödinger equation. J. Comput. Phys. **235**, 423–445 (2013)
Barrie, P.J.: NMR applications to porous solids. Annu. Rep. NMR Spectrosc. **30**, 37–96 (1995)
Basser, P.J.: Diffusion and diffusion tensor MR imaging: fundamentals. In: Magnetic Resonance Imaging of the Brain and Spine, 1752–1767 (2008)
Basser, P.J., Mattiello, J., LeBihan, D.: Estimation of the effective self-diffusion tensor from the NMR spin echo. J. Magn. Reson. Ser. B **103**(3), 247–254 (1994)
Basser, P.J., Jones, D.K.: Diffusion-tensor MRI: theory, experimental design and data analysis—a technical review. NMR Biomed. **15**, 456–467 (2002)
Berg-Sorenson, K., Castin, Y., Bonderup, E., Molmer, K.: Momentum diffusion of atoms moving in laser fields. J. Phys. B Mol. Opt. Phys. **25**(20), 4195 (1992)
Blanchet, K.D.: Redefining personalized medicine in the postgenomic era: developing bladder cancer therapeutics with proteomics. BJU Int. **105**(2), i–iii (2010)
Borgonovi, F., Gorshkov, V.N., Tsifrinovich, V.I.I.: Magnetic Resonance Force Microscopy and a Single-Spin Measurement. World Scientific, Singapore (2006)

Bransden, B.H., Joachain, C.J.: Quantum Mechanics, 2nd edn. Pearson Education, Essex, United Kingdom (2000)

Brix, G., Kolem, H., Nitz, W.R., Bock, M., Huppertz, A., Zech, C.J., Dietrich, O.: Basics of MRI and Magnetic Resonance Spectroscopy. Springer, Berlin, Heidelberg (2008)

Brix, G., Seebass, M., Hellwig, G., Griebel, J.: Estimation of heat transfer and temperature rise in partial-body regions during MR procedures: an analytical approach with respect to safety considerations. Magn. Reson. Imaging **20**, 65–76 (2002)

Bryant, D.J., Payne, J.A., Firmin, D.N., Longmore, D.B.: Measurement of flow with NMR imaging using a gradient pulse and phase difference technique. J. Comput. Assist. Tomogr. **8**(4), 588–593 (1984)

Burkhardt, C.E., Leventhal, J.J.: Foundations of Quantum Physics. Springer-Verlag, New York (2008)

Cabal, C., Darias, D., González, E., Musacchio, A.: Theranostics and molecular imaging: new concepts and technologies for drug development. Biotecnología Aplicada **30**(3) (2013)

Callaghan, P.T.: Principles of Nuclear Magnetic Resonance Microscopy. Oxford University Press, Oxford, United Kingdom (1994)

Chang, D.C., Rorschach, H.E., Nichols, B.L., Hazlewood, C.F.: Implications of diffusion coefficient measurements for the structure of cellular water. Ann. New York Acad. Sci. **204**, 434–443 (1973)

Chang, S., Minogin, V.: Density-matrix approach to dynamics of multilevel atoms in laser fields. Phys. Rep. **365**(2), 65–143 (2002)

Chen, C., Xu, L.X.: Tissue temperature oscillations in an isolated pig kidney during surface heating. Ann. Biomed. Eng. **30**, 1162–1171 (2002)

Chopra, R., Wachsmuth, J., Burtnyk, M., Haider, M.A., Bronskill, M.J.: Analysis of factors important for transurethral ultrasound prostate heating using MR temperature feedback. Phys. Med. Biol. **51**, 827–844 (2006)

Cline, H.E., Dumoulin, C.L., Hart, H.R., Lorensen, W.E., Ludke, S.: 3D reconstruction of the brain from magnetic resonance images using a connectivity algorithm. Magn. Reson. Imaging **5**(5), 345–352 (1987)

Cohen-Tannoudji, C., Diu, B., Laloe, F.: Quantum Mechanics, vol. 1. Wiley, New York, USA (1977)

Cohen-Tannoudji, C., Dupont-Roc, J., Grynberg, G.: Atom-Photon Interactions: Basic Processes and Applications, 2nd edn. Wiley-VCH Verlag GmbH and Co., New York, USA (1998)

Cooper, R.L., Chang, D.B., Young, A.C., Martin, C.J., Ancker-Johnson, B.: Restricted diffusion in biophysical systems: experiment. Biophys. J. **14**(3), 161 (1974)

Cowan, B.P.: Nuclear Magnetic Resonance and Relaxation, 1st edn. Cambridge University Press, Cambridge, United Kingdom (1997)

Dada, O.M., Awojoyogbe, O.B., Ukoha, A.C.: A computational analysis for quantitative evaluation of petrol-physical properties of rock fluids based on bloch NMR diffusion model for porous media. J. Petrol. Sci. Eng. **127**, 137–147 (2015)

Dada, O.M., Awojoyogbe, O.B., Adesola, O.A., Boubaker, K.: Magnetic resonance imaging derived flow parameters for the analysis of cardiovascular diseases and drug development. Magn. Reson. Insights **6**, 83–93 (2013a)

Dada, O.M., Awojoyogbe, O.B., Baroni, S., Aweda, M.A.: Application of bloch NMR equation and pennes bioheat equation to theranostics (Accepted for presentation at the world molecular imaging congress, Savannah, Georgia, USA; 18–21 Sept 2013b)

Dada, M.O., Awojoyogbe, B.O.: Computational Molecular Magnetic Resonance Imaging for Neuro-oncology. Springer (2021)

Dahlem, M.A., Müller, S.C.: Reaction-diffusion waves in neuronal tissue and the window of cortical excitability. Ann. der Phys. **516**(7–8), 442–449 (2004)

De Marsily, G.: Quantitative Hydrogeology: Groundwater Hydrology for Engineers. Academic Press, Inc, San Diego, California (1986)

Demir, H., Süngü, İ.Ç.: Numerical solution of a class of nonlinear Emden-Fowler equations by using differential transform method. Cankaya Univ. J. Sci. Eng. **12**(2) (2009)

Dumas, S., Jacques, V., Sun, W.C., Troughton, J.S., Welch, J.T., Chasse, J.M., Schmitt-Willich, H., Caravan, P.: High relaxivity MRI contrast agents part 1: impact of single donor atom substitution on relaxivity of serum albumin-bound gadolinium complexes. Invest. Radiol. **45**(10), 600 (2010)
Eberly, J.H., Stroud, C.R.: Coherent transients. In: Drake, G.W.F. (eds.) Springer Handbooks of Atomic, Molecular, and Optical Physics, pp. 1065–1076. Springer Science, New York, USA (2006)
Falk, J.: Epigenetics & sequencing-gene expression systems' second international meeting. Chromatin methylation to disease biology & theranostics. IDrugs Invest. Drugs J. **11**(9), 650 (2008)
Falk, M.H., Issels, R.D.: Hyperthermia in oncology. Int. J. Hyperth. **17**(1), 1–18 (2001)
Field, J.H.: Relationship of quantum mechanics to classical electromagnetism and classical relativistic mechanics. Eur. J. Phys. **25**(3), 385 (2004)
Fritsche, L., Haugk, M.: A new look at the derivation of the Schrödinger equation from Newtonian mechanics. Ann. Phys. **12**(6), 371–402 (2003)
Goose, J.E., Sansom, M.S.: Reduced lateral mobility of lipids and proteins in crowded membranes. PLoS Comput. Biol. **9**(4) (2013)
Gupta, A., Stait-Gardner, T., Ghadirian, B., Price, W.S., Dada, O.M., Awojoyogbe, O.B.: Theory, Dynamics and Applications of Magnetic Resonance Imaging-I. Science PG, New York, USA (2014)
Haacke, E.M., Lin, W., Li, D.: Whole body magnetic resonance angiography. In: Young, I.R. (ed.) Methods in Biomedical Magnetic Resonance Imaging and Spectroscopy, pp. 488–500. Wiley, New York (2000)
Haacke, E.M., Brown, R.W., Thompson, M.R., Venkatesan, R.: Magnetic Resonance Imaging: Physical Principles and Sequence Design, pp. 39–40. A John Wiley and Sons, New York (1999)
Haacke, E.M., Xu, Y., Cheng, Y.C.N., Reichenbach, J.R.: Susceptibility weighted imaging (SWI). Magn. Reson. Med. **52**(3), 612–618 (2004)
Hahn, E.L.: Heritage of the bloch equations in quantum optics. Felix Bloch Festschrist **75** (1980)
Hanusch, K.U., Janssen, C.H., Billheimer, D., Jenkins, I., Spurgeon, E., Lowry, C.A., Raison, C.L.: Whole-body hyperthermia for the treatment of major depression: associations with thermoregulatory cooling. Am. J. Psychiatry **170**(7), 802 (2013)
Hazlewood, C.F., Chang, D.C., Nichols, B.L., Woessner, D.E.: Nuclear magnetic resonance transverse relaxation times of water protons in skeletal muscle. Biophys. J . **14**(8), 583 (1974)
Hazlewood, U.F., Rorschach, H.E., Lin, C.: Diffusion of water in tissues and MRI. Magn. Reson. Med. **19**(2), 214–216 (1991)
Jacques, V., Dumas, S., Sun, W.C., Troughton, J.S., Greenfield, M.T., Caravan, P.: High relaxivity MRI contrast agents part 2: optimization of inner-and second-sphere relaxivity. Invest. Radiol. **45**(10), 613 (2010)
Jellison, B.J., Field, A.S., Medow, J., Lazar, M., Salamat, M.S., Alexander, A.L.: Diffusion tensor imaging of cerebral white matter: a pictorial review of physics, fiber tract anatomy, and tumor imaging patterns. Am. J. Neuroradiol. **25**(3), 356–369 (2004)
Khalil, H.K., Grizzle, J.W.: Nonlinear Systems, vol. 3. Prentice hall, New Jersey (1996)
Khanafer, K., Vafai, K.: The role of porous media in biomedical engineering as related to magnetic resonance imaging and drug delivery. Heat Mass Transf. **42**(10), 939–953 (2006)
Kimmich, R., Anoardo, E.: Field-cycling NMR relaxometry. Prog. Nucl. Magn. Reson. Spectrosc. **44**(3), 257–320 (2004)
Kreyszig, E.: Advanced Engineering Mathematics. Wiley, Singapore (1996)
Kuffler, S.W., Potter, D.D.: Glia in the leech central nervous system: physiological properties and neuron-glia relationship. J. Neurophysiol. **27**(2), 290–320 (1964)
Laub, G.: Time-of-flight method of MRA. In: Young, I.R. (ed.) Methods in Biomedical Magnetic Resonance Imaging and Spectroscopy, pp. 500–504. Wiley, New York (2000)
Laurent, S., Forge, D., Port, M., Roch, A., Robic, C., Vander Elst, L., Muller, R.N.: Magnetic iron oxide nanoparticles: synthesis, stabilization, vectorization, physicochemical characterizations, and biological applications. Chem. Rev. **108**(6), 2064–2110 (2008)

Le Bihan, D., Iima, M.: Diffusion magnetic resonance imaging: what water tells us about biological tissues. PLoS Biol. **13**(7), e1002203 (2015)

Le Bihan, D., Mangin, J.F., Poupon, C., Clark, C.A., Pappata, S., Molko, N., Chabriat, H.: Diffusion tensor imaging: concepts and applications. J. Magn. Reson. Imaging **13**(4), 534–546 (2001)

Li, S.G., Ruan, F., McLaughlin, D.: A space-time accurate method for solving solute transport problems. Water Resour. Res. **28**(9), 2297–2306 (1992)

Li, A., Wong, C.S., Wong, M.K., Lee, C.M., Yeung, M.A.: Acute adverse reactions to magnetic resonance contrast media–gadolinium chelates. Br. J. Radiol. (2014)

Li, W.H., Parigi, G., Fragai, M., Luchinat, C., Meade, T.J.: Mechanistic studies of a calcium-dependent MRI contrast agent. Inorg. Chem. **41**(15), 4018–4024 (2002)

Liu, E.H., Saidel, G.M., Harasaki, H.: Model analysis of tissue responses to transient and chronic heating. Ann. Biomed. Eng. **31**(8), 1007–1014 (2003)

Loman, A., Gregor, I., Stutz, C., Mund, M., Enderlein, J.: Measuring rotational diffusion of macromolecules by fluorescence correlation spectroscopy. Photochem. Photobiol. Sci. **9**(5), 627–636 (2010)

Martínez, M.C., de la Torre, J.G.: Brownian dynamics simulation of restricted rotational diffusion. Biophys. J . **52**(2), 303–310 (1987)

Merzbacher, E.: Quantum Mechanics. Wiley, New York (1998)

Metcalf, H.J., Straten, P.: Laser Cooling and Trapping of Neutral Atoms. Wiley-VCH Verlag GmbH & Co, KGaA (2007)

Meystre, P.: Light-matter interaction. In: Drake, G.W.F. (eds.) Springer Handbooks of Atomic, Molecular, and Optical Physics, pp. 997–1007. Springer Science, New York, USA (2006)

Mohanty, S.: Optically-actuated translational and rotational motion at the microscale for microfluidic manipulation and characterization. Lab Chip **12**(19), 3624–3636 (2012)

Mohanty, S.K., Sharma, M., Panicker, M.M., Gupta, P.K.: Controlled induction, enhancement, and guidance of neuronal growth cones by use of line optical tweezers. Opt. Lett. **30**(19), 2596–2598 (2005)

Mohanty, S.K., Uppal, A., Gupta, P.K.: Self-rotation of red blood cells in optical tweezers: prospects for high throughput malaria diagnosis. Biotech. Lett. **26**(12), 971–974 (2004)

Mohanty, S.K., Uppal, A., Gupta, P.K.: Optofluidic stretching of RBCs using single optical tweezers. J. Biophoton. **1**(6), 522–525 (2008)

Mori, S., Barker, P.B.: Diffusion magnetic resonance imaging: its principle and applications. Anat. Rec. **257**(3), 102–109 (1999)

Morriss, C.E., Freedman, R., Straley, C., Johnston, M., Vinegar, H.J., Tutunjian, P.N.: Hydrocarbon saturation and viscosity estimation from NMR logging in the belridge diatomite. In: SPWLA 35th Annual Logging Symposium. Society of Petrophysicists and Well-Log Analysts (1994)

Naik, D., Viswamitra, S., Kumar, A.A., Srinath, M.G.: Susceptibility weighted magnetic resonance imaging of brain: a multifaceted powerful sequence that adds to understanding of acute stroke. Ann. Indian Acad. Neurol. **17**(1), 58–61 (2014)

Nelson, D.L., Cox, M.M.: Lehninger Principles of Biochemistry. W. H. Freeman, New York, USA (2008)

Nelson, E.: Derivation of the Schrödinger equation from Newtonian mechanics. Phys. Rev. **150**(4), 1079 (1966)

Neuman, K.C., Block, S.M.: Optical trapping. Rev. Sci. Instrum. **75**(9), 2787–2809 (2004)

Nicholson, C., Rice, M.E.: Diffusion of ions and transmitters in the brain cell microenvironment. In: Flux, K., Agnati, L.F. (eds.) Volume transmission in the brain: novel mechanisms for neural transmission, pp. 279–294. Raven Press, New York (1991)

Nicholson, C., Syková, E.: Extracellular space structure revealed by diffusion analysis. Trends Neurosci. **21**(5), 207–215 (1998)

Panayotounakos, D.E., Kravvaritis, D.C.: Exact analytic solutions of the Abel, Emden-Fowler and generalized Emden-Fowler nonlinear ODEs. Nonlinear Anal. Real World Appl. **7**(4), 634–650 (2006)

Pape, H., Clauser, C., Iffland, J.: Permeability prediction for reservoir sandstones and basement rocks based on fractal pore space geometry. In: Proceedings of SEG Annual Meeting, New Orleans (1998)
Parker, S.P.: Encyclopedia of Science and Technology, vol. 21. McGraw-Hill (1997)
Paschotta, R.: Encyclopedia of Laser Physics and Technology, vol. 1. Wiley-VCH, Berlin (2008)
Peleg, Y., Pnini, R., Zaarur, E.: Schaum's outline of theory and problems of quantum mechanics, pp. 123–124. McGraw-Hill, New York, USA (1998)
Price, W.S., Kuchel, P.W.: Effect of nonrectangular field gradient pulses in the Stejskal and Tanner (diffusion) pulse sequence. J. Magn. Reson. **94**, 133–139 (1991)
Price, W.S.: Pulsed-field gradient nuclear magnetic resonance as a tool for studying translational diffusion: part 1. Basic theory. Concept. Magn. Reson. **9**(5), 299–336 (1997)
Price, W.S.: Pulsed-field gradient nuclear magnetic resonance as a tool for studying translational diffusion: part II. Experimental aspects. Concepts Magn. Reson. **10**(4), 197–237 (1998)
Price, W.S.: NMR Studies of Translational Motion: Principles and Applications. Cambridge University Press, Cambridge, United Kingdom (2009)
Prokopová-Kubinová, Š, Vargova, L., Tao, L., Ulbrich, K., Šubr, V., Syková, E., Nicholson, C.: Poly [N-(2-hydroxypropyl) methacrylamide] polymers diffuse in brain extracellular space with same tortuosity as small molecules. Biophys. J. **80**(1), 542–548 (2001)
Rai, A.J., Yee, J., Fleisher, M.: Biomarkers in the era of personalized medicine—a multiplexed SNP assay using capillary electrophoresis for assessing drug metabolism capacity. Scand. J. Clin. Lab. Invest. **70**(sup242), 15–18 (2010a)
Rai, P., Mallidi, S., Zheng, X., Rahmanzadeh, R., Mir, Y., Elrington, S., Khurshid, A., Hasan, T.: Development and applications of photo-triggered theranostic agents. Adv. Drug Deliv. Rev. **62**(11), 1094–1124 (2010b)
Redfield, A.G.: Relaxation theory: density matrix formulation. In: Grant, D.M., Harris, R.K. (eds.) The Encyclopaedia of Nuclear Magnetic Resonance. Wiley, Chichester, United Kingdom (1996)
Rohrer, M., Bauer, H., Mintorovitch, J., Requardt, M., Weinmann, H.J.: Comparison of magnetic properties of MRI contrast media solutions at different magnetic field strengths. Invest. Radiol. **40**(11), 715–724 (2005)
Rorschach, H.E., Chang, D.C., Hazlewood, C.F., Nichols, B.L.: The diffusion of water in striated muscle. Ann. New York Acad. Sci. **204**, 445–452 (1973)
Rungseevijitprapa, W., Siepmann, F., Siepmann, J.: Disperse systems. In: Modern Pharmaceutics, vol. 1, pp. 375–440. CRC Press. (2009)
Sakurai, J.J.: In: Tuan, S.F. (ed.) Modern Quantum Mechanics, Revised Edition. Addison-Wesley Publishing Company, Inc., Reading (1994)
Shih, T.C., Yuan, P., Lin, W.L., Kou, H.S.: Analytical analysis of the Pennes bioheat transfer equation with sinusoidal heat flux condition on skin surface. Med. Eng. Phys. **29**(9), 946–953 (2007)
Slichter, C.P.: Principles of Magnetic Resonance, vol. 1. Springer Science & Business Media, Berlin, Germany (2013)
Spiegel, M.R.: Fourier Analysis with Applications to Boundary Value Problems. McGraw-Hill, New York (1974)
Spiegel, M.R.: Schaum's Outline Series of Theory and Problems of Advanced Mathematics for Engineers and Scientists. McGraw-Hill, Singapore (1983)
Straten, P.V.D., Metcalf, H.: Atoms and Molecules Interacting with Light. Cambridge University Press, Cambridge (2016)
Sykova, E.: Activity-related ionic and volume changes in neuronal microenvironment. In: Flux, K., Agnati, L.F. (eds.) Volume Transmission in the Brain: Novel Mechanisms for Neural Transmission, pp. 217–336. Raven Press, New York (1991)
Syková, E., Nicholson, C.: Diffusion in brain extracellular space. Physiol. Rev. **88**(4), 1277–1340 (2008)
Tanner, J.E.: Transient diffusion in a system partitioned by permeable barriers. Application to NMR measurements with a pulsed field gradient. J. Chem. Phys. **69**(4), 1748–1754 (1978)

Taylor, D.G., Bushell, M.C.: The spatial mapping of translational diffusion coefficients by the NMR imaging technique. Phys. Med. Biol. **30**(4), 345 (1985)

Torrey, H.C.: Bloch equations with diffusion terms. Phys. Rev. **104**(3), 563 (1956)

Ungar, P.J., Weiss, D.S., Riis, E., Chu, S.: Optical molasses and multilevel atoms: theory. J. Opt. Soc. Am. B Opt. Phys. **6**, 2058–2071 (1989)

Velten, K., Schmidt, D.M., Kahlen, K.: Mathematical Modeling and Simulation: Introduction for Scientists and Engineers. John Wiley & Sons (2024)

Wust, P., Hildebrandt, B., Sreenivasa, G., Rau, B., Gellermann, J., Riess, H., Felix, R., Schlag, P.M.: Hyperthermia in combined treatment of cancer. Lancet Oncol. **3**(8), 487–497 (2002)

Wylie, C.R., Barrett, L.C.: Advanced Engineering Mathematics. McGraw-Hill, Tokyo, Japan (1982)

Xiao-Feng, P., Yuan-Ping, F.: Quantum Mechanics in Nonlinear Systems. World Scientific, Singapore (2005)

Yan, Z., Konotop, V.V.: Exact solutions to three-dimensional generalized nonlinear Schrödinger equations with varying potential and nonlinearities. Phys. Rev. Sect. E Stat. Nonlin. Soft Matter Phys. **80**(3), 36607 (2009)

Yang, M., Zhang, X., Vafai, K., Ozkan, C.S.: High sensitivity piezoresistive cantilever design and optimization for analyte-receptor binding. J. Micromechan. Microeng. **13**(6), 864 (2003)

Yang, N., Vafai, K.: Modelling of low-density lipoprotein (LDL) transport in the artery-effects of hypertension. Int. J. Heat Mass Transf. **49**(5), 850–867 (2006)

Zoppou, C., Knight, J.H.: Analytical solution of a spatially variable coefficient advection–diffusion equation in up to three dimensions. Appl. Math. Model. **23**(9), 667–685 (1999)

Chapter 4
Physics-Informed Neural Network for Addressing Spatial and Temporal Resolution Problems in Magnetic Resonance Molecular-Imaging of Adult Brain Tumours

Abstract Magnetic Resonance Molecular Imaging (MRMI) is a powerful technique for studying brain tumours at the molecular level. However, achieving high spatial and temporal resolution remains challenging. In this chapter, we develop a novel approach for enhancing the precision of molecular imaging in the diagnosis and treatment of adult brain tumours. This leveraged the power of artificial intelligence, specifically Physics-Informed Neural Networks (PINNs), to address spatial and temporal resolution problems in Magnetic Resonance Molecular-Imaging.

Keywords Magnetic resonance molecular imaging · Bloch NMR flow equation · Physics-informed neural networks · Spatial and temporal resolution · Machine learning models · Adult brain tumours

4.1 Introduction

Brain tumours can be broadly categorized into two types: primary and metastatic. Primary brain tumours originate from the tissue within the brain itself. These can be further classified based on the type of cells from which they arise. Metastatic brain tumours, on the other hand, are not native to the brain. Instead, they are the result of cancer cells that have travelled from another part of the body to the brain via the bloodstream.

Among the various types of brain tumours, gliomas are currently the main focus of brain tumour researchers. Gliomas are a type of tumour that starts in the glial cells, which support and protect the neurons in the brain. The term "glioma" is an umbrella term that encompasses different types of gliomas. These range from high-grade (HG) tumours, such as glioblastoma multiforme (GBM), which are fast-growing and often aggressive, to low-grade (LG) tumours, such as oligodendrogliomas or astrocytomas, which tend to grow more slowly (Rehman et al. 2020).

Treatment for gliomas and other types of brain tumours typically involves a combination of chemotherapy, radiotherapy, and surgery. The choice of treatment depends on several factors, including the type, size, and location of the tumour, as well as the

B. O. Awojoyogbe and M. O. Dada, *Digital Molecular Magnetic Resonance Imaging*, Series in BioEngineering, https://doi.org/10.1007/978-981-97-6370-2_4

patient's overall health. Despite the challenges, ongoing research and advancements in medical technology continue to improve the prognosis for individuals diagnosed with brain tumours. Surgery may be applied to treat gliomas.

4.2 Spatial and Temporal Resolution

a. **Spatial Resolution**: This refers to the smallest detail that can be detected in an image. Put differently, it is the capacity to discern between two objects that are near to each other. More detailed photos with a higher spatial resolution make it possible to see the tumour and the structures around it more clearly. Better spatial resolution, for instance, is a benefit of 3D-CTA (Computed Tomography Angiography) when imaging large heterogeneous lesions (Mona 2019).
b. **Temporal Resolution**: This refers to how often data is collected, or the smallest time difference that can be detected. Temporal resolution is essential for capturing dynamic changes in brain tumour imaging, such as the passage of a contrast agent through blood vessels. For instance, a trade-off between spatial and temporal resolution is necessary for DSC (Dynamic Susceptibility Contrast) perfusion MR (Magnetic Resonance) imaging in brain tumours. This trade-off results in less spatial coverage to meet the temporal resolution requirements for an accurate relative CBV (Cerebral Blood Volume) estimation (Chakhoyan et al. 2018).

4.2.1 *Importance of Spatial and Temporal Resolution*

i. **Identification and Characterization of Tumours**: High spatial resolution in imaging is crucial as it enables the detection of tiny details within an image. This is especially vital in the context of brain tumour imaging. The ability to identify and characterize small lesions is critical in this field. Thus, high spatial resolution plays a key role in the accurate diagnosis and treatment of brain tumours (Morin and Mahesh 2018).
ii. **Monitoring Treatment Efficacy**: High temporal resolution in imaging is crucial as it offers insights into the time intervals between images. This becomes particularly useful in monitoring the progress of treatments over time. By providing a time-lapse view of the treatment process, it aids in assessing the effectiveness of therapies. Therefore, high temporal resolution plays a significant role in the successful management of medical treatments (To et al. 2011).

4.3 Magnetic Resonance Molecular Imaging

Magnetic Resonance Molecular Imaging (MRMI) is an advanced imaging technique within the realm of magnetic resonance imaging (MRI) that specializes in visualizing and scrutinizing specific molecular attributes, behaviours, or markers within the biological structures of the body. This method is finely tuned to detect and characterize molecular changes, often focusing on key markers associated with various health conditions, prominently in cancer research and diagnosis (Gauberti et al. 2018).

The primary objective of MRMI is to generate high-resolution images that precisely delineate and highlight molecular alterations occurring at the cellular or tissue levels. This is accomplished by leveraging specialized contrast agents designed to selectively target and interact with distinct biomarkers linked to diseases, such as cancer. These contrast agents serve as molecular probes, allowing for a more refined and detailed imaging process that accentuates the presence, distribution, and activity of these molecular signatures.

By providing a nuanced understanding of molecular changes within the body, MRMI holds immense promise for early detection, comprehensive characterization, and monitoring of disease progression. This detailed molecular imaging technique contributes significantly to the potential for tailored and precise therapeutic interventions by offering comprehensive insights into the molecular landscape of diseases like cancer. Today, MRI is well established as the single most useful imaging modality available in radiological practice, especially for the detection and characterization of soft tissue pathologies such as solid tumours in cancer. Magnetic resonance imaging produces exquisite three-dimensional images noninvasively, with high spatial resolution and high contrast, and the quality and acquisition times of images continue to improve as technological innovations such as parallel imaging.

4.4 Artificial Intelligence in Medicine

Machine learning models are used in medicine to search medical data and unearth insights to help improve patient experiences and health outcomes. With recent advancements in computer science and informatics, artificial intelligence (AI) is now becoming an essential part of modern healthcare. Medical professionals are supported by AI algorithms and other applications powered by AI in clinical settings and ongoing research. Currently, clinical decision support and imaging analysis are the two most frequent uses of AI in medical settings. Clinical decision support tools give healthcare professionals quick access to data or studies that are pertinent to their patients, assisting them in making decisions about treatments, medications, mental health, and other patient needs. To analyse CT scans, X-rays, MRIs, and other images for lesions or other findings that a human radiologist might miss, AI tools are being used in the field of medical imaging.

Many healthcare organisations around the world have begun field-testing new AI-supported technologies, such as algorithms designed to support patient monitoring and AI-powered tools to screen COVID-19 patients, as a result of the difficulties that the COVID-19 pandemic created for many health systems.

The overall standards for the use of AI in medicine are still being established, as are the research and test results. However, there are more and more chances for AI to help doctors, scientists, and the patients they treat. There is currently little doubt that AI will play a significant role in the digital health platforms that will influence and support modern medicine.

4.4.1 Tkinter

Tkinter is a Python library used for creating graphical user interfaces (GUIs). While it is not directly involved in machine learning computations, it can be incredibly helpful for building interfaces to interact with machine learning models or tools. Tkinter can be used in machine learning projects by the following:

i. **Data Pre-processing Interfaces**: Tkinter can create GUIs to pre-process data before feeding it into machine learning models. Users can input data, clean it, normalize it, and perform other pre-processing tasks using a user-friendly interface.
ii. **Model Training and Testing**: You can create GUIs to control the training and testing of machine learning models. Users can select different hyperparameters, choose algorithms, and monitor the training process through interactive displays.
iii. **Visualization of Results**: Tkinter allows you to display model outputs visually. You can plot graphs, confusion matrices, or other visual representations of model performance for easier understanding.
iv. **Deployment of Models**: When you want to create an application that uses a trained machine learning model, Tkinter can help build the user interface for the application. It can provide a user-friendly experience for interacting with the model predictions.
v. **Hyperparameter Tuning**: Create interfaces for hyperparameter tuning where users can adjust parameters and immediately see the impact on model performance.

The research utilized MR fingerprinting (MRF) to measure T1 and T2 relaxation times in different brain tumour regions. By analysing these relaxation times in solid tumour and peritumoral white matter areas across glioblastomas, lower grade gliomas, and metastases, distinct tissue properties were identified. Significant differences in relaxation times were observed between lower grade gliomas and metastases as well as between peritumoral regions of glioblastomas and lower grade gliomas. Table 4.1 shows the data used in PINN models.

These differences suggested varying cellular content, water distribution, and structural organisation among the different tumour types. MRF-based quantification of

Table 4.1 Relaxometric data for solid tumour at 3.0 T (Badve et al. 2017)

Relaxometric properties of solid tumour at 3.0 T			
ROI	GBM	Metastasis	LGG
T_1 (ms)	1639 ± 247	1324 ± 273	1600 ± 197
T_2 (ms)	138 ± 22	105 ± 27	172 ± 53

tissue properties revealed unique characteristics, aiding in the differentiation of brain tumour types and potential implications for clinical diagnosis and treatment planning (Badve et al. 2017). In my work, I make use of T1 and T2 relaxation time of solid tumour region of different brain tumour as a parameter for my PINN model. And for the implementation, I generated 500 sample points between the highest and lowest values of T1 and T2 for the three types of brain tumours which were used for both the temporal and spatial classification of the three types of brain tumours, 500 (t) parameters were generated between 0 and 1 for temporal classification for the three types of brain tumours. Furthermore parameters (t1) and (t2) were gotten by dividing (t) by 10 and 100, respectively, for each value of (t) and M_y was gotten by substituting the actual parameter into the solution to the time dependent Bloch equation. 500 (x) parameter was generated between 0 and 2 for spatial classification for the three types of brain tumours. Furthermore, parameter (×1) and (×2) were gotten by dividing (x) by 10 and 100, respectively, for each value of (x) and M_y was gotten by substituting the actual parameter into the solution to the time independent Bloch equation (Dada and Awojoyogbe 2021).

4.4.2 Deep Learning

Deep learning is a subset of machine learning, which is a branch of artificial intelligence (AI). It involves the use of neural networks with multiple layers (deep neural networks) to model and solve complex problems. The term "deep" refers to the presence of multiple layers in the neural network architecture. While a single-layer neural network can still make approximate predictions, additional hidden layers can help optimize and refine for accuracy.

4.4.3 Physics-Informed Neural Network (PINN) Model Structure

The code I have written has a Physics-Informed Neural Network (PINN) that is organized as a fully connected network (FCN) featuring several layers and a distinct activation function. Figure 4.1 is the PINN model class and explanation of its structures.

```
class FCN(nn.Module):
    "Defines a connected network"

    def __init__(self, N_INPUT, N_OUTPUT, N_HIDDEN, N_LAYERS):
        super().__init__()
        activation = nn.Tanh
        self.fcs = nn.Sequential(*[
                        nn.Linear(N_INPUT, N_HIDDEN),
                        activation()])
        self.fch = nn.Sequential(*[
                        nn.Sequential(*[
                            nn.Linear(N_HIDDEN, N_HIDDEN),
                            activation()]) for _ in range(N_LAYERS-1)])
        self.fce = nn.Linear(N_HIDDEN, N_OUTPUT)

    def forward(self, x):
        x = self.fcs(x)
        x = self.fch(x)
        x = self.fce(x)
        return x
```

Fig. 4.1 PINN model structure

i. **Layers**: The network was constructed with multiple layers, including an input layer, several hidden layers, and an output layer. The dimensions of the input and output were determined by N_INPUT and N_OUTPUT, respectively. The number of hidden layers and the number of neurons in each hidden layer were set by N_LAYERS and N_HIDDEN, respectively.
ii. **Activation Function**: I chose the hyperbolic tangent function (nn.Tanh) as the activation function for the network. After each linear transformation in the hidden layers, I applied this activation function.
iii. **Loss Function**: To train the network, I used a loss function that was a combination of a data loss term and a physics loss term. The data loss term was the mean squared error between the network's predictions on the training data and the actual training data. The physics loss term was the mean squared error of the residual of the Bloch NMR flow equation when evaluated at certain points in the domain. I added these two terms together to form the total loss, which was then minimized during training as shown in Fig. 4.2.

```
# compute the "physics loss"
yhp = model(x_physics)
dx  = torch.autograd.grad(yhp, x_physics, torch.ones_like(yhp), create_gra          es dy/dx
dx2 = torch.autograd.grad(dx,  x_physics, torch.ones_like(dx),  create_gra flowers flower flows es d^2y/dx^2
physics = dx2 + mu*dx + k*yhp# computes the residual of the bloch NMR flow differential equation
loss2 = (1e-4)*torch.mean(physics**2)

# backpropagate joint loss
loss = loss1 + loss2# add two loss terms together
```

Fig. 4.2 Physics loss of the model

The Bloch NMR flow equation was incorporated into the network architecture through the physics loss term. I computed the first and second derivatives of the network's predictions with respect to the input using PyTorch's automatic differentiation capabilities. These derivatives, along with the network's predictions, were then used to compute the residual of the Bloch NMR flow equation. I calculated the mean squared error of this residual and added it to the data loss to form the total loss. By doing this, I enabled the network to learn from both the data and the underlying physics of the problem. This approach led to a model that was more accurate and generalizable, especially when the amount of available data was limited. This is the key idea behind Physics-Informed Neural Networks. The Bloch equations used are as shown below.

4.4.4 Equation for Temporal Resolution

The time-dependent Bloch NMR flow equation was used as shown in Eq. (3.113):

$$\frac{d^2M_y}{dt^2} + T_0\frac{dM_y}{dt} + (\gamma^2B_1^2 + T_g)M_y = \frac{M_0}{T_1}\gamma B_1 \tag{4.1}$$

where M_y is the transverse magnetization, M_0 is equilibrium magnetization, γ is gyromagnetic ratio, $T_0 = \frac{1}{T_1} + \frac{1}{T_2}$, $T_g = \frac{1}{T_1T_2}$, T_1 is spin–lattice relaxation time while T_2 is spin–spin relaxation time. Under non-adiabatic condition, $\gamma^2B_1^2$ (x, t)$\ll T_g$ and assuming $M_0 \approx 0$, Eq. 4.1 becomes:

$$\frac{d^2M_y}{dt^2} + T_0\frac{dM_y}{dt} + T_gM_y = 0 \tag{4.2}$$

$$M_y(t) = c_1 \exp\left(-\frac{T_0t}{2} - \frac{t\sqrt{T_0^2 - 4T_g}}{2}\right) + c_2 \exp\left(-\frac{T_0t}{2} + \frac{t\sqrt{T_0^2 - 4T_g}}{2}\right) \tag{4.3}$$

where $c_1 = c_2 = 0.1$. Equation 4.3 is the solution to the time-dependent NMR equation.

4.4.5 Equation for Spatial Resolution

The time-independent Bloch NMR flow equation was used as shown in Eq. (4.4):

$$v^2\frac{d^2M_y}{dx^2} + vT_0\frac{dM_y}{dx} + (\gamma^2B_1^2 + T_g)M_y = \frac{M_0}{T_1}\gamma B_1 \tag{4.4}$$

If $M_0 \approx 0$ and $T_g \gg \gamma^2 B_1^2$ Eq. (4.4) becomes:

$$v^2 \frac{d^2 M_y}{dx^2} + v T_0 \frac{dM_y}{dx} + T_g M_y = 0 \tag{4.5}$$

Simplifying Eq. 4.5 gives

$$\frac{d^2 M_y}{dx^2} + \left(\frac{T_0}{v}\right) \frac{dM_y}{dx} + \left(\frac{T_g}{v^2}\right) M_y = 0 \tag{4.6}$$

$$M_y(x) = c_1 \exp\left(-\frac{T_0 x}{2v^2} - \frac{x\sqrt{T_0^2 - 4T_g}}{2v^2}\right) + c_2 \exp\left(-\frac{T_0 x}{2v^2} + \frac{x\sqrt{T_0^2 - 4T_g}}{2v^2}\right) \tag{4.7}$$

where

$$\begin{aligned} T_0 &= \frac{1}{T_1} + \frac{1}{T_2} \\ T_g &= \frac{1}{T_1 T_2} \end{aligned} \tag{4.8}$$

and $c_1 = c_2 = 0.1$. Equation 4.7 is the solution to the time-independent NMR flow equation.

4.4.6 Training the PINN

The process of training the Physics-Informed Neural Network (PINN) involved several steps:

i. **Initialization**: I started by initializing the fully connected network (FCN) with the specified number of input nodes, output nodes, hidden nodes, and layers. The weights of the network were initialized randomly.
ii. **Defining the Loss Function**: The loss function was a combination of a data loss term and a physics loss term. The data loss term was the mean squared error between the network's predictions on the training data and the actual training data. The physics loss term was the mean squared error of the residual of the Bloch NMR flow equation when evaluated at certain points in the domain.
iii. **Optimization**: I used the Adam optimizer to minimize the total loss. Adam is a popular optimization algorithm in machine learning that combines the advantages of two other extensions of stochastic gradient descent: AdaGrad and RMSProp. It adapts the learning rate for each weight in the model individually and computes adaptive learning rates for different parameters.

iv. **Training Loop**: In each training step, I first zeroed out the gradients of the optimizer. Then, I computed the data loss and the physics loss, and added them together to form the total loss. I then performed back propagation on the total loss to compute the gradients, and used the optimizer to update the model's parameters.
v. **Hyper parameters**: The selection of hyper parameters, such as the learning rate, the number of hidden layers, and the number of hidden nodes, can significantly affect the performance of the model. In this case, the learning rate was set to 1e-4 for the Adam optimizer, and the network had 3 hidden layers with 32 nodes each. These values were likely chosen based on prior experience or through experimentation.
vi. **Use of Training Data**: The training data was used in two ways. First, it was used to compute the data loss term in the loss function. The network's predictions on the training data were compared to the actual training data to compute this term. Secondly, the training data was used to compute the physics loss term. The network's predictions were differentiated with respect to the input at the training data points, and these derivatives were used to compute the residual of the Bloch NMR flow equation.

4.4.7 Evaluating the PINN

Below is how the performance of the PINN was evaluated:

i. **Convergence with Exact Solution**: The convergence between the PINN's predictions and the exact solution indicates that the model is accurately capturing the underlying dynamics of the system. This is a key measure of the model's performance.
ii. **Loss Function**: The loss function, which includes both the data loss and the physics-informed loss, was another important metric. A decrease in the loss function value over the training period indicates that the model is improving its fit to the data and its compliance with physical laws.
iii. **Validation Set**: Although not explicitly mentioned in the provided code, typically a separate validation set is used to evaluate the model's performance during training. This set is not used to update the model's parameters, but rather to check how well the model generalizes to unseen data. If the model's performance on the validation set starts to deteriorate while its performance on the training set continues to improve, this is a sign of overfitting.
iv. **Overfitting Prevention Techniques**: The use of the physics loss term in the loss function can be seen as a form of regularization, as it encourages the model to find solutions that not only fit the data but also comply with underlying physical laws. This can help prevent the model from overfitting to the training data.

4.5 Transverse Magnetization Profiles

By putting the cancer tissue properties such as T_1, T_2 relaxation times and velocity of the three types of brain tumours. The network was able to fit the solution very closely in the vicinity of the training points, but did not learn an accurate solution outside them as shown in Fig. 4.3.

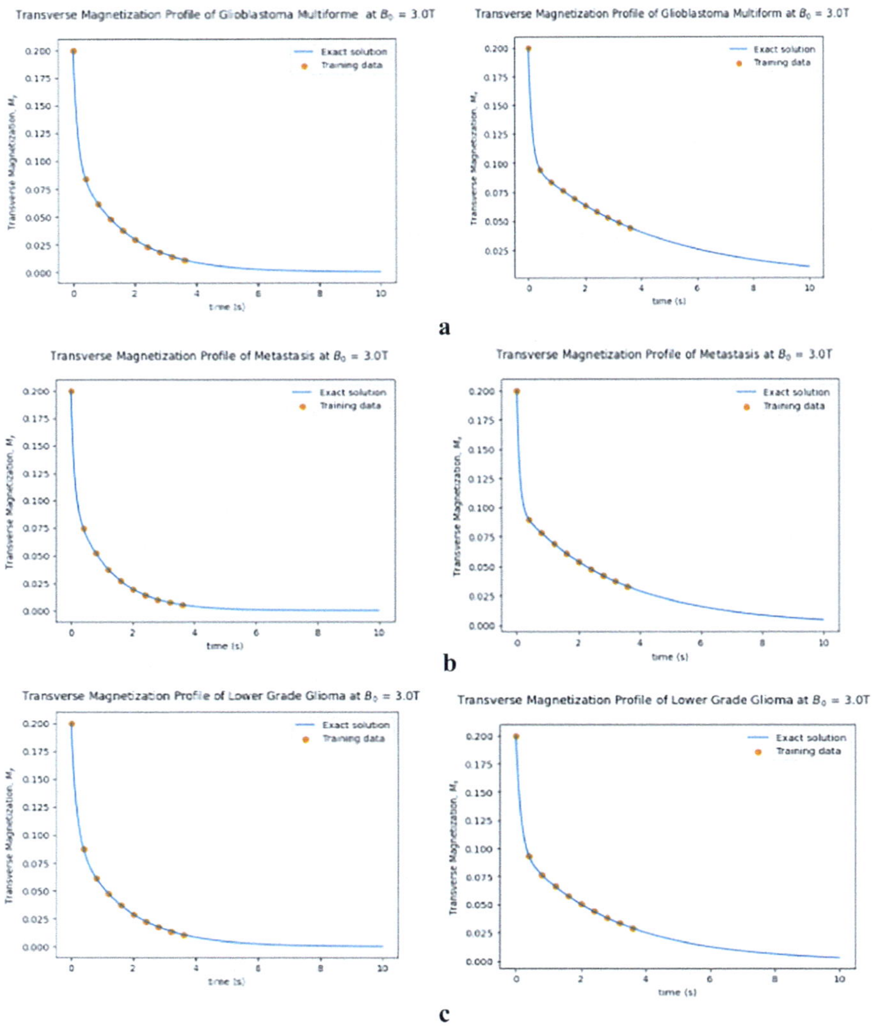

Fig. 4.3 Transverse magnetization profile for the three types of adult brain tumours (**a**) Glioblastoma Multiform (**b**) for Metastasis (**c**) for Lower Grade Glioma

4.5.1 PINN Model Training Result for Glioblastoma Multiform (GBM)

The Physics-Informed Neural Network (PINN) model predicts the Transverse Magnetization Profile of Glioblastoma Multiforme. It compares the exact solution with the neural network prediction at different training steps ranging from 1 to 18,000. The model's predictions improve as training progresses, demonstrating the effectiveness of the PINN model in enhancing the precision of molecular imaging in the diagnosis and treatment of adult brain tumours.

4.5.2 PINN Model (Training Result) for Metastasis.

The Physics-Informed Neural Network (PINN) model predicts the transverse magnetization profile of metastasis. It compares the exact solution with the neural network prediction at different training steps ranging from 1 to 18,000. The model's predictions improve as training progresses, demonstrating the effectiveness of the PINN model in enhancing the precision of molecular imaging in the diagnosis and treatment of adult brain tumours.

4.5.3 PINN Model (Training Result) for Lower Grade Glioma

The Physics-Informed Neural Network (PINN) model predicts the transverse magnetization profile of metastasis. It compares the exact solution with the neural network prediction at different training steps ranging from 1 to 18,000. The model's predictions improve as training progresses, demonstrating the effectiveness of the PINN model in enhancing the precision of molecular imaging in the diagnosis and treatment of adult brain tumours.

4.5.4 Classification of Adult Brain Tumours

Six data set was generated from Table 4.1 above and how additional parameters for the model were gotten was explained below that table. Moreover, I discovered that three out of the six models did not perform well.., I then went on to create three applications for the three models that had the better performance. The images below show my classifications result.

4.6 Conclusion

We have developed and implemented a Physics-Informed Neural Network (PINN) model. The model was trained to predict the transverse magnetization profile of the three types of adult brain tumours. At the initial training step, there was a noticeable difference between the exact solution and the neural network prediction as shown in Figs. 4.3, 4.4, 4.5 and 4.6. However, as training progressed, the model's predictions aligned more closely with the exact solution, indicating an improvement in accuracy.

The successful implementation of the PINN model signifies a major step forward in addressing spatial and temporal resolutions in magnetic resonance molecular

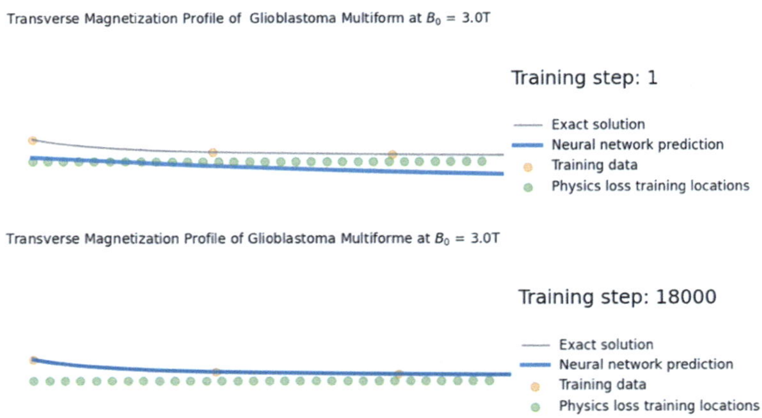

Fig. 4.4 Training Progress of PINN Model for Glioblastoma Multiforme

Fig. 4.5 Training progress of PINN model for metastasis

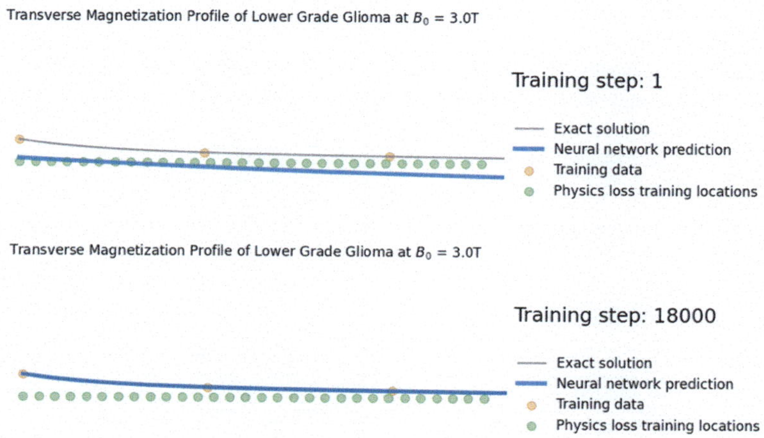

Fig. 4.6 Training progress of PINN model for low grade glioma

imaging of adult brain tumours. The model's ability to accurately capture the underlying dynamics of the system demonstrates its potential in enhancing the precision of molecular imaging. This chapter also demonstrates the potential of Tkinter, a Python library, for creating graphical user interfaces (GUIs), in various stages of the machine learning process, from data pre-processing to model deployment as shown in Fig. 4.7. Our next investigation will focus on: (i) improving the PINN model by incorporating more diverse data sets to enhance the model's ability to generalize and make accurate predictions across a wider range of scenarios; (ii) **refining** the training process to improve the model's performance. This could involve tuning hyperparameters, experimenting with different loss functions, or exploring other optimization algorithms; (iii) the use of a separate validation set to evaluate the model's performance during training. This can provide valuable insights into the model's ability to generalize to unseen data, and could help in early detection of overfitting, and improve the robustness of the model; (iv) further exploration of overfitting prevention techniques, such as the use of the physics loss term in the loss function to ensure the model finds solutions that not only fit the data but also comply with underlying physical laws; (v) the challenges of implementing PINNs, such as computational cost and model interpretability, and developing strategies to mitigate these issues. This could involve leveraging more efficient computing resources or developing methods for better understanding and interpreting the model's predictions; and (vi) exploring the application of PINNs in other areas of medical imaging.

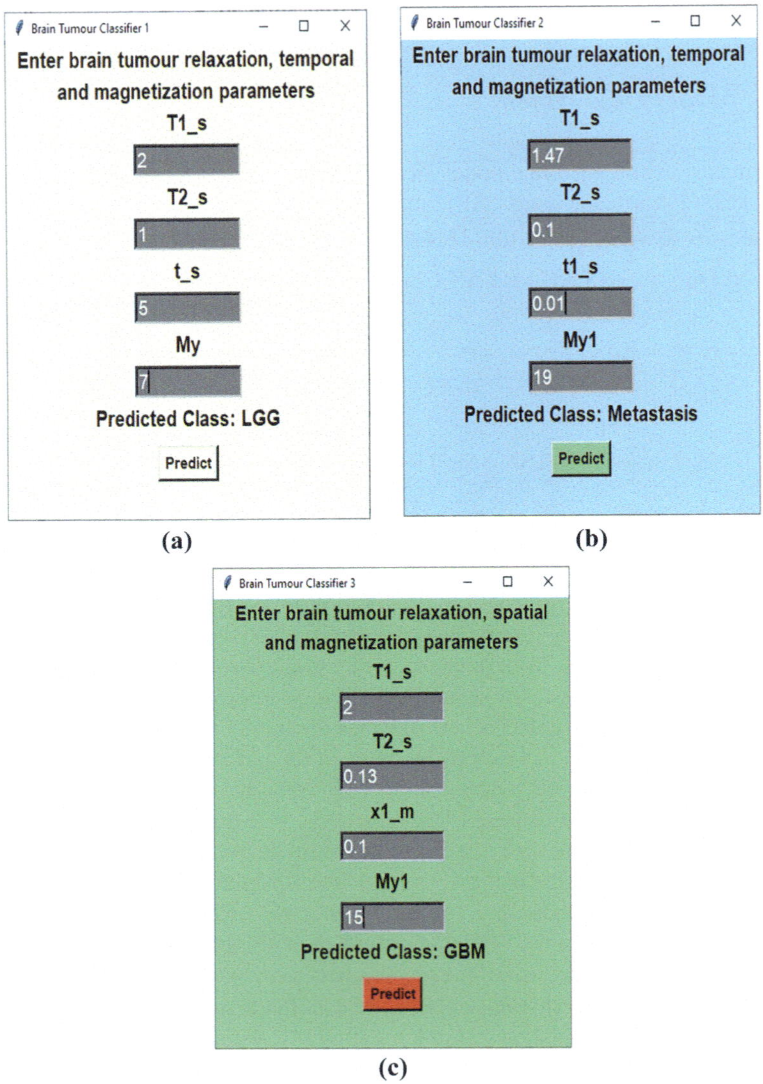

Fig. 4.7 Brain tumour classification deployment

Acknowledgements The authors acknowledge the motivated research efforts of Taiwo John Majebi as a promising young researcher.

References

Badve, C., Yu, A., Dastmalchian, S., Rogers, M., Ma, D., Jiang, Y., Gulani, V.: MR fingerprinting of adult brain tumors: initial experience. Am. J. Neuroradiol. **38**(3), 492–499 (2017)

Chakhoyan, A., Leu, K., Pope, W.B., Cloughesy, T.F., Ellingson, B.M.: Improved spatiotemporal resolution of dynamic susceptibility contrast perfusion MRI in brain tumors using simultaneous multi-slice echo-planar imaging. Am. J. Neuroradiol. **39**(1), 43–45 (2018)

Dada, M.O., Awojoyogbe, B.O.: Fundamental physics of nuclear magnetic resonance. In: Computational Molecular Magnetic Resonance Imaging for Neuro-oncology. Biological and Medical Physics, Biomedical Engineering. Springer, Cham (2021)

Gauberti, M., Fournier, A.P., Vivien, D., Martinez de Lizarrondo, S.: Erratum to: molecular magnetic resonance imaging (mMRI). Preclin. MRI: Methods Protoc. E1–E1 (2018)

Mona, P.: (2019). https://practicalneurology.com/articles/2019-nov-dec/brain-tumor-imaging/pdf

Morin, R.L., Mahesh, M.: The importance of spatial resolution to medical imaging. J. Am. Coll. Radiol. **15**(8), 1127 (2018)

Rehman, Z.U., Zia, M.S., Bojja, G.R., Yaqub, M., Jinchao, F., Arshid, K.: Texture based localization of a brain tumor from MR-images by using a machine learning approach. Med. Hypotheses **141**, 109705 (2020)

To, A.C., Flamm, S.D., Marwick, T.H., Klein, A.L.: Clinical utility of multimodality LA imaging: assessment of size, function, and structure. JACC: Cardiovasc. Imaging **4**(7), 788–798 (2011)

Chapter 5
Machine Learning Model for Diagnosis of Pulmonary Arterial Hypertension and Severe Aortic-Valve Stenosis Using Magnetic Resonance Relaxometry Data

Abstract This chapter develops and implements a Machine Learning model for the diagnosis of pulmonary arterial hypertension (PAH) and aortic-valve stenosis (AS) from Magnetic resonance relaxometry data that classifies the selected human diseases. The goal is to (i) develop machine learning models to improve the accuracy of predictions and decision-making in different models, (ii) use machine learning techniques for sorting across big datasets to identify hidden trends, patterns, and insights, and (iii) explore the data in the developed models for the diagnosis of human medical conditions.

Keywords Pulmonary arterial hypertension · Aortic-valve stenosis · Magnetic resonance relaxometry data · Machine learning models · Diagnosis of human medical conditions

5.1 Introduction

Pulmonary arterial hypertension (PAH) is a progressive disease of the pulmonary vasculature that is characterized by elevated pressures within the pulmonary vascular tree. Over the past several decades, there has been a significant increase in our comprehension of the pathobiology and epidemiology of PAH, along with enhancements in therapy alternatives. Between 48 and 55 cases of PAH are thought to occur for per million adult population. A diagnosis of PAH now requires evidence of a mean pulmonary artery pressure > 20 mmHg, pulmonary vascular resistance > 2 Wood units, and a pulmonary artery wedge pressure $\leq$ 15 mmHg at right heart catheterization, according to a recent amendment to the criteria. However, the structure and function of the pulmonary circulation may be further affected by several mechanisms potentially leading to pulmonary arterial and venous remodeling. In heart failure, recent data even suggest that the severity of PH correlates most strongly with venous and small arteriolar intimal thickening. In addition, the function of the right ventricle is often affected independently from the afterload increase, leading to uncoupling of the right ventricle/pulmonary artery unit with further exercise limitation and adverse

B. O. Awojoyogbe and M. O. Dada, *Digital Molecular Magnetic Resonance Imaging*, Series in BioEngineering, https://doi.org/10.1007/978-981-97-6370-2_5

outcome. This is especially true in heart failure with preserved ejection fraction. Over the past 5 years, since the 5th World Symposium on Pulmonary Hypertension (WSPH) in 2013, significant advances have improved our understanding of PH-LHD. Possible treatments include chemotherapy and radio therapy surgery. Survival depends on the stage of the disease, overall health, and other factors. International Journal of Applied Engineering Research indicates that significant progress has been made, resulting in a greater knowledge of PH-LHD, its problems, and its evidence gaps. The most complicated scenario is PH with cardiac failure with intact ejection fraction, since it can be mistakenly identified as group 1 PH. We suggest a revised hemodynamic criterion for PH caused by LHD and a three-step practical method to differential diagnosis based on the most recent data. This involves determining a non-invasive probability of PH-LHD as well as a certain "left heart" phenotype. Accurate pulmonary arterial wedge pressure measurements, and in patients with a high probability, provocative testing to establish the diagnosis are the foundations for invasive confirmation of PH-LHD.

The early metabolomics revolution effectively began with *De Statica Medicina*, published by Santorio Santorio in 1614. His quantitative approach was described to modern medicine and the first systematic study of basal metabolism. Both mass-spectrometric (MS) and nuclear magnetic resonance (NMR) strategies, two of the most common modern analytical techniques for metabolic measurements, were first described in the early part of twentieth century: mass-spectrometric by J. J. Thomson, nuclear magnetic resonance by F. Aston in 1913, and by Bloch and Purcell in 1946. Although not described as such, the first analytical measurement of metabolites was published by Pauling in 1971, before the pioneering work of Nicholson and Sadler in the early 1980s and Willmitzer in 1987 (International Journal of Applied Engineering Research). Metabolic profiling now includes the measurement of a range of biofluids, including blood plasma and serum, saliva, urine, knee-joint synovial fluid, semen and cerebrospinal fluid (Percival et al. 2018). Indeed, biological media will vary in suitability for each disease/condition investigated, with a wide range and high volume of metabolic data being extracted from each (Visentin et al. 2017). For example, common high-field (HF) NMR metabolomics studies performed at operating frequencies of 600 MHz and above can detect more than 150 metabolites in human urine, and quantify almost 70 metabolites in human blood serum. Moreover, with these practices becoming commonplace, tools and repositories, such as the Human Metabolome database (HMDB) and Metaboanalyst (Chong et al. 2018), have been developed to serve the omics community.

With the rapidly expanding functional improvements in both NMR and MS techniques, additional avenues for the incorporation of metabolite profiling in medicine have arisen, based on miniaturization of these technologies, biomarker discovery, and prognostic monitoring (Weljie et al. 2006). Although MS and NMR are both considered standard techniques for metabolomics and metabonomic studies (Blekherman et al. 2011), this investigation will focus on the latter, and more specifically, low-field (LF) compact, or mobile, bench top NMR facilities. In general, these instruments operate at frequencies below 100 MHz, and are based on permanent magnets as opposed to the large, HF superconducting magnets commonly found in analytical characterization suites. These facilities operating at LF suffer from the same issues of sensitivity which plagued early designs of NMR spectrometers; however, with the employment of approaches such as solvent suppression, and magnet arrays, it is possible to simultaneously observe and monitor 20 or more metabolites in saliva and urine at LF, as demonstrated here for the first time. Although this limits the detection capabilities of NMR when compared to those of larger, high-field instruments, the advantages of more compact, mobile NMR instruments, incorporating applications of chemometric/metabolomics approaches to the multivariate analysis of complex mixtures, have been demonstrated in fields such as materials science (Schäler et al. 2015), forensic chemistry (Singh and Blümich 2017), chemical education, and biomedical sciences.

Herein, we present an updated protocol for the analysis of biofluids through compact, benchtop NMR measurements. Comprehensive protocols for metabolite measurement and profiling are currently available, and these rely on HF NMR or LC/MS techniques, such as that described in the pioneering works of Nicholson and others (Chang et al. 2006). We describe a complete procedural development for the analysis of biofluids by LF NMR analysis, including validation and quantification experiments, experimental guidelines, and metabolomics data analysis, and these are experimentally demonstrated by an appropriate example based in the area of diabetes (preprints).

Implementation of machine learning algorithms in R for predicting selected human diseases from nuclear magnetic resonance (NMR) relaxometry data is teaching the machine to predict the disease of the user with respect to the collected data. Disease prediction system has data sets collected from different health related sites. With the help of this system the user will be able to know the probability of the disease with the given data.

To diagnose a patient's disease, one needs an expert in laboratory and Magnetic Resonance Imaging (MRI) scanner. But experts are few couple with the fact that equipment are not available. So, there is the need to teach a machine how to read NMR Relaxometry Data and diagnose a patient's disease efficiently (Computational diagnosis).

5.2 Magnetic Resonance Relaxometry (MRR)

Magnetic resonance imaging relaxometry is the measuring of relaxation times in a nuclear magnetic resonance experiment. If one knows and decouples the contrast mechanisms that contribute to the overall MR signal, the quantitative character of the MR signal can be exploited to obtain highly important absolute measurements of biophysical parameters and respond by producing an electromagnetic signal with a frequency characteristic of the magnetic field at the nucleus. This process occurs near resonance, when the oscillation frequency matches the intrinsic frequency of the nuclei, which depends on the strength of the static magnetic field, the chemical environment, and the magnetic properties of the isotope involved. In practical applications with static magnetic fields up to ca. 20 T, the frequency is similar to VHF and UHF television broadcasts (60–1000 MHz). NMR results from specific magnetic properties of certain atomic nuclei. Nuclear magnetic resonance spectroscopy is widely used to determine the structure of organic molecules in solution and study molecular physics, crystals as well as non-crystalline materials. NMR is also routinely used in advanced medical imaging techniques, such as in magnetic resonance imaging (MRI). All isotopes that contain an odd number of protons and/or neutrons have an intrinsic nuclear magnetic moment and angular momentum. The most commonly used nuclei are 1H and 13C, although isotopes of many other elements can be studied by high-field NMR spectroscopy as well. A key feature of NMR is that the resonance frequency of a particular simple substance is usually directly proportional to the strength of the applied magnetic field. It is this feature that is exploited in imaging techniques; if a sample is placed in a non-uniform magnetic field then the resonance frequencies of the sample's nuclei depend on spot of their location in the field. Since the resolution of the imaging technique depends on the magnitude of the magnetic field gradient, many efforts are made to develop increased gradient field. The principle of NMR usually involves strength and invloving three sequential steps are (Chizhik et al. 2014):

i. The alignment (polarization) of the magnetic nuclear spins in an applied, constant magnetic field (B_0).
ii. The perturbation of this alignment of the nuclear spins by a weak oscillating magnetic field, usually referred to as a radio-frequency (RF) pulse. The oscillation frequency required for significant perturbation is dependent upon the static magnetic field (B_0) and the nuclei of observation.
iii. The detection of the NMR signal during or after the RF pulse, due to the voltage induced in a detection coil by precession of the nuclear spins around B_0. After an RF pulse, precession usually occurs with the nuclei's intrinsic Larmor frequency, and in itself, does not involve transitions between spin states or energy levels.

The two magnetic fields are usually chosen to be perpendicular to each other as this maximizes the NMR signal strength. The frequencies of the time-signal response by the total magnetization (M) of the nuclear spins are analyzed in NMR spectroscopy and magnetic resonance imaging. Both use applied magnetic fields (B_0) of great strength, often produced by large currents in superconducting coils, in order to achieve dispersion of response frequencies and of very high homogeneity and stability in order to deliver spectral resolution. The details of these are described by chemical shifts, the Zeeman effect, and Knight shifts (in metals). The information provided by NMR can also be increased using hyperpolarization, and/or using two-dimensional, three-dimensional and higher-dimensional techniques. NMR phenomena are also utilized in low-field NMR, NMR spectroscopy and MRI in the Earth's magnetic field (referred to as Earth's field NMR), and in several types of magnetometers.

5.3 Relaxation Process

The relaxation rate is highly influenced by the microscopic environment's mobility (fluctuations, diffusion) and the strength of the applied magnetic field. As a general rule, large magnetic fields improve sensitivity to fast dynamics, while low fields increase sensitivity to slow dynamics. Thus, the relaxation rate as a function of magnetic field strength is a microdynamic fingerprint. Key features of materials science are frequently defined in several domains using the terms mobility/dynamics/stiffness/viscosity/rigidity of the sample. These properties are typically affected by atomic and molecular mobility in the sample, which can be quantified using time-domain NMR and fast field cycling relaxometry (Bortolotti et al. 2024). The process of population relaxation refers to nuclear spins that return to thermodynamic equilibrium in the magnet. This process is also called T1, "spin–lattice" or "longitudinal magnetic" relaxation, where T1 refers to the mean time for an individual nucleus to return to its thermal equilibrium state of the spins. After the nuclear spin population has relaxed, it can be probed again, since it is in the initial, equilibrium (mixed) state. T2 relaxometry involves measuring the T2 relaxation rate in the area of interest. The relaxation rate is best described by a combination of exponential decay curves in part due to different compartments of water in tissue but for simplicity a first approximation using a single curve is used. The minimum number of points needed to define an exponential curve is three though more is ideal. This can be done for each pixel and the T2 rates averaged or the signal from pixels in the selected volume can be averaged and the calculation performed from this. The relaxation curve was initially generated by a series of single spin echo "Hahn" sequences repeating the sequence to generate each new point on the curve. To save time, the Carr-Purcell multi-echo sequence was proposed, resulting in a shortening of the time as occurs in imaging. An accumulation of slight errors in the length of the 180° RF pulse can cause an artificial decrease in the T2 time so a modification of this sequence by Meiboom and Gill changing the phase of alternating 180° pulses by 90° results in correction of the error on the even echoes. A relatively long repetition rate is needed to prevent T1 effects from causing errors. Also, the echo time should be kept to a minimum to limit diffusion effects that can also falsely reduce the T2 time (Fig. 5.1).

The above data type contains information about T1 and T2 relaxation times taken from the interventricular septum (sp) and the left ventricular lateral (lvl) wall. This relaxometric data has been taken from an experimental measurement. A total of 42 subjects were included in the study by Wang et al. (2018) which included 18 subjects with PAH, 19 with severe AS, and 5 healthy controls. However, since these data is quite small for building machine learning models, we have assume a large sample point of 1500 with 500 data points (patients) per class. A script written in R language has been used to simulate a larger dataset as shown in Fig. 5.1.

Step 3: The simulated dataset was saved into .csv (comma separated value file) format for further processing. It can also be saved into .txt (tab-delimited text file).

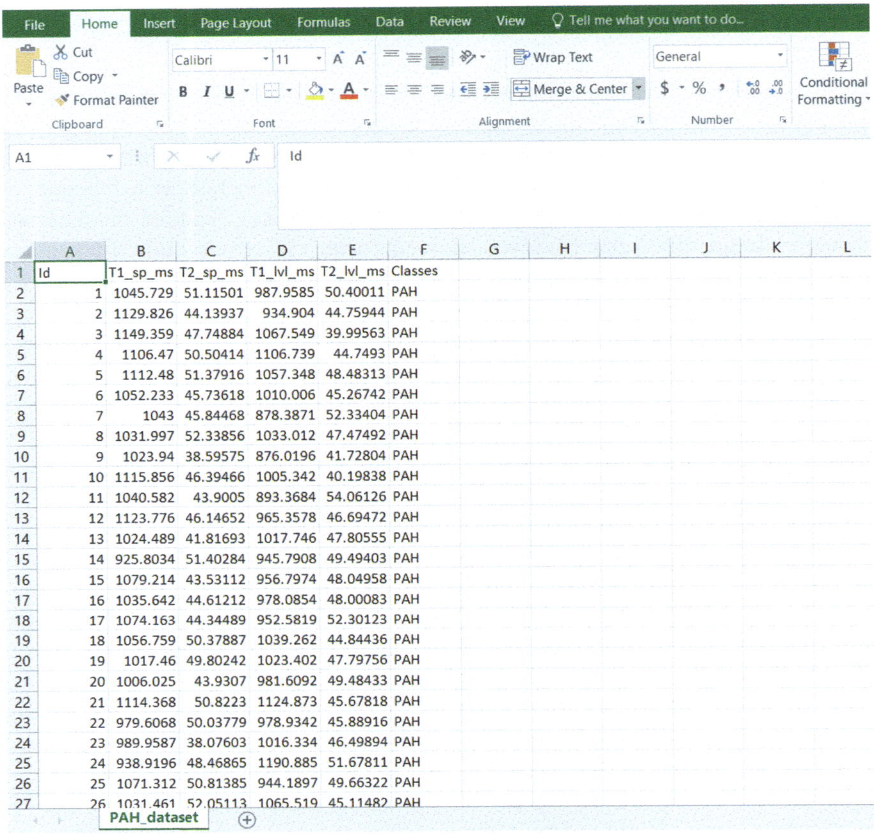

Id	T1_sp_ms	T2_sp_ms	T1_lvl_ms	T2_lvl_ms	Classes
1	1045.729	51.11501	987.9585	50.40011	PAH
2	1129.826	44.13937	934.904	44.75944	PAH
3	1149.359	47.74884	1067.549	39.99563	PAH
4	1106.47	50.50414	1106.739	44.7493	PAH
5	1112.48	51.37916	1057.348	48.48313	PAH
6	1052.233	45.73618	1010.006	45.26742	PAH
7	1043	45.84468	878.3871	52.33404	PAH
8	1031.997	52.33856	1033.012	47.47492	PAH
9	1023.94	38.59575	876.0196	41.72804	PAH
10	1115.856	46.39466	1005.342	40.19838	PAH
11	1040.582	43.9005	893.3684	54.06126	PAH
12	1123.776	46.14652	965.3578	46.69472	PAH
13	1024.489	41.81693	1017.746	47.80555	PAH
14	925.8034	51.40284	945.7908	49.49403	PAH
15	1079.214	43.53112	956.7974	48.04958	PAH
16	1035.642	44.61212	978.0854	48.00083	PAH
17	1074.163	44.34489	952.5819	52.30123	PAH
18	1056.759	50.37887	1039.262	44.84436	PAH
19	1017.46	49.80242	1023.402	47.79756	PAH
20	1006.025	43.9307	981.6092	49.48433	PAH
21	1114.368	50.8223	1124.873	45.67818	PAH
22	979.6068	50.03779	978.9342	45.88916	PAH
23	989.9587	38.07603	1016.334	46.49894	PAH
24	938.9196	48.46865	1190.885	51.67811	PAH
25	1071.312	50.81385	944.1897	49.66322	PAH
26	1031.461	52.05113	1065.519	45.11482	PAH

Fig. 5.1 A cross-section generated dataset for Pulmonary Arterial hypertension (PAH), severe Aortic Stenosis (AS) and healthy patients

5.4 Visualization and Analysis of the Projection Data

As such, Fig. 5.2 shows the display of R-Studio when it is opened for the first time.

The MRI Relaxometry data is obtained from 1500 projections across the original object of three different medical conditions with the relaxation rate and relaxation time. The number of projections is saved in a solitary file in column-major order with the name of the file giving in a .csv file. The .csv was prepared using a Microsoft Excel program. It consists of 1500 rows which is the number of projections with 6 columns. The tables below show the file which contains 500 projections.

i. T1_sp_ms: This shows the name of the file that contains all the first time at which the detection, classification and visualization of the condition was classified.
ii. T2_sp_ms: This show the name of the file that contains all the second time at which the detection, classification and visualization of the condition was classified.

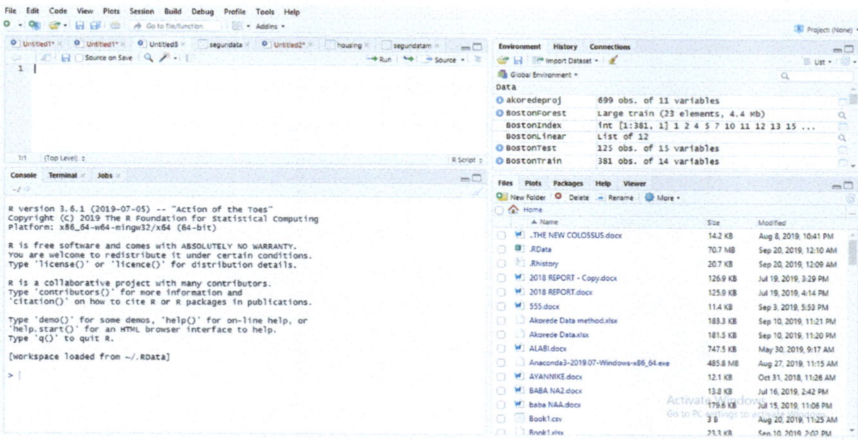

Fig. 5.2 R-Studio screen

iii. T1_lvl_ms: This shows the name of the file that contains all the first time at which the detection, classification and visualization of the condition was classified.
iv. T1_lvl_ms: This shows the name of the file that contains all the first time at which the detection, classification and visualization of the condition was classified.
v. Classes: This shows the three types of conditions considered in this chapter (Dada and Awojoyogbe 2021) in which each one of them contains 500 instances which give total instances of 1500. Based on Eq. (4.8), we can write:

$$T_0 = \frac{1}{T1} + \frac{1}{T2} \tag{5.1}$$

$$T_g = \frac{1}{T1T2} \tag{5.2}$$

5.5 Reading the Input File

See Table 5.1.

5.5.1 Distribution of Classes

We will use data visualization to explore a dataset prepared as a spreadsheet. The dataset contains information about the different types of classes and counting the total number of unique classes:

From Fig. 5.3, the distribution of classes in a pie chart form showing the classes are balanced. The number of observations in the chart is roughly the same. PAH and AS involve the medical condition of patients while control represents individuals without

Table 5.1 Reading the impute file

Id		T1_sp_ms	T2_sp_ms	T1_lvl_ms	T2_lvl_ms	Classes
0	1	1045.728984	51.115006	987.958459	50.400106	PAH
1	2	1129.825764	44.139366	934.904028	44.759437	PAH
2	3	1149.358552	47.748844	1067.548759	39.995627	PAH
3	4	1106.470145	50.504141	1106.738658	44.749295	PAH
4	5	1112.479794	51.379162	1057.348218	48.483128	PAH
...	...	...	...	...	...	...
1495	1496	1048.213562	46.141391	985.050808	43.354780	Control
1496	1497	1073.214609	43.520676	968.069279	44.638845	Control
1497	1498	962.353010	48.074221	967.604789	42.349506	Control
1498	1499	1001.929691	45.850867	931.481973	40.494836	Control
1499	1500	1020.168228	49.267723	1037.037413	43.779646	Control

Total 1500 rows × 6 columns

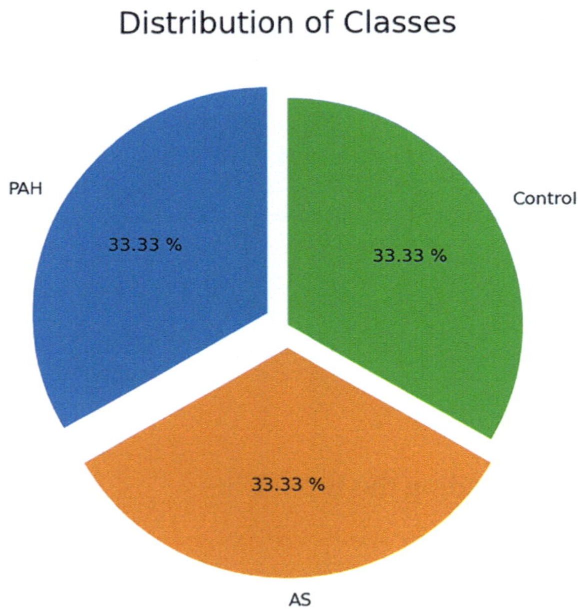

Fig. 5.3 Visualized class distribution

this medical conditions. The pie chart shows equal representation of individuals with PAH, AS and Control.

Box Plot

Figure 5.4 is a graphical representation of the distribution of dataset. The line inside the box represents the median within the box which gives an idea of the central tendency of the data. By having two boxes side by side, the median of T1_Septum is higher than T2_Septum, it suggests a shift in the central tendency.

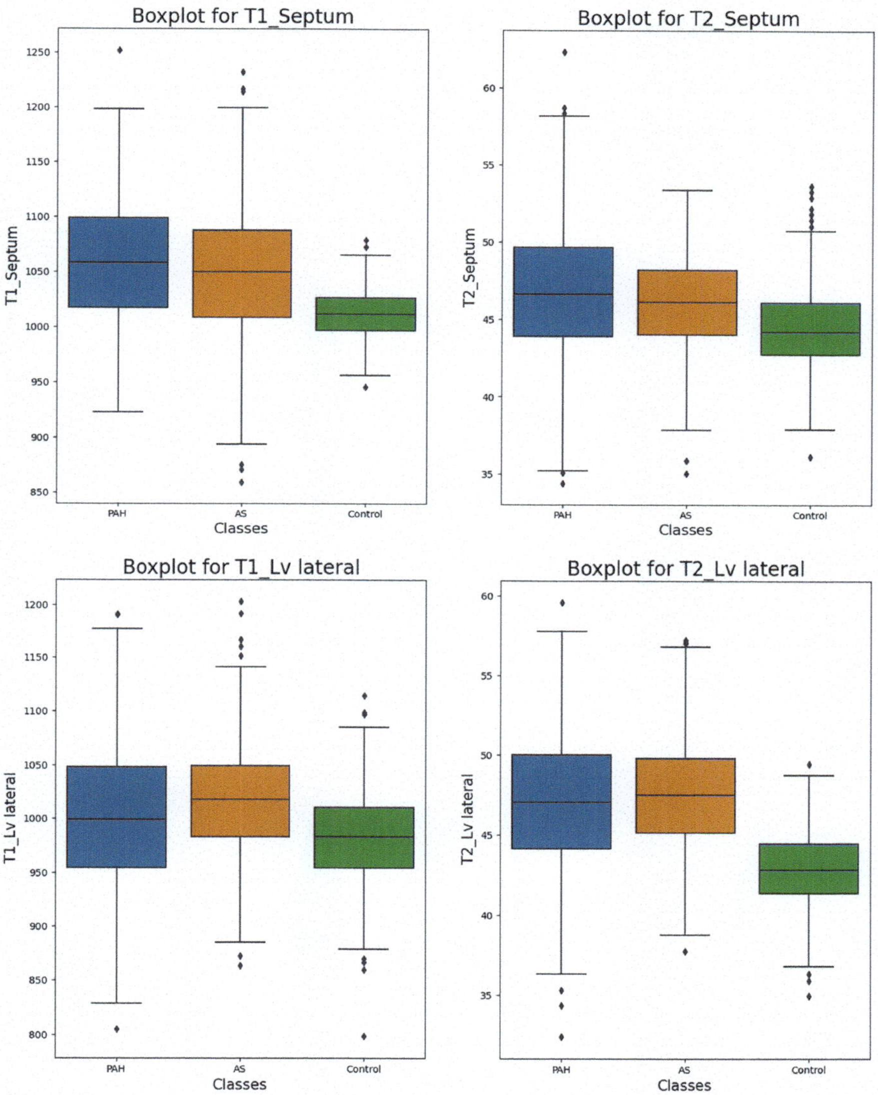

Fig. 5.4 Box plot

Violin Plot

The widened areas represent kernel density estimates of the distribution data which show that the data is spread out along the y-axis. A violin plot is a hybrid of a box plot and a kernel density plot, which shows peaks in the data. It is used to visualize the distribution of numerical data. Unlike a box plot that can only show summary statistics, violin plots depict summary statistics and the density of each variable. On each side of the gray line is a kernel density estimation to show the distribution shape of the data (Fig. 5.5).

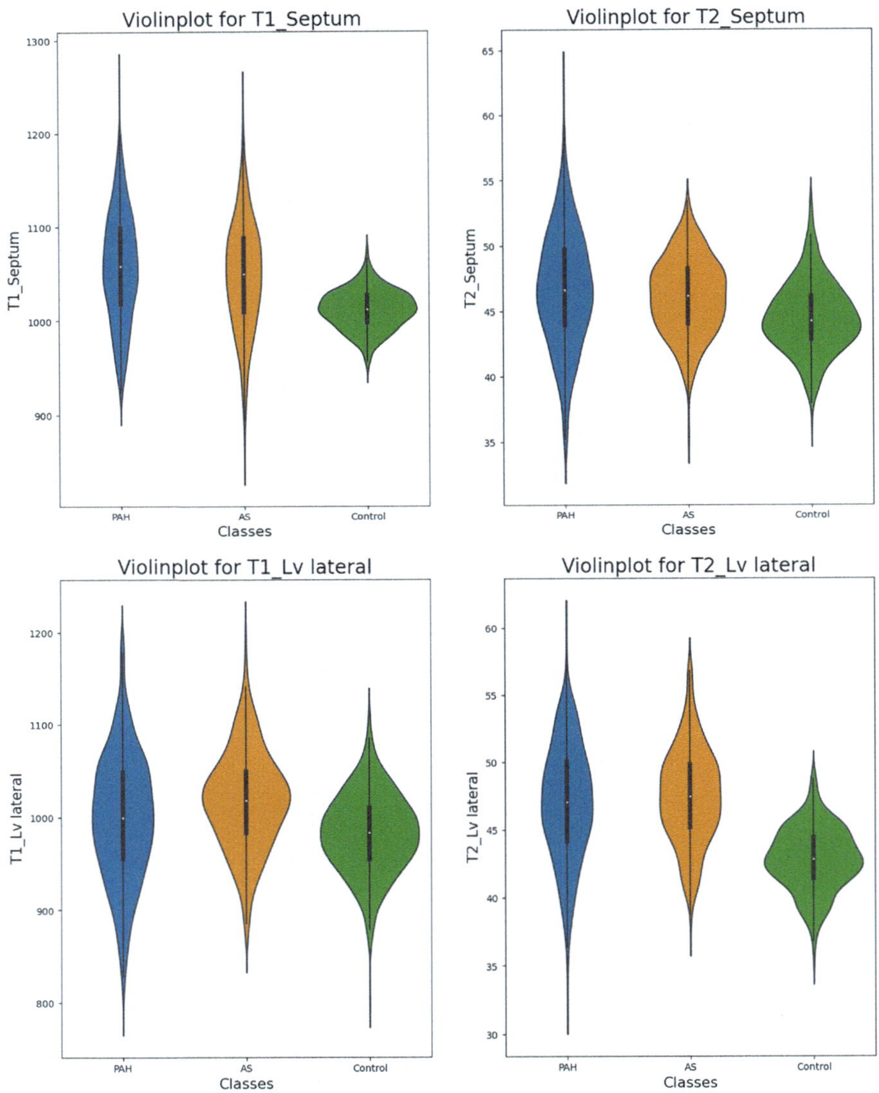

Fig. 5.5 Violin plot

5.5.2 Bivariate Analysis

Bivariate analysis is a form of statistical analysis. It involves the analysis of two variables for the purpose of determining the relationship between them (Fig. 5.6).

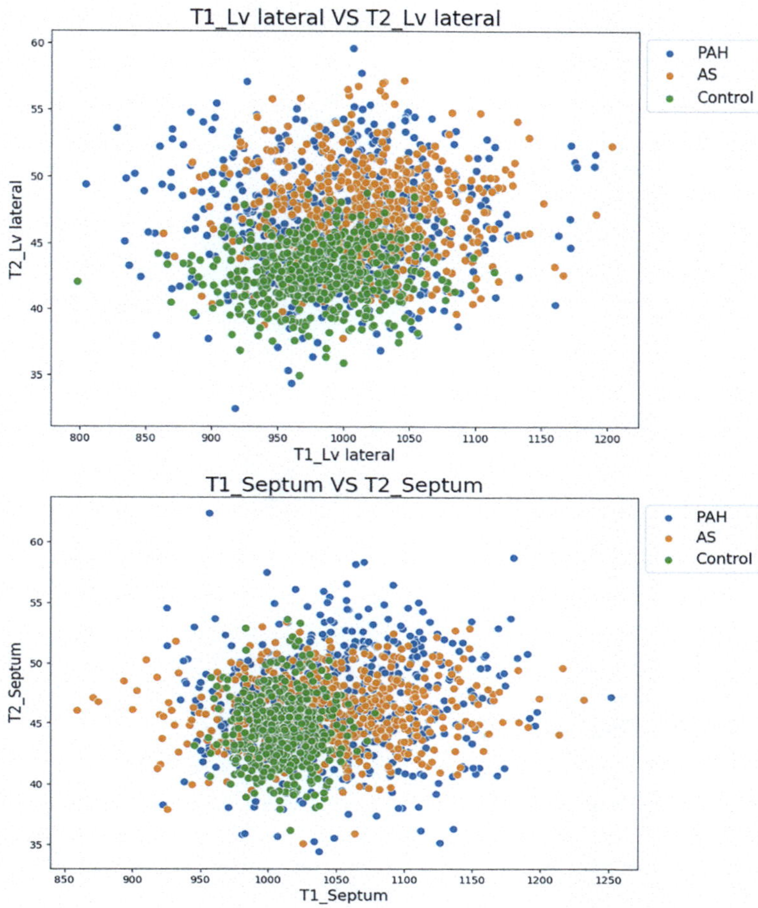

Fig. 5.6 Bivariate analysis

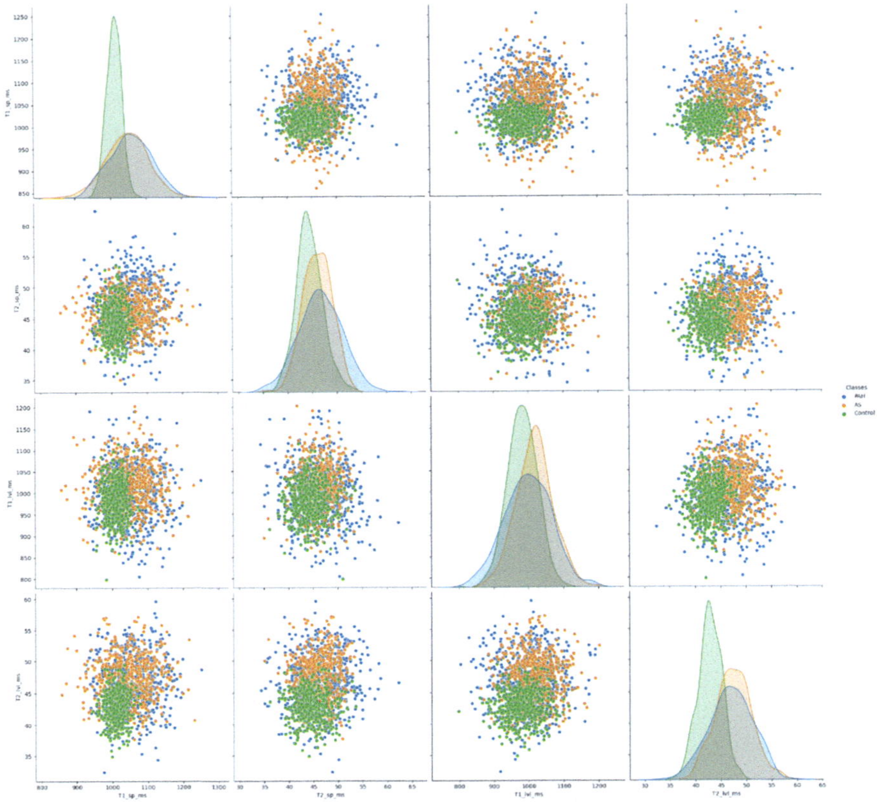

Fig. 5.7 Multivariate analysis

5.5.3 Multivariate Analysis

It refers to an analysis involving multiple dependent variables resulting in one outcome. Creating different graphs for all the features to perform bivariate analysis would be extremely tedious. Seaborn provides a convenient way to perform multivariate analysis using the 'pair plot' function (Fig. 5.7).

5.5.4 3D Visualization

3D scatter plots are used to plot data points on three axes in the attempt to show the relationship between three variables. Each row in the data table is represented by a marker whose position depends on its values in the columns set on the X, Y, and Z axes (Fig. 5.8).

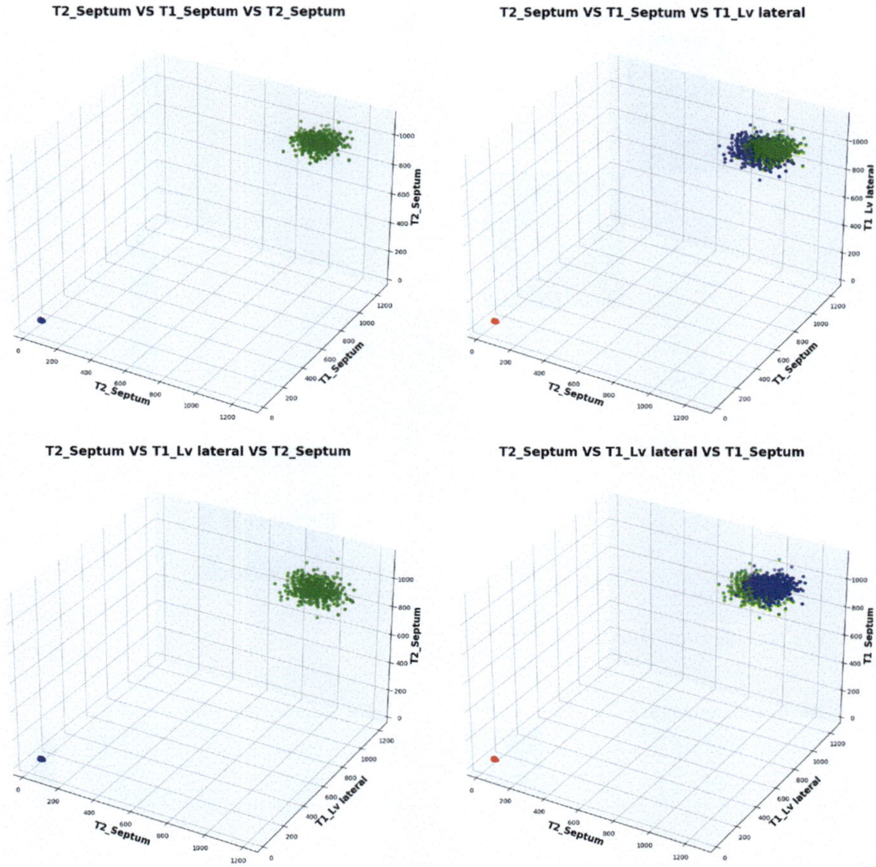

Fig. 5.8 3D visualization

Feature Correlations

See Fig. 5.9.

Confusion Matrix

From Fig. 5.10, confusion matrix for logistic regression, despite its name, is a classification model rather than regression model. It is a process of modeling the probability of a discrete outcome given input variables.

KNN Model Performance

See Fig. 5.11.

Different Models Trained

See Fig. 5.12.

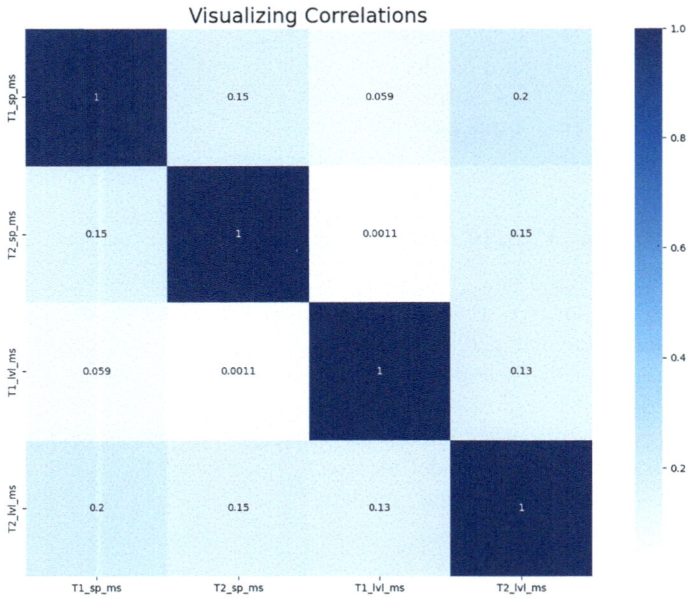

Fig. 5.9 Checking correlations

Fig. 5.10 Confusion matrix

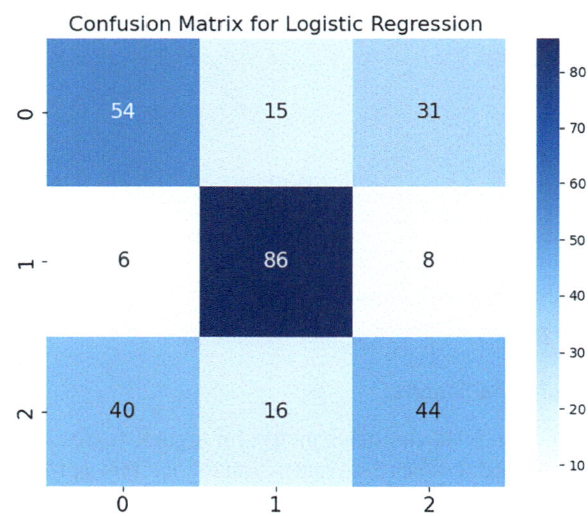

PAH data of the first 25 entries after training machine learning model for PAH predictions is shown in Table 5.2 and Table 5.3 shows AS data of the first 25 entries after training machine learning model for AS predictions. These datasets would be used for prediction on Django App.

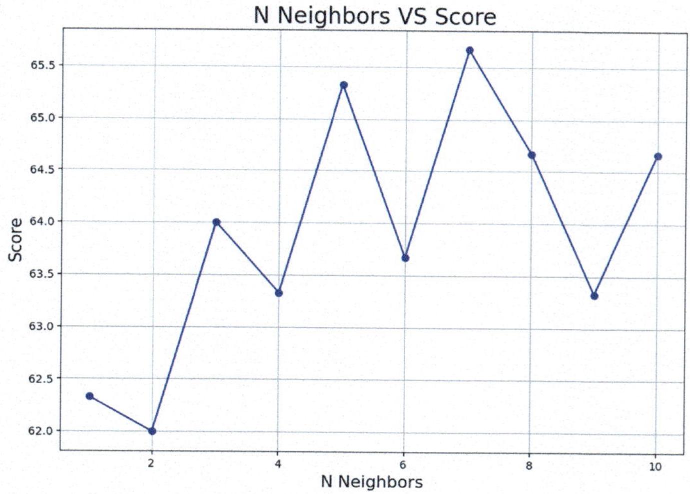

Fig. 5.11 Different values for N-neighbors

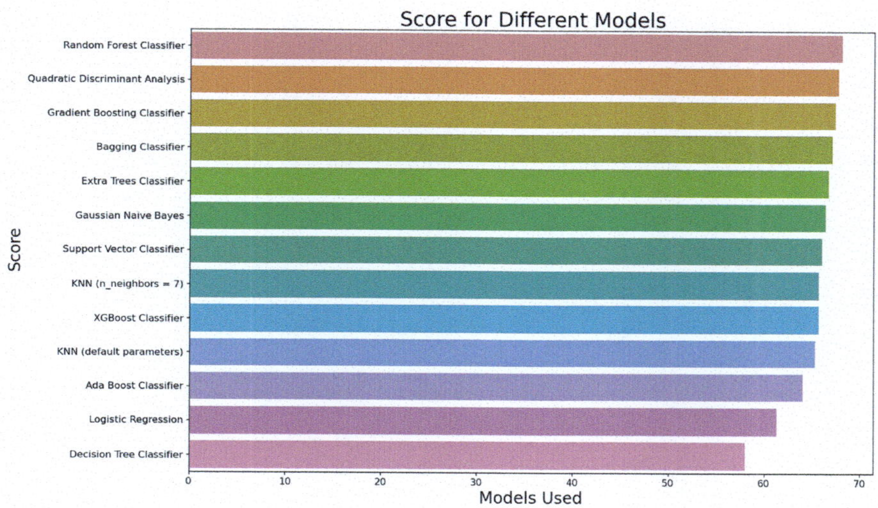

Fig. 5.12 Final result for different models

Table 5.4 is the control data of the first 25 entries after training the machine learning model for Control predictions. These datasets were used for prediction on and app developed with Django framework.

Table 5.2 PAH data of the first 25 entries

Id	T1_sp_ms	T2_sp_ms	T1_lvl_ms	T2_lvl_ms	Classes
1	1045.729	51.11501	987.9585	50.40011	PAH
2	1129.826	44.13937	934.904	44.75944	PAH
3	1149.359	47.74884	1067.549	39.99563	PAH
4	1106.47	50.50414	1106.739	44.7493	PAH
5	1112.48	51.37916	1057.348	48.48313	PAH
6	1052.233	45.73618	1010.006	45.26742	PAH
7	1043	45.84468	878.3871	52.33404	PAH
8	1031.997	52.33856	1033.012	47.47492	PAH
9	1023.94	38.59575	876.0196	41.72804	PAH
10	1115.856	46.39466	1005.342	40.19838	PAH
11	1040.582	43.9005	893.3684	54.06126	PAH
12	1123.776	46.14652	965.3578	46.69472	PAH
13	1024.489	41.81693	1017.746	47.80555	PAH
14	925.8034	51.40284	945.7908	49.49403	PAH
15	1079.214	43.53112	956.7974	48.04958	PAH
16	1035.642	44.61212	978.0854	48.00083	PAH
17	1074.163	44.34489	952.5819	52.30123	PAH
18	1056.759	50.37887	1039.262	44.84436	PAH
19	1017.46	49.80242	1023.402	47.79756	PAH
20	1006.025	43.9307	981.6092	49.48433	PAH
21	1114.368	50.8223	1124.873	45.67818	PAH
22	979.6068	50.03779	978.9342	45.88916	PAH
23	989.9587	38.07603	1016.334	46.49894	PAH
24	938.9196	48.46865	1190.885	51.67811	PAH
25	1071.312	50.81385	944.1897	49.66322	PAH

Classes

PAH:	500
AS:	500
CONTROL:	500

Table 5.3 AS data of the first 25 entries

501	1080.124	45.61205	1030.736	45.88131	AS
502	900.2674	46.11532	1009.81	51.72695	AS
503	1092.111	48.12337	967.1612	55.86551	AS
504	1013.692	48.07485	1001.703	41.66186	AS
505	1037.097	41.15307	998.1337	44.99245	AS
506	1139.873	49.63765	1008.945	47.96009	AS
507	1115.174	47.34628	1063.797	47.16851	AS
508	1015.813	47.98292	1031.723	49.61473	AS
509	1113.337	43.60191	1137.416	45.6555	AS
510	1139.74	48.69063	1003.046	52.77958	AS
511	1047.545	47.89874	954.686	48.29525	AS
512	1015.241	45.47541	1106.156	49.89388	AS
513	1100.749	44.50449	994.3662	47.88911	AS
514	1011.308	49.94917	991.4552	45.73753	AS
515	963.5378	43.14863	1006.698	42.80421	AS
516	1058.665	45.68337	1063.09	47.42412	AS
517	1054.063	45.71305	987.1033	41.755	AS
518	967.6798	42.41719	1031.375	45.45512	AS
519	949.269	45.81261	989.2386	48.74601	AS
520	1047.336	45.34882	1096.78	49.74584	AS
521	1148.338	52.87158	986.0519	52.84463	AS
522	1044.177	43.62283	981.1972	53.09804	AS
523	1116.374	49.96889	1056.512	50.57866	AS
524	1122.064	42.79958	1020.583	40.69787	AS

5.5.5 Model Deployment

Deploying a machine learning model for predicting Pulmonary Arterial hypertension (PAH), severe Aortic Stenosis (AS), and control on a Django application involves integrating the trained model into Django project.

Deployed Diagnosed PAH and AS Classification App

See Fig. 5.13.

Table 5.4 Control data of the first 25 entries

1001	1020.743	43.4642	1076.065	46.61981	Control
1002	999.7208	43.59673	1036.036	43.74541	Control
1003	1009.956	44.37567	961.4238	42.83991	Control
1004	1016.279	51.85617	998.3656	42.73037	Control
1005	990.1333	39.22985	945.2066	43.95047	Control
1006	995.7267	41.18131	976.0061	42.72023	Control
1007	994.9269	44.11016	956.0426	42.74529	Control
1008	1005.456	43.75639	988.0904	44.82187	Control
1009	1005.444	46.3515	940.5363	44.74245	Control
1010	1026.363	42.31937	1002	43.95142	Control
1011	1014.076	44.08211	974.4461	43.83748	Control
1012	1032.745	41.44998	990.8755	44.81195	Control
1013	1013.896	43.35377	994.357	45.38671	Control
1014	1019.455	43.37326	939.6589	44.20698	Control
1015	1009.607	44.25152	970.0097	42.52237	Control
1016	1000.922	45.37501	1020.955	45.28317	Control
1017	982.9786	47.74162	947.5558	40.33058	Control
1018	958.045	48.80403	942.219	41.59793	Control
1019	1035.03	45.39131	945.6755	40.46074	Control
1020	983.5704	42.87042	959.716	41.18929	Control
1021	1019.759	45.51564	1047.099	45.41182	Control
1022	977.6409	42.78967	970.6695	43.73412	Control
1023	1021.562	39.68894	959.9616	42.60079	Control
1024	1016.459	42.23853	958.4186	44.99907	Control

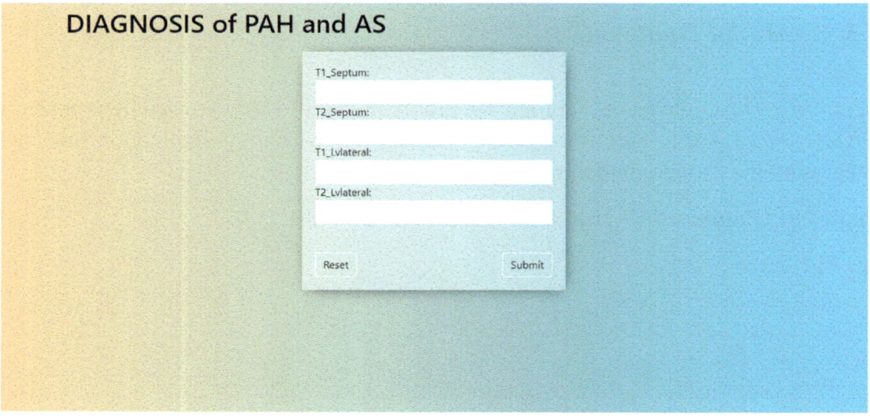

Fig. 5.13 Deployment on Django

5.6 Conclusion

In this chapter, we have developed a computational diagnosis technique to forecast specific human diseases. While lung transplantation remains the only curative intervention for PAH, there are a number of promising therapies under investigation which may further reduce morbidity and improve outcomes. The method is set up so that the user's data is used as input by the system to predict the chosen human ailment as an output. The NMR relaxometry data framework was used to successfully develop a machine learning method.

References

Blekherman, G., Laubenbacher, R., Cortes, D. F., Mendes, P., Torti, F. M., Akman, S., Torti, S.V., Shulaev, V.: Bioinformatics tools for cancer metabolomics. Metabolomics, **7**, 329–343 (2011)

Bortolotti, V., Conte, P., Landi, G., Lo Meo, P., Nagmutdinova, A., Spinelli, G. V., Zama, F: Robust Algorithms for the Analysis of Fast-Field-Cycling Nuclear Magnetic Resonance Dispersion Curves. Computers **13**(6), 129 (2024)

Chang, W. H., Chen, J. H., Hwang, L. P.: Single-sided mobile NMR with a Halbach magnet. Magn. Reson. Imaging **24**(8), 1095–1102 (2006)

Chizhik, V. I., Chernyshev, Y. S., Donets, A. V., Frolov, V. V., Komolkin, A. V., Shelyapina, M. G.: Magnetic resonance and its applications (No. 15506). Cham: Springer International Publishing (2014)

Chong, J., Soufan, O., Li, C., Caraus, I., Li, S., Bourque, G., Wishart, D.S., Xia, J.: MetaboAnalyst 4.0: towards more transparent and integrative metabolomics analysis. Nucleic Acids Res. **46**(W1), W486–W494 (2018)

Dada, M.O., Awojoyogbe, B.O.: Fundamental physics of nuclear magnetic resonance. Computational Molecular Magnetic Resonance Imaging for Neuro-oncology, pp. 11–114 (2021)

Percival, B. C., Grootveld, M., Gibson, M., Osman, Y., Molinari, M., Jafari, F., Sahota, T., Martin, M., Casanova, F., Mather, M.L., Wilson, P. B.: Low-field, benchtop NMR spectroscopy as a potential tool for point-of-care diagnostics of metabolic conditions: Validation, protocols and computational models. High-throughput, **8**(1), 2 (2018)

Schäler, K., Roos, M., Micke, P., Golitsyn, Y., Seidlitz, A., Thurn-Albrecht, T., Schneider, H., Hempel, G., Saalwächter, K.: Basic principles of static proton low-resolution spin diffusion NMR in nanophase-separated materials with mobility contrast. Solid State Nucl. Magn. Reson. **72**, 50–63 (2015)

Singh, K., Blümich, B.: Desktop NMR for structure elucidation and identification of strychnine adulteration. Analyst **142**(9), 1459–1470 (2017)

Visentin, S., Crotti, S., Donazzolo, E., D'Aronco, S., Nitti, D., Cosmi, E., Agostini, M.: Medium chain fatty acids in intrauterine growth restricted and small for gestational age pregnancies. Metabolomics **13**(5), 54 (2017)

Wang, J., Zhao, H., Wang, Y., Herrmann, H. C., Witschey, W. R., Han, Y.: Native T1 and T2 mapping by cardiovascular magnetic resonance imaging in pressure overloaded left and right heart diseases. J. thorac. dis. **10**(5), 2968 (2018)

Weljie, A.M., Newton, J., Mercier, P., Carlson, E., Slupsky, C.M.: Targeted profiling: quantitative analysis of 1H NMR metabolomics data. Anal. Chem. **78**(13), 4430–4442 (2006)

Chapter 6
A Convolution Neural Network for Artificial Intelligence-Based Classification of Alzheimer's Diseases

Abstract The purpose of this chapter is to construct a high-accuracy deep learning algorithm with the best performing model for the diagnosis and classification of medical pictures related to Alzheimer's disease. The CNN model will be used to train MRI brain images. In order to construct and deploy the model, this chapter makes use of specialized application software and libraries including Jupyter notebook, Python 3, Keras, and TensorFlow. The trained model will be distributed as an application known as an APK, which will make it accessible as an android application for personal diagnosis. It will be a web service for diagnosing Alzheimer's disease and determining the stage of the disease remotely.

Keywords Convolution neural network · Artificial intelligence-based classification · Alzheimer's diseases · Jupyter notebook · Python 3 · Keras · TensorFlow

6.1 Introduction

Alzheimer's disease is the most general term of a large class of disorders which are clinically known as "dementias", diseases. In 1906, Dr. Alois Alzheimer published his description of senile dementia. After his female patient died in her early fifties with an unusual mental illness, he used microscope to study her brain tissue, and observed the "tangles" (now called neurofibrillary) and "plaques" (abnormal clumps) which are now the hallmarks of the disease (Diamond 2008). Dementia is a wide class of brain diseases. They were initially called senile dementia, in his honor, senile dementia was renamed Alzheimer's disease.

Alzheimer's disease progresses in transitional stages. They are characterized by a progressive decline of thinking ability and of memory. When this disease is present in the brain, the nerve cells die and their connection with other nerve cells are lost. Alzheimer's disease can be classified into:

B. O. Awojoyogbe and M. O. Dada, *Digital Molecular Magnetic Resonance Imaging*, Series in BioEngineering, https://doi.org/10.1007/978-981-97-6370-2_6

i. Very mildly demented: This is the stage where patient starts to forget where they put their stuff, other people's names recently, etc. It is hard to detect through cognitive ability test.
ii. Mildly demented: This is the stage where patients don't remember words, can't find their way to the destination, loss focus and work-abilities. This is also the stage where patients even forget that they are losing memory. From this stage, with cognitive testing, it can be found.
iii. Moderately demented: Starts to forget the recent activities, important old histories, have hard time calculating the budget, hard to go outside alone, and loss of empathy.
iv. And non-demented, if the disease is not detected (Kruthika 2019).

It is important to discover the stage in which the disease is present in a patient to enable the treatment have higher effect of slowing down the disease process. Thus, having an artificial intelligence-based classification that detects Alzheimer's disease in their early stages can help elongate the life expectancy of a patient, reduce the cost of caring for Alzheimer's patients as well as an overall improvement of life quality.

6.2 Signs and Symptoms of Alzheimer's Disease

The symptom of Alzheimer's disease can originally appear after age of 60 and the danger will rise with age (Jiang 2020). The first sign of Alzheimer's disease is the failure to remember new information over time. Symptoms vary in many people with Alzheimer's disease but the most common symptoms are:

i. Memory loss that disorders daily life activities.
ii. Challenges in scheduling or solving problems
iii. Difficulty carrying out familiar tasks at home, at work or at leisure
iv. Misperception with time or place
v. Trouble understanding visual images and spatial relationships
vi. Difficulties with words when speaking or writing
vii. Misplacing things and losing the ability to retrace steps
viii. Decreased or poor judgment
ix. Withdrawal from work or social activities
x. Changes in disposition and personality including boredom and depression (National Institute on Aging 2022a, b).

AD is a degenerative multi-faceted neurological brain disorder. People who have Mild Cognitive Impairment are more prone to develop Alzheimer's disease than other people. People don't notice the effects of Alzheimer's disease (AD) until years after the disease has already caused changes in the brain. This is because AD starts at least two decades before the symptoms appear. According to Alzheimer's Disease International (ADI), more than 50 million people across the globe are living with dementia. It is anticipated that this proportion will rise to 152 million individuals by the year 2050, which indicates that one person will develop dementia every three seconds (Helaly 2022).

The expense of care for AD patients is predicted to climb considerably, necessitating early and precise computer-aided diagnosis (Wen et al. 2020). Therefore, this study is highly significant to caregivers, hospitals and Alzheimer disease patients in classifying the various stages of the disease which will be useful for treatment administration.

6.3 Jupyter Notebook

Jupyter notebook was created to create open-source software, open standards, and services for interactive computing in dozens of programming languages. The jupyter notebook can be run on a local desktop without the need for internet connectivity, or it can be installed on a remote server (for example, Anaconda) and accessed via the internet. They are excellent for demonstrating work flows since you can see both the code and the results.

6.4 Python Libraries

Python libraries are collections of pre-combined code that can be used iteratively to reduce coding time. They are significant because they eliminate the need to write code from scratch. Some of the fundamental Python libraries are: Matplotlib, which aids in data analysis and numerical plotting, pandas, which provides fast, expressive, and flexible data structures for working with structured and time-series data, and numpy, one of the fundamental Python libraries that supports large multidimensional arrays and matrices as well as a collection of high-level mathematical functions for easy execution. There are approximately 137,000 Python libraries available.

6.5 Data

Dataset consists of two files—Training and Testing both containing a total of 6400 images each segregated into the severity of Alzheimer's Classes obtained from Kaggle. Kaggle is a crowd-sourced platform to attract, nurture, train and challenge data scientists from all around the world to solve data science, machine learning and predictive analytics problems. It has over 536,000 active members from 194 countries and it receives close to 150,000 submissions per month.

6.6 Criteria for Exclusion from the Study

Individuals or subject who have a contraindication to MR scanning implanted metal clips or wires of the type that may concentrate radiofrequency fields or cause tissue damage from twisting in a magnetic field were excluded from the study. Here are some examples:

1. Clipping an aneurysm (A ballooning or weakened area in an artery that can occur in the brain, aorta etc.)
2. Neural stimulator implant
3. Pacemaker or defibrillator implanted in the heart
4. Cochlear (hearing device) implantation
5. A foreign body in the eye (e.g., metal shavings)
6. Any implanted gadget (pumps, infusion devices, etc.)
7. Shrapnel wounds

6.7 Definition of Moderate, Mild, Very and Non-demented

Alzheimer's disease can be classified into:

i. Very mildly demented: This is the stage where patient starts to forget where they put their stuff, other people's names recently, etc. It is hard to detect through cognitive ability test.
ii. Mildly demented: This is the stage where patients don't remember words, can't find their way to the destination, loss focus and work-abilities. This is also the stage where patients even forget that they are losing memory. From this stage, with cognitive testing, it can be found.
iii. Moderately demented: Starts to forget the recent activities, important old histories, have hard time calculating the budget, hard to go outside alone, and loss of empathy.
iv. And non-demented if the disease is not detected (Jeong 2022).

6.8 Spyder IDE

Spyder is a free and open-source scientific environment written in Python, for Python, and designed by and for scientists, engineers and data analysts. It features a unique combination of the advanced editing, analysis, debugging, and profiling functionality of a comprehensive development tool with the data exploration, interactive execution, deep inspection, and beautiful visualization capabilities of a scientific package.

6.9 Machine Learning

The objective of machine learning (ML) is to create software that can create new knowledge from data. Machine learning's inductive inference, or the generalizations drawn from a collection of observed instances, can be compared to early Artificial Intelligence (AI) methods, which mostly dealt with deductive inference (Ławrynowicz and Tresp 2014).

6.10 Convolutional Neural Network

The disease is detected using a convolutional neural network with multiple layers, including pooling, softmax regression, and completely interconnected layers. A CNN expands the length and width of the images while reducing the complexity of the image. The overfitting issue is lessened by a pooling process because it reduces the amount of computation and parameters (Patil et al. 2022).

One of the most prevalent deep neural networks is the CNN. The convolutional layer, as its name suggests, is a layer in the CNN technique and a mathematical operation for feature extraction. Convolutional, non-linearity, pooling, and fully connected layers are just a few of the many layers that make up CNN. However, the pooling and non-linearity layers lack parameters while the convolutional and fully connected layers have factors (Al-Adhaileh 2022). The convolution for one pixel can be calculated through the convolutional layer using the equation below:

$$\mathrm{Net}(i,j) = (x * w)[i,j] = \sum\sum mn[m,n] \times [i-m, j-n] \tag{6.1}$$

where x denotes the input data, which is made up of a collection of images, w denotes the kernel or filter matrix, and the asterisk denotes the convolution process. Net(i, j) is the yielding of the convolutional layer that forwards it to the following layer. The input and kernel's element-by-element product is computed, grouped, and then expressed as the analogous point in the succeeding layer (Al-Adhaileh 2022).

6.11 MaxPooling2D

For reducing dimensionality and extracting spatial features, MaxPooling with a 22 stride is used. Utilizing maxpooling also increases accuracy by providing the convolution feature map's pooling with the highest possible average values. We used common CNN hyperparameters with model fine-tuning to speed up the process of choosing hyperparameters. With accuracy as the loss function and a learning rate of 0.0001, the network is optimized to produce the best classification results with a smaller batch size and less memory usage. Due to the multi-label input data, we used the categorical cross-entropy loss function.

CNN employs pooling for two reasons. First, the classification requires a fixed size for the output feature map of pooling. No matter how big the filters are, using maxpooling on each of 256 filters will produce a 256-dimensional output. For instance, in order to reduce the data dimensionality and reduce the amount of time needed for data training for subsequent layers in the network, downsampling is a crucial step in the layer pooling process. The pooling layer can be used in image handling systems (Mosleh 2021). Secondly, space features are substantial. The spatial information is made less redundant and the size of the feature map is reduced thanks to this layer. Sample reduction aids in lowering computation costs and incorporating meaningful data into feature maps (Mosleh 2021).

6.12 Nonlinearity

The output of the mathematical operation was carried out through the convolutional layer and then passed on to the non-linearity layer, which is the following layer. This layer can be used to modify or remove the output that was formed. This layer, however, is employed to saturate or restrict the output. The convolution layer has a permanent embedding of the nonlinearity layer. Two activation functions, sigmoid and tanh, have been primarily used in deep learning techniques over the past 20 years. However, as demonstrated by the following two equations, the rectified linear unit (ReLU) has simpler descriptions of both the functions and gradient (Mosleh 2021).

$$\mathrm{ReLU}(x) = \max(0, x) \tag{6.2}$$

$$d/dx\,\mathrm{ReLU}(x) = \{1 \text{ if } x > 0;\ 0 \text{ otherwise}\} \tag{6.3}$$

6.13 Fully Connected Layer

The fully connected layer follows the pooling layer and is used to connect and arrange all of the neurons in a convolutional neural network. As a result, every neuron in a layer that is fully connected is directly connected to every neuron in the layer above it and the layer below it. The CNN technique's most frequently used parameter is this layer. The time required for the CNN to train on the data can be reduced with the aid of this layer. A fully connected layer's main drawback is that it necessitates a number of factors that necessitate a more involved computation of training samples. As a result, we disregarded the quantity of connections and neurons. Thus, the dropout layer can be used to perform the excluded neurons and connections (Mosleh 2021).

The names of these layers serve as descriptions. Here, every layer from before is connected. All of the earlier learning layers have been combined at this point. The model's classification is provided by the topmost layer. Equal to the number of classes found in the image dataset, there are the same number of outputs. There are four classes in the Fully Connected Deep Layer (Mosleh 2021).

6.14 Softmax

There is no normalization applied to the output produced by the fully connected layer. An output Softmax layer is used for normalization. This layer's output, is a positive integer that can be used for categorization.

The softmax layer is the final layer in the model that is being presented, and it is responsible for calculating the probability distribution of N-dimensional vectors for the input images. Multiclass classification in deep learning-based models is the main purpose of utilizing softmax at the output layer in this model. The SoftMax function, which returns the likelihood of a data point belonging to each class, is utilized for the suggested multi-classifier and is calculated as:

$$\mathrm{f(xi)} = e^{xi/\sum_{j=1}^{k} e^{xj}} \quad \text{for i} = 1, 2, \ldots, \text{k} \quad \text{and} \quad \text{x} = [\text{x1}, \ldots, \text{xk}] \tag{6.4}$$

where k is the number of classes in the multi-class classifier, x is the input vector, e^{xi} is the standard exponential function for the input vector, and e^{xj} is the standard exponential function for the output vector.

CNN Model

The model architecture of the convolution neural network employed in this study is illustrated in Fig. 6.1.

Training and Testing Result

Figure 6.2 shows the epoch training networks for 100 iterations. The training result is the result gotten from the data used to create the model, while the validation result is used to evaluate the final performance of the machine learning model, it is also known as the test set result.

The loss for the training is 0.0132 and the validation loss is 0.1967, the training accuracy is 0.9963 and the validation accuracy is 0.9541, the training Area Under Curve (AUC) is 0.9997 and the validation AUC is 0.9885 and the f1-score is 0.9963 and the validation f1-score is 0.9539.

Model Accuracy

An accuracy metric is used to measure the algorithm's performance in an interpretable way. It is calculated in the form of a percentage. It is the measure of how accurate your model's prediction is, compared to the true data i.e. how often the classifier

```
Model: "cnn_model"
_________________________________________________________________
 Layer (type)                Output Shape              Param #
=================================================================
 conv2d (Conv2D)             (None, 176, 176, 16)      448

 conv2d_1 (Conv2D)           (None, 176, 176, 16)      2320

 max_pooling2d (MaxPooling2D  (None, 88, 88, 16)       0
 )

 sequential (Sequential)     (None, 44, 44, 32)        14016

 sequential_1 (Sequential)   (None, 22, 22, 64)        55680

 sequential_2 (Sequential)   (None, 11, 11, 128)       221952

 dropout (Dropout)           (None, 11, 11, 128)       0

 sequential_3 (Sequential)   (None, 5, 5, 256)         886272

 dropout_1 (Dropout)         (None, 5, 5, 256)         0

 flatten (Flatten)           (None, 6400)              0

 sequential_4 (Sequential)   (None, 512)               3279360

 sequential_5 (Sequential)   (None, 128)               66176

 sequential_6 (Sequential)   (None, 64)                8512

 dense_3 (Dense)             (None, 4)                 260

=================================================================
Total params: 4,534,996
Trainable params: 4,532,628
Non-trainable params: 2,368
_________________________________________________________________
```

Fig. 6.1 CNN model architecture

makes the correct predictions. From the epoch training networks in Fig. 6.3, at the 100th iteration, the training accuracy is 0.9963 and the validation accuracy is 0.9541. This implies that the accuracy is very high.

Model AUC

The AUC tells how much the model is capable of distinguishing between classes. The higher the AUC, the better the model is at predicting 0 classes as 0 and 1 classes as 1. By analogy, the higher the AUC, the better the model is at distinguishing between patients with the disease and no disease (Narkhede 2018). The epoch training networks in Fig. 6.2 show that at the last (100th) iteration, the training AUC is 0.9997 and the validation is 0.9541. The model area under curve is very high with a value of 0.9997 (Fig. 6.4).

```
Epoch 1/100
WARNING:tensorflow:From C:\Users\Chima\anaconda3\lib\site-packages\tensorflow\python\autograph\pyct\static_analysis\liveness.
py:83: Analyzer.lamba_check (from tensorflow.python.autograph.pyct.static_analysis.liveness) is deprecated and will be remove
d after 2023-09-23.
Instructions for updating:
Lambda fuctions will be no more assumed to be used in the statement where they are used, or at least in the same block. http
s://github.com/tensorflow/tensorflow/issues/56089
256/256 [==============================] - 1310s 5s/step - loss: 1.6544 - acc: 0.3112 - auc: 0.5760 - f1_score: 0.3079 - val_
loss: 1.8671 - val_acc: 0.2402 - val_auc: 0.5192 - val_f1_score: 0.0969
Epoch 2/100
256/256 [==============================] - 1238s 5s/step - loss: 1.1290 - acc: 0.4940 - auc: 0.7681 - f1_score: 0.4805 - val_
loss: 5.2493 - val_acc: 0.2402 - val_auc: 0.5078 - val_f1_score: 0.0969
Epoch 3/100
256/256 [==============================] - 3171s 12s/step - loss: 0.8153 - acc: 0.6293 - auc: 0.8791 - f1_score: 0.6227 - val
_loss: 2.0945 - val_acc: 0.4766 - val_auc: 0.7089 - val_f1_score: 0.3790
Epoch 4/100
256/256 [==============================] - 1962s 8s/step - loss: 0.7084 - acc: 0.6763 - auc: 0.9073 - f1_score: 0.6730 - val_
loss: 3.3236 - val_acc: 0.2446 - val_auc: 0.5744 - val_f1_score: 0.1091
Epoch 5/100

oss: 0.2083 - val_acc: 0.9502 - val_auc: 0.9883 - val_f1_score: 0.9501
Epoch 96/100
256/256 [==============================] - 667s 3s/step - loss: 0.0126 - acc: 0.9952 - auc: 0.9999 - f1_score: 0.9952 - val_l
oss: 0.2123 - val_acc: 0.9517 - val_auc: 0.9883 - val_f1_score: 0.9516
Epoch 97/100
256/256 [==============================] - 669s 3s/step - loss: 0.0142 - acc: 0.9956 - auc: 0.9997 - f1_score: 0.9956 - val_l
oss: 0.2198 - val_acc: 0.9478 - val_auc: 0.9878 - val_f1_score: 0.9476
Epoch 98/100
256/256 [==============================] - 672s 3s/step - loss: 0.0134 - acc: 0.9952 - auc: 0.9998 - f1_score: 0.9952 - val_l
oss: 0.2409 - val_acc: 0.9468 - val_auc: 0.9857 - val_f1_score: 0.9467
Epoch 99/100
256/256 [==============================] - 667s 3s/step - loss: 0.0229 - acc: 0.9926 - auc: 0.9995 - f1_score: 0.9926 - val_l
oss: 0.2404 - val_acc: 0.9409 - val_auc: 0.9880 - val_f1_score: 0.9402
Epoch 100/100
256/256 [==============================] - 668s 3s/step - loss: 0.0132 - acc: 0.9963 - auc: 0.9997 - f1_score: 0.9963 - val_l
oss: 0.1967 - val_acc: 0.9541 - val_auc: 0.9885 - val_f1_score: 0.9539
```

Fig. 6.2 Epoch training networks

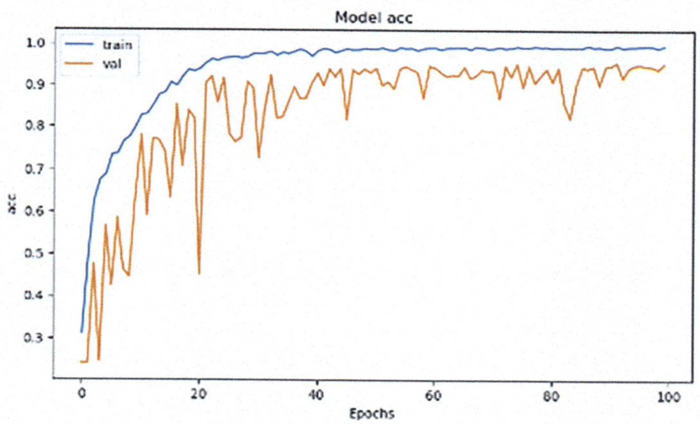

Fig. 6.3 The model accuracy

A poor model has an AUC near 0 which means it has the worst measure of separability. An excellent model has AUC near to the 1 which means it has a good measure of separability. This means that this model is able to separate between MildDemented, VeryMildlyDemented, ModerateDemented and the Non-Demented cases of Alzheimer's diseases.

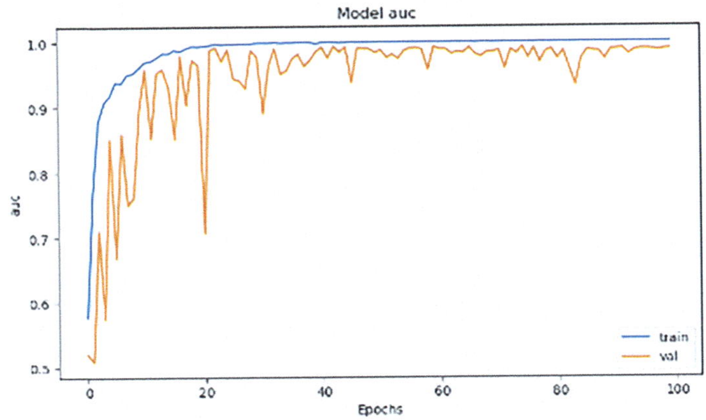

Fig. 6.4 The model AUC

Model Loss

The loss value implies how poorly or well a model behaves after each iteration of optimization. This loss is calculated on training and validation and its interpretation is how well the model is doing for these two sets. From Fig. 6.5, the training and validation has a low model loss. As the training iteration continues, the loss keeps reducing. A high value of loss shows how bad the model is doing and a low loss show how good the model is doing.

The epoch training networks in Fig. 6.2, show that at the last (100th) iteration, the training loss is 0.0132 and the validation loss is 0.1967. These indicates low loss.

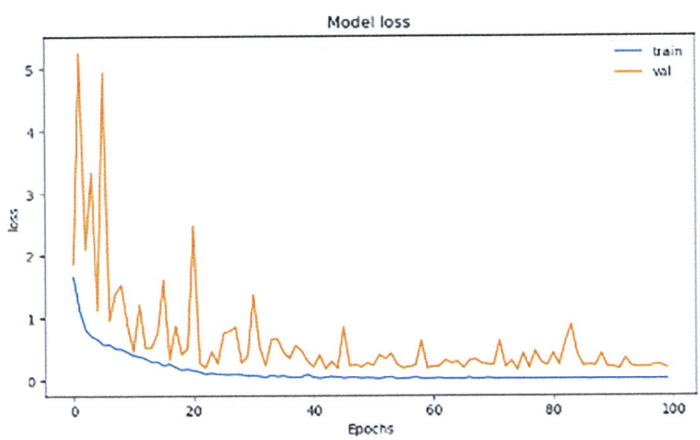

Fig. 6.5 The model loss

```
80/80 [==============================] - 44s 545ms/step - loss: 0.2174 - acc: 0.9516 - auc: 0.9873 - f1_score: 0.9519
Testing Accuracy: 95.16%
```

Fig. 6.6 Model evaluation

Model Evaluation

See Fig. 6.6.

6.15 Evaluation of Result

Model evaluation is the process of using various evaluation metrics to comprehend the performance and strengths and weaknesses of a machine learning model. In the early stages of research, it is crucial to evaluate a model's effectiveness. An evaluation metric measures the performance of a model after training. As seen from Fig. 6.2, the model evaluation is 95.16%.

6.15.1 Classification Report on Tested Data

The precision report answers the question of what percent of all predictions were correct. It is the percentage of genuine positive and true negative events. This demonstrates the classifier's ability for handling favorable observations while making few comments about unfavorable ones. It is determined by:

$$\text{Precision} = \text{True Positive}/(\text{True Positive} + \text{False Positive}) \tag{6.5}$$

The precision result for this project for different classes of Alzheimer's disease is given in Table 6.1.

Table 6.1 Classification report table

	Precision	Recall	F1-score	Support
NonDemented	0.99	0.97	0.98	639
VeryMildDemented	1.00	1.00	1.00	635
MildDemented	0.94	0.89	0.91	662
ModerateDemented	0.88	0.95	0.91	624
Micro avg	0.95	0.95	0.95	2560
Micro avg	0.95	0.95	0.95	2560
Weighted avg	0.95	0.95	0.95	2560
Weighted avg	0.95	0.95	0.95	2560

6.15.2 *Recall*

Recall (also referred to as sensitivity) is the frequency at which positive cases are actually anticipated to be positive. Due to the precision with which the observations are evaluated, this metric is appealing, particularly in the clinical setting. During this examination, it is more crucial to correctly identify a mildly demented than it is to wrongly identify a non-demented one. It is determined by:

$$\text{Recall} = \text{True Positive}/(\text{True Positive} + \text{False Negative}) \quad (6.6)$$

$$\text{Recall} = \text{TP}/(\text{TP} + \text{FN}) \quad (6.7)$$

where TP = True Positive, FN = False Negative.

The recall result for this project for different classes of Alzheimer's disease is given in Table 6.1.

6.15.3 *F1-Score*

F1-score defines what percentage of positive predictions were correct. The weighted harmonic mean of recall and precision is the f1 score. Its range is between [0, 1]. The model performance improves with increasing F1 Score. As a result, this score takes into account both false positives and negatives. It is calculated by:

$$\text{F1} = 2\text{TP}/(2\text{TP} + \text{FP} + \text{FN}) \quad (6.8)$$

or

$$\text{F1} = 2 * (\text{recall} + \text{precision})/(\text{recall} + \text{precision}) \quad (6.9)$$

The f1-score result for this project for different classes of Alzheimer's disease is given in Table 6.1.

6.15.4 *Support*

The number of occurrences of the class in the given dataset is known as support. Insufficient support in the training data may point to structural flaws in the classifier's reported scores and may call for stratified sampling or rebalancing. The support result for this project for different classes of Alzheimer's disease is given in Table 6.1 which shows the precision, recall, f1-score and support values for Non-Demented, Very Mild Demented, mild Demented and moderately Demented.

```
Balanced Accuracy Score: 95.22 %
Matthew's Correlation Coefficient: 93.58 %
```

Fig. 6.7 Other classification metrics result

In Table 6.1, the micro average (averages the total true positives, false negatives, and false positives) is only displayed for multi-label or multi-class with a subset of classes. This is because it corresponds to accuracy otherwise it would be the same for all metrics.

Weighted avg

The weighted avg essentially refers to the proportion of each class's support relative to the sum of all support values.

Other Classification Metrics

Balanced accuracy is used when working with imbalanced data, or when one of the target classes shows up much more frequently than the other. It is the arithmetic mean of sensitivity and specificity. The balance accuracy score for this model is 95.22% (Fig. 6.7).

Mathew's Correlation Coefficient is a statistical tool used for model evaluation. It measures the difference between the predicted values and actual values. It helps to summarize the confusion matrix or an error matrix. It is calculated as shown below using the values from the confusion matrix.

$$MCC = \frac{TN \times TP - FN \times FP}{\sqrt{(TP + FP)(TP + FN)(TN + FP)(TN + FN)}} \tag{6.10}$$

The MCC is more reliable statistical rate that produces a high score only if the prediction obtained good results in all of the four confusion matrix categories (true positives, false negatives, true negatives, and false positives). The Mathew's Correlation Coefficient for this model is 93.58%.

6.16 Confusion Matrix on the Test Data

From the confusion matrix, it was observed that based on 6,400 tested images, the following results were obtained: 622 true-positive (the model correctly predicted the NonDemented class), 635 true-positive (the model correctly predicted the VeryMild-Demented class), 588 true-positive (the model correctly predicted the MildDemented class), 591 true-positive (the model correctly predicted the ModerateDemented class), 36 false-positive (the model incorrectly predicted all the classes), and 88 false-negative (the model incorrectly predicted all the classes) (Fig. 6.8).

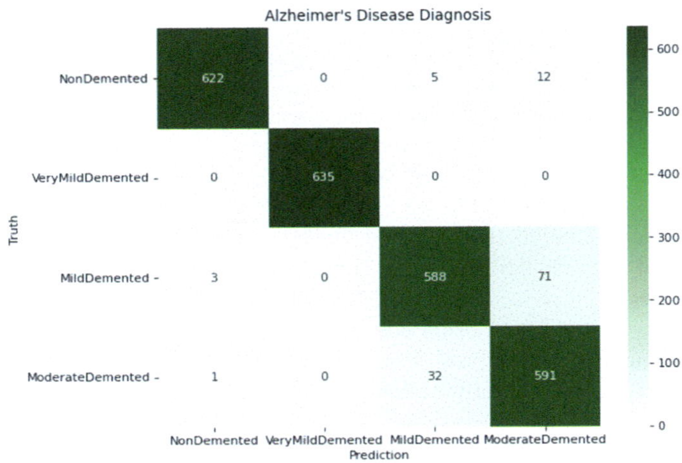

Fig. 6.8 Confusion matrix

It is observable that the number of false predictions is low compared with the total number of predictions made, indicating a high-performance trained model.

6.17 Model Deployment

For our machine learning model to be used for more practical decisions, there is a need to deploy it into production. Machine Learning Deployment is the method of joining a machine learning model into an existing production. It is one of the last stages in machine learning, there exist different python libraries for deployment of machine learning models like voila, streamlit, flask, tkinter, Django etc. For this model deployment, Streamlit library was used.

6.17.1 Deployment Process with Streamlit

I. On the anaconda prompt run "pip install Streamlit", this is the installation of Streamlit library. Streamlit is an open-source Python library that makes easy to create and share beautiful, custom web apps for machine learning and data science. In just few minutes you can build and deploy powerful data apps, we then wrote a python code where the model saved was called and widgets to make our GUI web application interactive were included.

II. Write a Python program of the model with best accuracy score then save it as a.pkl file using the pickle library. A PKI is a file created by pickle, a Python module that enables objects to be serialized to files on disk and desterilized back into the program at runtime. It contains byte stream that represents the object (https://fileinfo.com/extension/pkl).

III. Create a Streamlit app. This can be done by writing a python code that creates the user interface and allows that machine learning model to take data and give feedbacks from the model. Save the model as a.py file.
IV. On anaconda prompt change the environment directory to where your files are saved. Then run the line, Streamlit run then paste the of your.py file. Example Streamlit run alzheimer'sDiseaseClassification.py
V. Register Github then create a repository and upload the files which includes a profile, requirement.txt, app.py, setup.sh, runtime.txt, classifier.pkl and README.md
VI. Sign up on Heroku cloud platform, login and create an app then navigate to the deploy button and click for deployment.

6.17.2 Deployed Alzheimer's Disease Classification App

The graphic user interface of the deployed Alzheimer's disease classification algorithm is illustrated in Figs. 6.9, 6.10 and 6.11.

6.18 Conclusion

Alzheimer's disease is distinguished by the formation of abnormal protein bundles in brain tissue, but experts are unsure what causes the condition. Due to the overwhelming number of people who appear with the ailment, manual analysis by medical professionals is troubled with the possibility of making mistakes and is a time-consuming process (Al Shehri 2022). Understanding AD, mild cognitive impairment (MCI) and cognitive normal (CN) manifestation is one of the most challenging tasks faced by neurologists over the past few years.

Physicians use different clinical methodologies to classify AD. Clinically, cerebrospinal fluid (CSF) concentration is an indicator of AD. The norepinephrine level increases in the CSF as the disease progresses. The CSF sample is collected using a ventricular puncture, wherein the physician makes a hole in the skull and collects CSF directly from one of the brain ventricles. It is a laborious procedure and may confer a risk of bleeding in the brain (Mosleh 2021).

In this project, we proposed a CNN approach for the classification of AD using imaging data. The CNN model achieved high accuracy and precision in classifying AD patients from healthy controls, indicating that it may be a valuable tool for early detection and diagnosis of AD. To compare experimental results, the measures of accuracy, precision, recall, f1-score and support are used. The current research contributes to the healthcare paradigm, involving the realization of data-driven and

Fig. 6.9 Welcome page of the web app

patient-centric health systems wherein advanced applications can be deployed to provide personalized assistance.

The proposed model can be utilized for real-time study and categorization of Alzheimer's disease. It is possible that policy makers and stakeholders may take decisive action to encourage improvement in access to AD biomarker testing, which would give the greatest possible service in the intermediate term. Further research is needed to validate the results of this study on larger datasets and to investigate the potential of the CNN model for use in clinical settings.

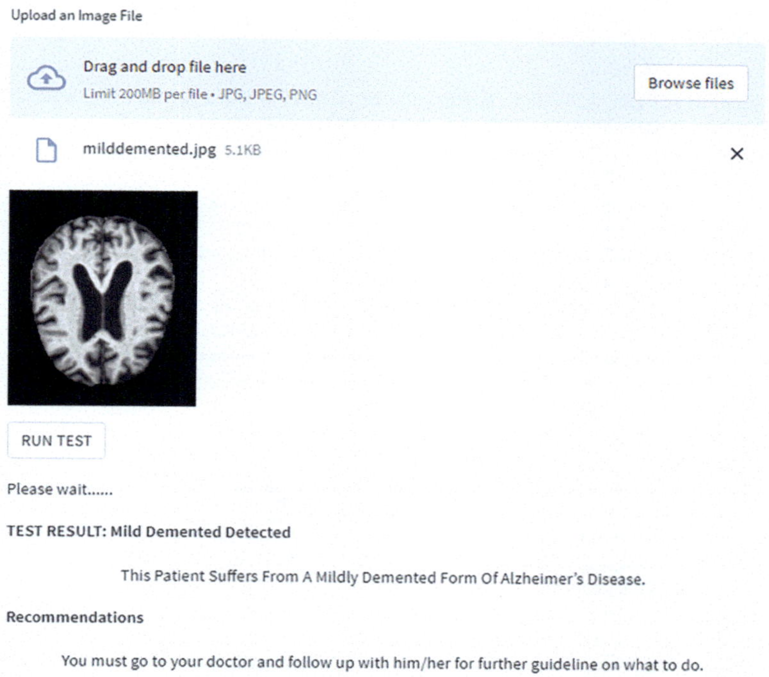

Fig. 6.10 A test page detecting that there is no presence of disease

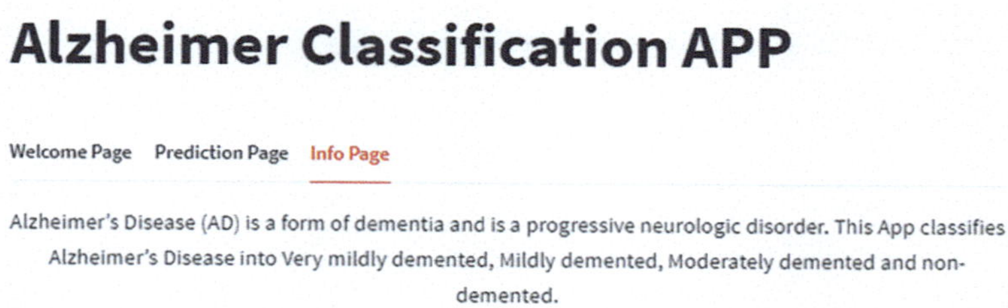

Fig. 6.11 A section of the web app. showing the Alzheimer disease classification web app

Acknowledgements The authors acknowledge the motivated research efforts of Chima Bartholomew Akpa (Department of Physics, Federal University of Technology, Minna, Nigeria) as a promising young researcher.

References

Al-Adhaileh, M.H.: Diagnosis and classification of Alzheimer's disease by using a convolution neural network algorithm. Soft. Comput. **26**(16), 7751–7762 (2022)

Al Shehri, W.: Alzheimer's disease diagnosis and classification using deep learning techniques. PeerJ Comput. Sci. **8**, e1177 (2022)

Diamond, J.: A Report on Alzheimer's Disease and Current Research, p. 1–11. Alzheimer Society of Canada (2008)

Helaly, H.A., Badawy, M., Haikal, A.Y.: Deep learning approach for early detection of Alzheimer's disease. Cogn. Comput. **14**(5), 1711–1727 (2022)

Jiang, T.: Deep learning application in Alzheimer disease diagnoses and prediction. In: 2020 4th International Conference on Artificial Intelligence and Virtual Reality, pp. 62–65 (2020)

Kruthika, K. R., Maheshappa, H. D.: Alzheimer's Disease Neuroimaging Initiative. Multistage classifier-based approach for Alzheimer's disease prediction and retrieval. Inform. Med. Unlocked **14**, 34–42 (2019)

Ławrynowicz, A., Tresp, V.: Introducing machine learning. Perspectives on Ontology Learning, **18**, 35–50 (2014)

MRI|Make softwares, change the world| Computer History Museum (2023). https://www.computerhistory.org/makesoftware/search/. Retrieved 25 Jan 2022

Narkhede, S.: Understanding auc-roc curve. Towards data sci. **26**(1), 220–227 (2018)

National Institute on Aging: Alzheimer's Disease & Related Dementias (2022). https://www.nia.nih.gov/health/alzheimers/symptoms. Retrieved 29 Dec 2022

National Institute on Aging: What Is Alzheimer's Disease? National Institute on Aging, Alzheimer's Disease Education and Referral Center, Alzheimer's Association (2022). Retrieved 29 Dec 2022

Patil, V., Madgi, M., Kiran, A.: Early prediction of Alzheimer's disease using convolutional neural network: a review. Egypt. J. Neurol. Psychiatry Neurosurg. **58**(1), 130 (2022)

Wen, J., Thibeau-Sutre, E., Diaz-Melo, M., Samper-González, J., Routier, A., Bottani, S., Dormont, D., Durrleman, S., Burgos, N., Colliot, O., Alzheimer's Disease Neuroimaging Initiative: Convolutional neural networks for classification of Alzheimer's disease: overview and reproducible evaluation. Med. Image Anal. **63**, 101694 (2020)

Chapter 7
Physics-Informed Neural Networks for Nuclear Magnetic Resonance-Guided Clinical Hyperthermia

Abstract This chapter addresses a critical need for improved precision in the application of hyperthermia for clinical purposes. The developed approach advances our understanding of the physical mechanisms behind medical interventions and opens the door to more effective and personalized clinical hyperthermia, which could lead to improved patient outcomes and advancements in medical management. The goal is to design and implement a Physics-Informed Neural Network (PINN) to control clinical hyperthermia procedures based on Nuclear Magnetic Resonance (NMR). These involve (i) acquiring a comprehensive understanding of the relevant physical principles underlying clinical hyperthermia and ascertain the most effective means of integrating them into the neural network model; (ii) developing a robust neural network architecture and implement it to effectively integrate physics-based knowledge while maintaining the adaptability and learning capability of neural networks; (iii) creating a model that uses NMR data to guide clinical hyperthermia procedures in real time, enabling precise targeting and temperature monitoring of the treatment area; (iv) verifying the developed model in details against accepted practices and assess its effectiveness relative to traditional methods. This entails evaluating the accuracy, efficiency, and dependability of the Physics-Informed Neural Network in treating clinical hyperthermia.

Keywords Physics-informed neural network · Nuclear magnetic resonance-guided clinical hyperthermia · Deep learning

7.1 Introduction

Hyperthermia is one type of cancer treatment that uses heat to destroy cancer cells. It frequently functions in concert with other medical interventions such as radiation therapy and chemotherapy (Awojoyogbe and Dada 2018). Interstitial heating, external beam radiation, and magnetic resonance-guided focused ultrasound (MRgFUS) are a few methods for inducing hyperthermia. People have been using heat as a cancer treatment since ancient times. The Edwin Smith Papyrus, an Egyptian

B. O. Awojoyogbe and M. O. Dada, *Digital Molecular Magnetic Resonance Imaging*, Series in BioEngineering, https://doi.org/10.1007/978-981-97-6370-2_7

medical text from 1700 BCE, describes the use of heat cauterization in the treatment of breast cancer. Because he believed heat could kill cancerous tissue while leaving healthy tissue intact, Hippocrates, the Greek physician of antiquity, advocated using heat to treat tumors (Rutgers 2023). The development of modern medicine in the nineteenth century led to advancements in the treatment of hyperthermia. In the early 1900s, American surgeon William Coley developed a technique called "Coley's toxins," wherein patients received injections of bacterial toxins to increase body temperature and strengthen the immune system's capacity to fight cancer. Around the same time, German physicians began using a technique called diathermy-which involves using electromagnetic waves to produce heat-to treat cancer. The development of non-invasive nuclear magnetic resonance imaging (NMR) in the 1970s revolutionized the management of hyperthermia by providing a way to monitor the body's temperature distribution. As a result, NMR-guided hyperthermia was developed, which allowed for more precise and effective treatment (Cai et al. 2021).

In recent years, there has been a growing interest in the use of Physics-Informed Neural Networks (PINNs) for medical applications, including therapy for hyperthermia. Physical information is combined with traditional neural network architecture in a specific type of machine learning model known as PINNs. Because of this, they are perfect for modeling complex physical systems, like the heat transfer process involved in treating hyperthermia (Wang et al. 2020).

7.2 Clinical Hyperthermia

The expense of identifying and treating diseases has recently gone up. Unfortunately, disease-related mortality has not decreased to correspond with these rising costs. The traditional "one drug fits all" approach to illness management is gradually giving way to customized medicine, which administers drugs to patients exactly when they need them. Although personalized medicine has a lot of promising future applications, its full potential is still untapped. Recent research has shown that creating customized treatment methods, also referred to as theranostics or personalized treatments, can successfully treat a variety of illnesses. Theranostics is the field that combines medicine and diagnostics. Therapy can be started first to find people who are more likely to respond to a specific treatment plan, or therapy can be started after a diagnosis to monitor early treatment response and predict treatment effectiveness. It is also possible that diagnostics and treatments are developed simultaneously.

The ultimate objective of theranostics is to visualize and track sick tissue, delivery timing, and medication efficacy; the long-term goal is to be able to precisely control therapy and dosage (Awojoyogbe and Dada 2018). Treatment for hyperthermia involves keeping the patient at a temperature between 40 and 48 °C for a minimum of an hour. Because high temperatures can cause damage to tissues, temperatures above 50 °C are regarded as coagulation, 60–90 °C as thermal ablation, and 200 °C as charring.

Ablation, also referred to as high-temperature hyperthermia, is defined as the direct application of thermal or chemical therapies to a tumor with the goal of achieving substantial tumor destruction or annihilation.

The three main approaches to treating clinical hyperthermia are localized hyperthermia, regionalized hyperthermia, and whole-body hyperthermia. These three forms of hyperthermia apply heat to localized, advanced, or deep-seated tumors, depending on the extent, depth, and stage of the tumor. For localized hyperthermia, which mainly uses microwaves, radio waves, or ultrasonography to transfer heat, small tumors (diameters between 3 and 6 cm) are particularly well-suited.

One area of the body where regional hyperthermia is used for thermal treatment is the abdominal cavity, which houses large body parts such as organs and limbs. Advanced tumors such as colorectal, ovarian, bladder, and prostate cancers, as well as soft tissue sarcomas, are usually treated with this technique. Many heat sources are used to apply heat to the tumor, including infrared radiators, ferromagnetic seeds, nanoparticles, resistive implants, ultrasound, microwaves, and radiofrequency energy.

7.3 The Use of Nuclear Magnetic Resonance

When a weak, oscillating magnetic field disturbs nuclei in a strong, constant magnetic field, it results in the physical phenomenon known as nuclear magnetic resonance (NMR). In response, the nucleus releases an electromagnetic signal at a frequency unique to the local magnetic field. Nuclear magnetic resonance is a useful tool for studying crystals, non-crystalline materials, molecular physics, and the structure of organic molecules in solution. It has broad applications in the physical sciences as well as the industry.

NMR is used in many different applications, such as relaxometry, diffusometry, magnetic resonance imaging (MRI), and high-resolution liquid and solid-state NMR spectroscopy. By virtue of the orientation of their magnetic moments with respect to an external magnetic field, atoms possessing magnetic properties are arranged into distinct energy levels. These techniques appear to differ in many aspects, such as instrumentation, sample preparation, and intended results, but they actually share the same concept, which is known as the magnetic resonance principle. Whether this field is homogeneous or inhomogeneous is determined by the NMR methodology. After reaching what is known as thermal equilibrium, nuclei are subjected to two weak radiofrequency fields. After two relaxation processes, the excited nuclei return to low energy levels and release their excess energy, either by interacting with the surrounding environment (lattice) or exchanging energy with nearby nuclei at lower energy levels. The first type of relaxation process is called spin–lattice relaxation (T1), while the second type is called spin–spin relaxation (T2). There is a time constant that describes the former. The time scale of the relaxing nuclei in any NMR application is critical and depends on the molecular weight as well as the physical state (liquid or solid) of the compound. In relaxation mechanisms, molecular motion affects the local magnetic fields generated by the atomic and electronic environment of the nucleus.

7.4 Python Modules/Packages

The standard library in Python programming is very large and provides a wide range of facilities. The library contains built-in modules that gives access to system functionality like file input and output that programmers would not have had access to. Also modules that offer standardized solutions for many issues that occur in routine programming are written. These modules promote that portability of python program and form the computational backbone of the project, facilitating the development, training, and evaluation of the PINN model. The choice of modules is driven by the need for efficient numerical computation, machine learning capabilities, and seamless integration with NMR data. Some of these modules and packages include Pytorch, TensorFlow, Matplotlib, Numpy, Pandas, etc.

7.4.1 Jupyter Notebook

The Jupyter notebook serves as the interactive computing environment for the development, experimentation, and documentation phases of the project. This open-source tool provides a web-based interface that seamlessly integrates code, visualizations, and explanatory text. It can be run on a desktop without need for internet connectivity too. It offers an easy, efficient and cost-effective document-focused interface. Its versatility makes it a good platform for carrying out these projects.

7.4.2 Google Colab

Google Colab, short for Collaboratory, is a cloud-based platform provided by Google that enables users to write and execute Python code in a collaborative environment. It is particularly popular in the field of data science, machine learning, and artificial intelligence due to its integration with popular libraries and its provision of free access to GPU (Graphics Processing Unit) and TPU (Tensor Processing Unit) resources. Key features of Google Colab include:

(i) Free Access to GPU and TPU: Users can leverage Google Colab's computing resources, including GPUs and TPUs, without incurring additional costs. This is especially valuable for tasks that require significant computational power, such as training complex machine learning models.

(ii) Cloud-Based Environment: it operates entirely in the cloud, eliminating the need for users to set up and manage local computing environments. This cloud-based approach allows for easy accessibility and collaboration from various devices.
(iii) Integration with Google Drive: Colab integrates seamlessly with Google Drive, enabling users to save and share Colab notebooks directly from their Google Drive accounts. This simplifies collaboration and provides an organized storage solution for projects.
(iv) Pre-installed Libraries and Packages: it comes pre-installed with many popular Python libraries and packages commonly used in data science and machine learning, such as NumPy, Pandas, Matplotlib, and TensorFlow. This saves users the effort of manually installing dependencies.
(v) Real-Time Collaboration: Multiple users can collaborate on the same Colab notebook in real-time. Changes made by one user are immediately visible to others, facilitating collaborative work on projects.
(vi) Notebook Format: Colab uses a notebook format, similar to Jupyter Notebooks. This format allows users to intersperse code with formatted text, images, and visualizations, making it an excellent tool for both coding and documentation.
(vii) Easy Sharing: notebooks can be easily shared with others, either by providing a link or by sharing directly through Google Drive. This makes it a convenient platform for sharing work with collaborators or the broader community. Google colab was basically used for this project when model training requires computational power higher than the local desktop's CPU can offer; leveraging on its GPU performance and cloud computing.

7.4.3 Deep Learning

Deep learning is a subset of machine learning, which is a branch of artificial intelligence (AI). It involves the use of neural networks with multiple layers (deep neural networks) to model and solve complex problems (Fig. 7.1).

The term "deep" refers to the presence of multiple layers in the neural network architecture. While a single-layer neural network can still make approximate predictions, additional hidden layers can help optimize and refine for accuracy.

7.5 Partial Differential Equation (PDE)

A partial differential equation (PDE) is a mathematical equation that involves multiple independent variables, and their partial derivatives with respect to those variables. PDEs are used to describe a wide range of physical phenomena in fields such as physics and engineering, where the behavior of a system depends on multiple variables and their rates of change. To uniquely determine the solution to the PDEs,

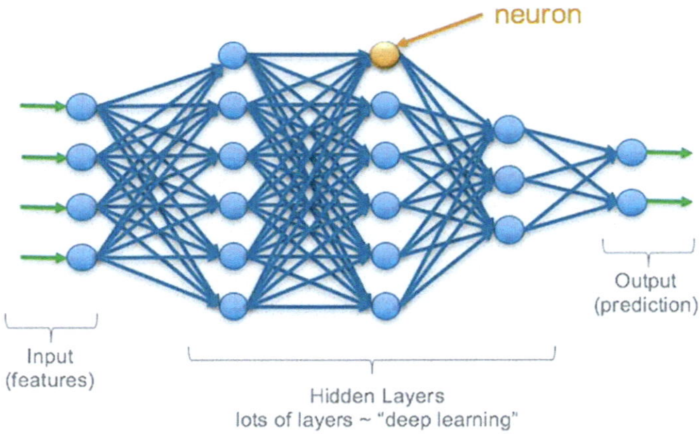

Fig. 7.1 Architecture of deep neural network

appropriate boundary and initial conditions must be specified. These conditions reflect the constraints imposed by the system's geometry, external influences, or the initial state of the system. Careful consideration and accurate representation of these conditions are essential for obtaining meaningful results.

The effective implementation of nuclear magnetic resonance-guided clinical hyperthermia demands a robust modeling framework that accurately captures the underlying physics of heat transfer within biological tissues. Partial Differential Equations (PDEs) serve as a mathematical cornerstone for formulating the spatiotemporal dynamics of temperature distribution during hyperthermia treatments.

For this study, the Pennes Bioheat Equation was utilized. This PDE provides a quantitative framework essential for improving therapeutic approaches and deepening our knowledge of heat transfer in biological contexts. It takes into account factors like blood perfusion, metabolic heat generation, and thermal conductivity. The Pennes's bio-heat equation is defined as follows:

$$\rho_t c_t \frac{\partial T}{\partial t} = k \frac{\partial^2 T}{\partial x^2} + W_b \rho_b c_b (T - T_b) + Q_m \tag{7.1}$$

where

ρ_t is the tissue density,
c_t is the tissue specific heat,
T is the tissue temperature,
t is time,
k is the tissue thermal conductivity,
Q_m denotes the metabolic heat generation,
W_b is the blood perfusion rate,
c_b is the blood specific heat,

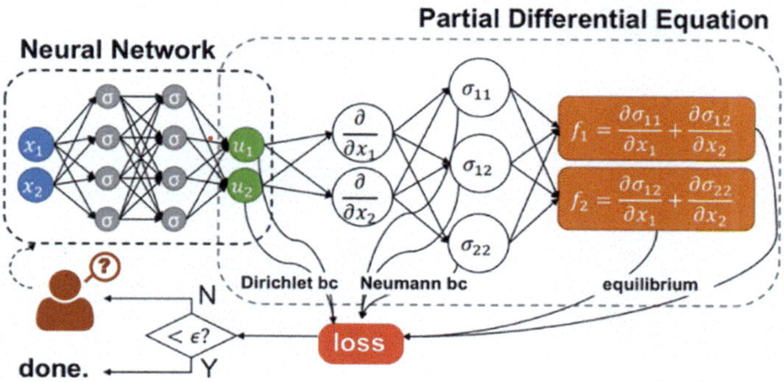

Fig. 7.2 Neural Network and partial differential equation

T_a is the core body temperature,
ρ_b is the blood density.

Based on earlier studies (Awojoyogbe and Dada 2018), the Pennes's bio-heat equation in the presence of a source of RF heating has the Specific Absorption Rate (SAR) term. Therefore, setting $Q_m = \rho SAR$, Eq. 7.1 becomes:

$$\rho_t c_t \frac{\partial T}{\partial t} = k \frac{\partial^2 T}{\partial x^2} + W_b \rho_b c_b (T - T_b) + \rho SAR \tag{7.2}$$

The SAR term accounts for the additional heat generated within the tissues due to the absorption of electromagnetic energy. In hyperthermia treatments, external sources such as radiofrequency or microwave fields are applied to raise the temperature of targeted tissues. The SAR term quantifies the power absorbed per unit mass of the tissue and is often measured in units of watts per kilogram (W/kg).

Including the SAR term in the Pennes Bioheat Equation allows for a more comprehensive modeling of hyperthermia scenarios where external energy deposition significantly contributes to the overall heat distribution within the biological tissues. This consideration is crucial for accurately predicting temperature changes during treatments involving electromagnetic heating. An example of a Physics-informed neural network for solving partial differential equations is illustrated in Fig. 7.2.

7.6 Physics-Informed Neural Network (PINN)

The structure of the PINN employed in the model developed in this study is shown in Fig. 7.3.

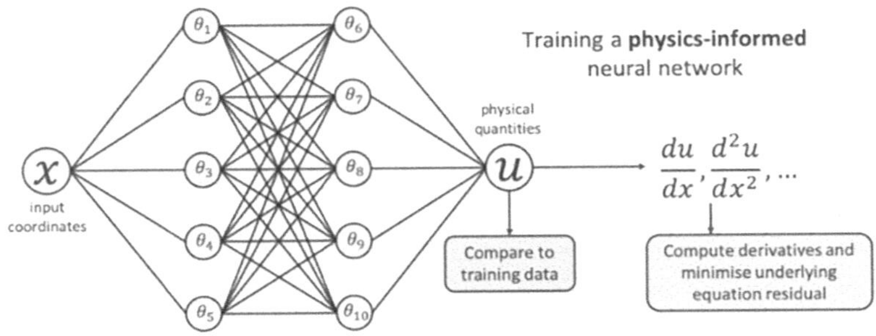

Fig. 7.3 Schematic of PINNs (Moseley 2021)

7.6.1 *Tanh (Activation) Function*

The hyperbolic tangent function, often abbreviated as tanh, is an activation function commonly used in neural networks and other machine learning models. It is a mathematical function that maps input values to output values in the range of -1 to 1. This property is useful in neural networks as it introduces non-linearity to the model, allowing it to learn complex relationships in the data. One key advantage of tanh over the sigmoid activation function is that it is zero-centered. This characteristic helps mitigate issues related to vanishing gradients during the training of deep neural networks. A zero-centered activation function enables the network to learn more effectively, especially in scenarios where the data has been normalized (Sharma et al. 2023).

It is often used in hidden layers of neural networks. The zero-centered nature of tanh can be particularly beneficial in preventing neurons from saturating too quickly, aiding in the learning process. The tanh function is smooth and differentiable everywhere, making it suitable for optimization algorithms such as back propagation. Back propagation relies on the ability to compute derivatives for updating the weights during training. This is because the smoothness ensures that small changes in the input result in continuous and predictable changes in the output, facilitating the learning process during the training of neural networks (Table 7.1).

7.7 PINN Model (Training Result) for Kidney

The training process for the PINN model for kidney is illustrated in Fig. 7.5. This shows no GPU was used and the model trained fast on a CPU.

The model trained for more than 5000 epochs, giving it more chance to learn and adjust after each iteration. The provided data comprise the overall loss, the mean squared error (MSE) for initial conditions (IC), boundary conditions (BC), and the mean squared error for the PDE residual (f). These losses helps in fine-tuning of the deep learning PINN model, adjusting it to obey the physical laws of the Pennes

Table 7.1 Table of values for Tissues used for the modeling (Awojoyogbe and Dada 2018)

Tissue	ρ (kg m^{-3})	k (W m^{-1} K^{-1})	c (J kg^{-1} K^{-1})	W_0 (kg m^{-3} s^{-1})	α (m^2 s^{-1})	T_1 (s)	T_2 (s)	T_0 (s^{-1})	T(°C)	SAR (W m^{-3})
Skeletal muscle	1050	0.50	3465	0.0009	1.37×10^{-7}	1.030	0.060	17.63754	38.6939	0.655351
Kidney	1050	0.54	3700	0.061	1.39×10^{-7}	0.830	0.082	13.39994	38.50342	0.783260
Liver	1060	0.52	3600	0.015	1.36×10^{-7}	0.610	0.057	19.18320	38.51901	1.060860
Adipose tissue	950	0.27	3100	0.0013	9.17×10^{-8}	0.250	0.080	16.50000	39.65281	2.392499
Cortical bone	1920	0.79	1300	0.0013	3.17×10^{-7}	0.400	0.060	19.16667	38.81808	0.849240
Tumour	920	0.42	3000	0.000009	1.52×10^{-7}	0.926	0.120	9.413247	58.70538	0.779418

$c_b = 3720$ J kg^{-1} K^{-1}, $T_b = 37.0$ °C and $\rho_b = 1060$ kg m^{-3}

```
500, Loss : 8.21024e-01, ic : 5.933e-01, bc : 1.044e-01, f : 1.233e-01
1000, Loss : 5.86976e-01, ic : 4.409e-01, bc : 6.695e-02, f : 7.917e-02
1500, Loss : 7.31394e-01, ic : 4.236e-01, bc : 2.273e-01, f : 8.055e-02
2000, Loss : 1.10343e+00, ic : 5.353e-01, bc : 5.147e-01, f : 5.350e-02
2500, Loss : 2.87310e+00, ic : 1.308e+00, bc : 1.524e+00, f : 4.081e-02
3000, Loss : 4.30851e-01, ic : 3.189e-01, bc : 4.666e-02, f : 6.525e-02
3500, Loss : 5.86908e-01, ic : 4.270e-01, bc : 9.281e-02, f : 6.706e-02
4000, Loss : 5.99646e-01, ic : 5.130e-01, bc : 3.430e-02, f : 5.232e-02
4500, Loss : 6.00904e+00, ic : 5.063e+00, bc : 9.293e-01, f : 1.709e-02
5000, Loss : 2.84233e+00, ic : 2.257e+00, bc : 5.442e-01, f : 4.125e-02
5044, Loss : 5.70006e+01, ic : 1.600e+01, bc : 4.100e+01, f : 6.486e-04tensor(2.8404, device='cuda:0', grad_fn=<AddBackward0>)
```

Fig. 7.5 Training losses for the kidney

Bioheat Equation. The first number in the bracket of the last line gives the final loss after 5044 iterations which is 2.8404.

7.7.1 Biothermal Profile for Kidney

Figure 7.6 shows the variation of tissue temperature during hyperthermia. It is a plot of the period for which the tissue is exposed to heat (t) against the depth of penetration of the RF-magnetic field.

7.8 PINN Model (Training Result) for Cortical Bone

The model trained for 5053 iterations and learning after each iteration is shown in Fig. 7.7, the final loss = 3.4370.

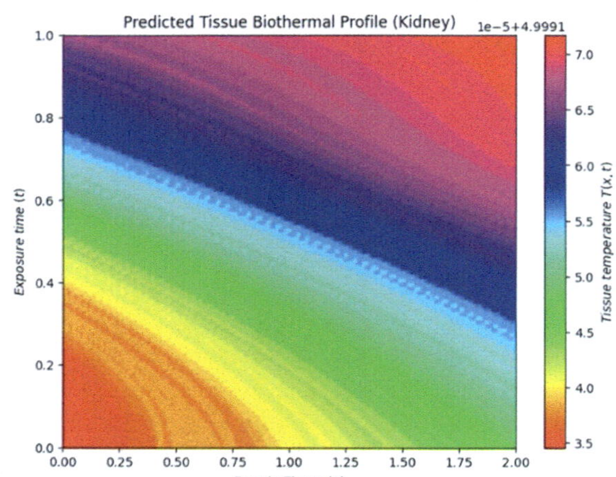

Fig. 7.6 Predicted biothermal profile of the kidney

```
500, Loss : 9.35182e-01, ic : 7.269e-01, bc : 6.986e-02, f : 1.384e-01
1000, Loss : 5.79529e-01, ic : 4.306e-01, bc : 6.507e-02, f : 8.385e-02
1500, Loss : 2.51437e+00, ic : 2.141e+00, bc : 2.864e-01, f : 8.736e-02
2000, Loss : 5.06780e+00, ic : 4.220e+00, bc : 7.409e-01, f : 1.070e-01
2500, Loss : 9.22940e-01, ic : 7.130e-01, bc : 1.264e-01, f : 8.349e-02
3000, Loss : 6.07907e-01, ic : 1.957e-01, bc : 3.427e-01, f : 6.951e-02
3500, Loss : 2.13717e+00, ic : 1.917e+00, bc : 1.367e-01, f : 8.400e-02
4000, Loss : 1.44190e+00, ic : 1.228e+00, bc : 1.417e-01, f : 7.200e-02
4500, Loss : 1.09898e+00, ic : 8.583e-01, bc : 1.839e-01, f : 5.680e-02
5000, Loss : 3.56323e+00, ic : 3.263e+00, bc : 2.814e-01, f : 1.926e-02
5053, Loss : 2.50004e+01, ic : 5.163e-09, bc : 2.500e+01, f : 4.643e-04tensor(3.4370, device='cuda:0', grad_fn=<AddBackward0>)
```

Fig. 7.7 Training losses for cortical bone

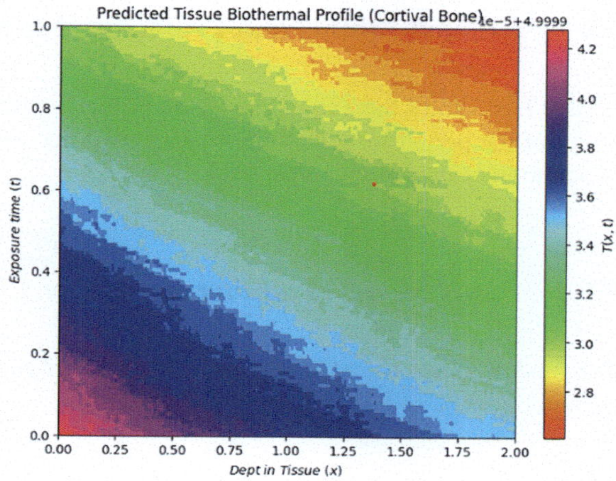

Fig. 7.8 Predicted biothermal profile of the cortical bone tissue

7.8.1 Biothermal Profile for the Cortical Bone

Figure 7.8 shows the variation of cortical bone (tissue) temperature during hyperthermia. It is a plot of the period for which the tissue is exposed to heat (t) against the depth of penetration of the RF-magnetic field.

7.8.2 PINN Model (Training Result) for Tumour

The model trained for 5019 iterations, learning after each iteration is shown in Fig. 7.9, the final loss = 3.9186.

```
500, Loss : 1.25985e+00, ic : 1.091e+00, bc : 7.161e-02, f : 9.711e-02
1000, Loss : 2.83643e+00, ic : 1.723e+00, bc : 8.664e-01, f : 2.469e-01
1500, Loss : 5.58204e-01, ic : 4.192e-01, bc : 6.037e-02, f : 7.859e-02
2000, Loss : 4.30335e+00, ic : 3.709e+00, bc : 5.382e-01, f : 5.586e-02
2500, Loss : 1.14105e+00, ic : 4.767e-01, bc : 5.150e-01, f : 1.494e-01
3000, Loss : 1.09397e+00, ic : 5.246e-01, bc : 5.163e-01, f : 5.305e-02
3500, Loss : 4.49988e-01, ic : 3.646e-01, bc : 4.123e-02, f : 4.413e-02
4000, Loss : 4.83311e-01, ic : 3.889e-01, bc : 4.511e-02, f : 4.934e-02
4500, Loss : 2.76976e+00, ic : 2.245e+00, bc : 3.304e-01, f : 1.939e-01
5000, Loss : 3.99827e+00, ic : 3.003e+00, bc : 9.843e-01, f : 1.113e-02
5019, Loss : 2.50001e+01, ic : 1.330e-10, bc : 2.500e+01, f : 7.348e-05tensor(3.9186, device='cuda:0', grad_fn=<AddBackward0>)
```

Fig. 7.9 Training losses for tumour

7.9 Predicted Biothermal Profile for Tumour

Figure 7.10 shows the variation of tumour temperature during hyperthermia. It is a plot of the period for which the tissue is exposed to heat (t) against the depth of penetration of the RF-magnetic field.

7.9.1 PINN Model (Training Result) for Liver

The model trained for 5172 epochs, learning after each iteration. The final loss as seen in Fig. 7.11 is 4.1359.

7.9.2 Predicted Biothermal Profile for Liver

Figure 7.12 shows the variation of liver temperature during hyperthermia. It is a plot of the period for which the tissue is exposed to heat (t) against the depth of penetration of the RF-magnetic field.

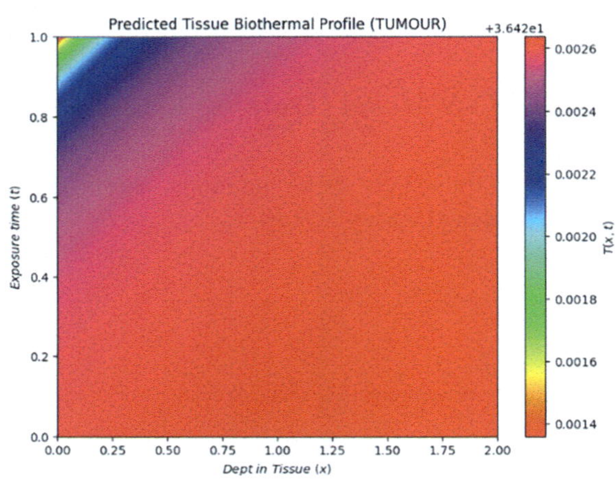

Fig. 7.10 Predicted biothermal profile of tumour

```
500, Loss : 8.48650e-01, ic : 5.385e-01, bc : 1.875e-01, f : 1.227e-01
1000, Loss : 8.28260e-01, ic : 5.528e-01, bc : 1.648e-01, f : 1.106e-01
1500, Loss : 1.75307e+00, ic : 2.978e-01, bc : 1.344e+00, f : 1.110e-01
2000, Loss : 5.72558e-01, ic : 4.154e-01, bc : 6.470e-02, f : 9.248e-02
2500, Loss : 2.76724e+00, ic : 2.491e+00, bc : 1.912e-01, f : 8.478e-02
3000, Loss : 1.11269e+00, ic : 7.997e-01, bc : 2.202e-01, f : 9.280e-02
3500, Loss : 4.75517e+00, ic : 3.574e+00, bc : 9.074e-01, f : 2.739e-01
4000, Loss : 4.42749e+00, ic : 3.576e+00, bc : 7.698e-01, f : 8.185e-02
4500, Loss : 5.35669e+00, ic : 4.479e+00, bc : 2.946e-01, f : 5.833e-01
5000, Loss : 3.75726e+00, ic : 3.305e+00, bc : 1.133e-01, f : 3.386e-01
5172, Loss : 2.50225e+17, ic : 1.859e+17, bc : 6.427e+16, f : 8.514e+09tensor(4.1359, device='cuda:0', grad_fn=<AddBackward0>)
```

Fig. 7.11 Training losses for liver

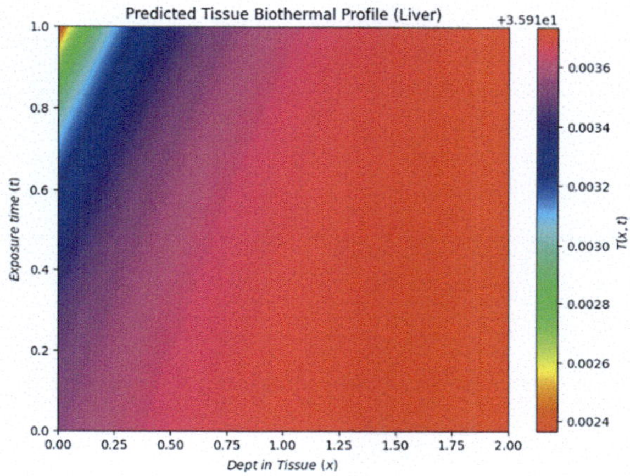

Fig. 7.12 Predicted biothermal profile of the tissue liver

7.9.3 *PINN Model (Training Result) for Skeletal Muscle*

The model trained for 5154 iterations, learning after each. As seen from Fig. 7.13, the final loss = 0.4558.

```
500, Loss : 8.61466e-01, ic : 6.206e-01, bc : 1.155e-01, f : 1.253e-01
1000, Loss : 7.81293e-01, ic : 3.308e-01, bc : 3.806e-01, f : 6.984e-02
1500, Loss : 1.70964e+00, ic : 1.481e+00, bc : 1.082e-01, f : 1.203e-01
2000, Loss : 5.59021e-01, ic : 4.428e-01, bc : 3.868e-02, f : 7.750e-02
2500, Loss : 1.67504e+00, ic : 9.338e-01, bc : 6.703e-01, f : 7.092e-02
3000, Loss : 5.20790e-01, ic : 3.925e-01, bc : 5.964e-02, f : 6.864e-02
3500, Loss : 9.84499e-01, ic : 6.120e-01, bc : 2.937e-01, f : 7.881e-02
4000, Loss : 9.06834e+00, ic : 8.265e+00, bc : 3.271e-02, f : 7.709e-01
4500, Loss : 6.60052e-01, ic : 5.585e-01, bc : 4.914e-02, f : 5.240e-02
5000, Loss : 4.57541e-01, ic : 3.547e-01, bc : 5.299e-02, f : 4.990e-02
5154, Loss : 5.85702e+18, ic : 2.500e+01, bc : 5.857e+18, f : 1.758e+14tensor(0.4558, device='cuda:0', grad_fn=<AddBackward0>)
```

Fig. 7.13 Training losses for skeletal muscle

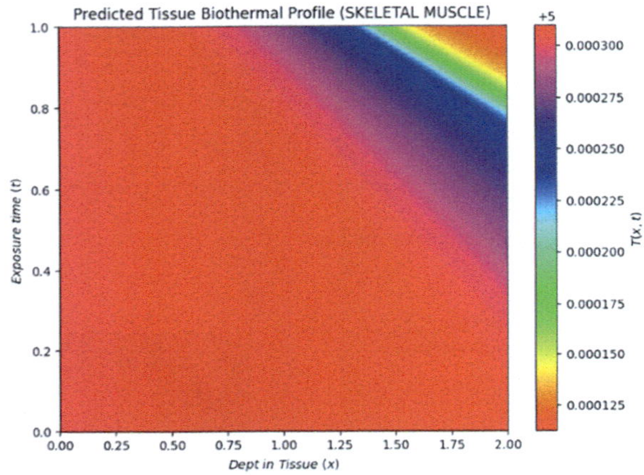

Fig. 7.14 Predicted biothermal profile of the skeletal muscle

```
500, Loss : 1.48954e+00, ic : 8.727e-01, bc : 4.505e-01, f : 1.664e-01
1000, Loss : 1.70521e+00, ic : 5.858e-01, bc : 1.008e+00, f : 1.109e-01
1500, Loss : 3.00531e+00, ic : 2.159e+00, bc : 5.152e-01, f : 3.307e-01
2000, Loss : 6.30093e-01, ic : 5.177e-01, bc : 5.463e-02, f : 5.774e-02
2500, Loss : 4.40558e+00, ic : 3.651e+00, bc : 5.555e-01, f : 1.990e-01
3000, Loss : 3.72519e+00, ic : 3.058e+00, bc : 6.396e-01, f : 2.776e-02
3500, Loss : 8.59493e-01, ic : 5.759e-01, bc : 1.601e-01, f : 1.234e-01
4000, Loss : 1.18514e+00, ic : 4.614e-01, bc : 6.278e-01, f : 9.590e-02
4500, Loss : 6.58627e-01, ic : 4.027e-01, bc : 1.693e-01, f : 8.661e-02
5000, Loss : 3.05415e+00, ic : 2.458e+00, bc : 3.471e-01, f : 2.491e-01
5229, Loss : 2.50001e+01, ic : 6.930e-07, bc : 2.500e+01, f : 4.919e-05tensor(3.0615, device='cuda:0', grad_fn=<AddBackward0>)
```

Fig. 7.15 Training losses for adipose tissue

7.9.4 Predicted Biothermal Profile for Skeletal Muscle

Figure 7.14 shows the variation of skeletal muscle (tissue) temperature during hyperthermia. It is a plot of the period for which the tissue is exposed to heat (t) against the depth of penetration of the RF-magnetic field.

7.9.5 PINN Model (Training Result) for Adipose Tissue

The model trained for 5229 epochs, learning after each iteration. The final loss as seen in Fig. 7.15 is 3.0615.

7.9.6 Predicted Biothermal Profile for Adipose Tissue

Figure 7.16 shows the variation of adipose tissue temperature during hyperthermia. It is a plot of the period for which the tissue is exposed to heat (t) against the depth of penetration of the RF-magnetic field.

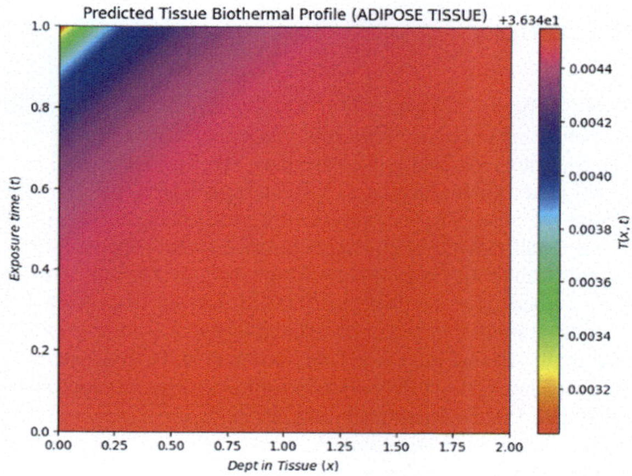

Fig. 7.16 Showing the biothermal profile of adipose tissue

7.10 Conclusion

In this chapter, the Pennes bioheat equation, NMR-guided temperature measurements and sophisticated computational modeling have been integrated to enable a new method of directing clinical hyperthermia procedures. The performance benefits of the new technique are demonstrated through simulations suitable to perform hyperthermia treatment of tumours and monitoring malignant tissue temperature changes without the usual concern for safety of normal tissues. The new theranostics approach can provide real time monitoring of tissues during hyperthermia treatment of tumours so that the surrounding tissues are not overheated in the process. All utilization of PINNs has demonstrated significant advancements in predicting temperature distributions during NMR-guided hyperthermia. The model's ability to leverage physics-informed constraints has not only enhanced accuracy but has also provided a deeper understanding of the intricate dynamics of heat transfer within biological tissues. The zero-centered nature of the tanh activation function employed in the PINN architecture has proven effective in mitigating issues related to vanishing, gradients, contributing to the stability and efficiency of the learning process.

Building on the success of this chapter, several recommendations for future work and enhancements emerge: (i) expanding the model to incorporate additional physics and biological factors that might influence hyperthermia outcomes. This could include variations in tissue composition, tumor heterogeneity, and other patient-specific parameters. (ii) Exploring advanced optimization techniques to fine-tune the model parameters and improve its efficiency. This could involve the exploration of different optimization algorithms and hyper parameter tuning strategies.

Acknowledgements The authors acknowledge the motivated research efforts of Enenche Oche Solomon as a promising young researcher.

References

Awojoyogbe, B.O., Dada, M.O.: Computational design of an RF controlled theranostic model for evaluation of tissue biothermal response. J. Med. Biol. Eng. **38**, 993–1013 (2018)

Cai, S., Wang, Z., Wang, S., Perdikaris, P., Karniadakis, G.E.: Physics-informed neural networks for heat transfer problems. J. Heat Transf. **14**(6) (2021). https://doi.org/10.1115/1.4050542

Moseley, B.: So, what is a physics-informed neural network? (2021) https://benmoseley.blog/my-research/so-what-is-a-physics-informed-neural-network/?ref=jousefmurad.com

Rutgers Cancer Institute of New Jersey Rutgers: Cancer as ancient Egyptians knew and understood it. Cancer History Project (2023). Retrieved December 2, 2023, from https://cancerhistoryproject.com/article/cancer-as-ancient-egyptians-knew-and-understood-it/

Sharma, P., Evans, L., Tindall, M., Nithiarasu, P.: Hyperparameter selection for physics-informed neural networks (PINNs)—application to discontinuous heat conduction problems. Numer. Heat Transf. Part B: Fundam. 1–15 (2023)

Wang, H., Burgei, W.A., Zhou, H.: Analytical solution of one-dimensional Pennes' bioheat equation. Open Phys. **18**(1), 1084–1092 (2020)

Chapter 8
Simulation of Temperature Distribution in Biological Tissues Using Physics-Informed Neural Networks

Abstract The simulation in this chapter sheds light on the complicated thermal behavior of tissues, such as their thermal conductivity, blood perfusion rate, and thermal diffusivity. To streamline the analysis, metabolic heat generation was assumed to be zero in both tissues. The distribution of temperature in biological tissues offers several significant benefits across various fields including healthcare and medical research. The goal is to demonstrate the prospective synergy between physics-based knowledge and machine learning approaches, which contributes to improved understanding of temperature distribution in biological tissues and its implications for a variety of healthcare applications. Additionally the use of the Pennes equation which describes how heat is generated within tissues due to metabolic activities transferred through thermal conduction, and dissipated by blood perfusion was used to model the simulation.

Keywords Physics-informed neural networks · Temperature distribution · Biological tissues · Pennes bioheat equation

8.1 Introduction

The regulation of temperatures is a homeostatic process that maintains a constant internal body temperature in the face of changing environmental variables. Maintaining a body temperature within a narrow range (between 36.5 and 37.5 °C) permits the body's enzymes and immunological systems to work properly (Osilla et al. 2023). The study of temperature distribution in medical applications is very important in hyperthermia treatment where elevated temperatures are purposely delivered to target tissues for therapeutic objectives such as cancer treatment. Heat is transferred in living tissues through a variety of physical mechanisms such as convection, conduction, blood perfusion, and advection across micro-vascular networks (Aijaz et al. 2023). By replicating temperature distribution under diverse settings, mathematical modeling complements experimental findings. Over the last few decades, major research in bioheat transfer has led to an understanding of the governing dynamics of thermal

B. O. Awojoyogbe and M. O. Dada, *Digital Molecular Magnetic Resonance Imaging*, Series in BioEngineering, https://doi.org/10.1007/978-981-97-6370-2_8

transport in a tissue. Several models for controlling heat flow in tissues have been proposed. Pennes presented a classical model in 1948, which was followed by various refinements and related models.

8.2 The Pennes Bioheat Equation

The Pennes equation accounts for blood flow through an advection term that includes the thermophysical properties of blood as well as the distinction between blood temperature and local tissue temperature (Sarkar et al. 2015). The distribution of temperature in biological tissues is a diverse field that spans fundamental physiological processes to advanced medical applications, integrating experimental studies, thermal imaging tools; and mathematical modeling plays a pivotal role in advancing our understanding of temperature distribution in biological tissues. Computational models, grounded in the principles of heat transfer and thermodynamics, enable researchers to simulate and predict temperature profiles within complex biological structures. These models consider factors such as blood flow, tissue composition, and metabolic heat generation, providing a quantitative framework for interpreting experimental findings, our understanding of how temperature affects biological systems, opening up possibilities for better medical treatments and biological research.

The Pennes bioheat equation is a mathematical model that describes the distribution of heat within biological tissues, taking into account blood perfusion as a mechanism for heat dissipation (Ostadhossein and Hoseinzadeh 2022). The bioheat equation is a partial differential equation in its most basic form, accounting for heat conduction, metabolic heat generation, blood perfusion, and other variables within a tissue. Depending on the assumptions and parameters considered in a particular application, the equation can be expressed in a number of ways.

The Pennes bioheat equation is given in Eq. (3.298) as follows;

$$\rho_t c_t \frac{\partial T}{\partial t} = k \frac{\partial^2 T}{\partial x^2} + W_b \rho_b c_b (T - T_b) + Q_m \tag{8.1}$$

where

ρ_t is the tissue density,
c_t is the tissue specific heat,
T is the tissue temperature,
t is time,
k is the tissue thermal conductivity,
Q_m denotes the metabolic heat generation,
W_b is the blood perfusion rate,
c_b is the blood specific heat,
T_a is the core body temperature,
ρ_b is the blood density.

8.3 Physics-Informed Neutral Network (PINN)

Physics-informed neural networks (PINNs) are a sort of universal function approximator that can incorporate knowledge of any physical rules governing a given dataset in the learning process and can be expressed using partial differential equations (PDEs). The goal of PINNs is to discover an unknown solution to a PDE or a physical system directly from data while remaining true to the underlying physics (Fig. 8.1).

The architecture of the neural network in a PINN can vary based on the problem. For this research, it includes input layers that correspond to spatial or temporal coordinates, hidden layers for feature extraction, and output layers for predicting the dependent variables. PINNs employ a loss function that includes terms relevant to data fitting and problem physics.

8.3.1 Activation Function

The activation function is a mathematical approach used to infuse non-linearity into the output of a neural network node, or neuron. The activation function assists neural networks in describing complex connections between inputs and outputs, allowing them to learn and predict on a diverse set of data. For this research, the activation function used is the hyperbolic tangent function (tanh). The hyperbolic tangent function, commonly denoted as tanh, is an activation function that squashes its input to the range of $(-1, 1)$ and mathematically, it is defined as:

$$F(x) = e^{x} - e^{-x}/e^{x} + e^{-x} \tag{8.2}$$

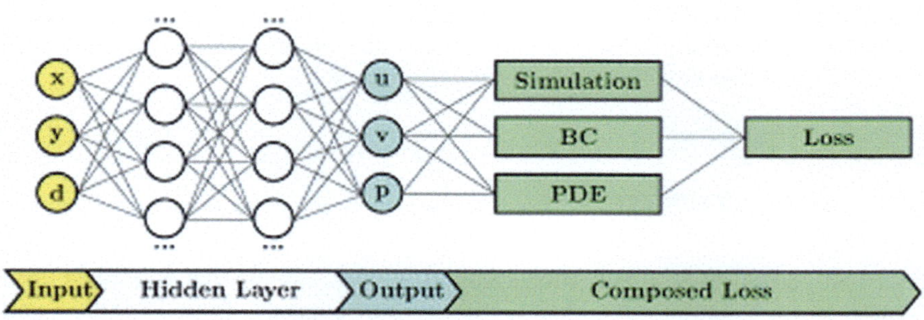

Fig. 8.1 Architecture of PINN (Pioch et al. 2023)

8.3.2 The Initialization or Constructor

The initialization refers to the process of setting up the initial parameters of a neural network before training. The constructor is typically the method in a programming class that is called when an instance of the class is created. The initialization of a neural network's weights and biases is critical because it influences the network's ability to learn from data during training. Proper initialization can help prevent issues like vanishing or expanding gradients, which may prevent the training process's convergence.

8.3.3 Residual Function (F Method)

The 'f' method uses the PINN architecture to calculate the residual (PDE) at supplied points. It entails calculating the derivatives of the projected answer in terms of geographical and temporal coordinates. The main benefit of residual learning is that it aids in the training of very deep neural networks by solving challenges like disappearing gradients. The network can learn to alter the residual rather than the complete mapping, making the optimization process easier.

8.3.4 Closure Method (Closure)

The closure method is utilized for optimization. It computes the mean squared errors for initial conditions, boundary conditions, and the PDE residual. These losses are then added together to generate the total loss. The method also prints the progress of optimization. When dealing with sophisticated optimization methods that require additional computations during the optimization process, using a closure allows for a more modular and flexible approach to optimization.

8.3.5 Adam Optimizer

The Adam optimizer is a well-known optimization technique that is used in deep learning to train neural networks. It incorporates concepts from RMSprop and Momentum, two additional optimization techniques. Adam is notable for its flexible learning rate capabilities and is particularly effective in a wide range of deep learning applications.

8.3.6 *L-BFGS Optimizer*

The L-BFGS (Limited-memory Broyden-Fletcher-Goldfarb-Shanno) optimizer is an iterative optimization algorithm used for solving unconstrained nonlinear optimization problems. It is particularly well-suited for optimizing smooth, differentiable functions where the computation of the full Hessian matrix is impractical. The use of both the Adam and L-BFGS optimizers is typical in PINN training. Adam is effective for the initial training phase, whereas L-BFGS is frequently employed for fine-tuning and refining the solution.

8.4 Tissue Parameters Used in the First Simulation

Table 8.1 showing the approximated properties for the kidney tissue.

8.4.1 *Tissue Parameter Used in the Second Simulation*

Table 8.2 showing the approximated properties for the Skin tissue:

Table 8.1 Approximated thermal properties for the kidney tissue (IT'IS Foundation 2023)

Kidney thermal conductivity (k)	0.53 W/m/°C
Kidney density (ρ_t)	1066 kg/m^3
Kidney specific heat capacity (c_t)	3763 J/kg/k
Thermal diffusivity of the kidney (α)	1.32×10^{-7} m^2/s
Blood temperature (T_a)	37 °C
Blood density (ρ_b)	1060 kg/m^3
Blood specific heat capacity (c_b)	3893 J/kg/k
Blood perfusion rate (W_b)	7.6 kg/m^3/s

Table 8.2 Approximated thermal properties for the skin tissue (IT'IS Foundation 2023)

Skin thermal conductivity (k)	0.37 W/m/°C
Skin density(ρ_t)	1109 kg/m^3
Skin specific heat capacity (c_t)	3391 J/kg/k
Thermal diffusivity of the skin (α)	9.84×10^{-8} m^2/s
Blood temperature (T_a)	37 °C
Blood density (ρ_b)	1060 kg/m^3
Blood specific heat capacity (c_b)	3893 J/kg/k
Blood perfusion rate (W_b)	7.6 kg/m^3/s

8.5 PINN Training for the Kidney Tissue

Using the parameters in Table 8.1, a PINN was trained and the process is shown in Fig. 8.2.

After more than 5000 iterations, the output indicates the training progress. The provided data comprise the overall loss, the mean squared error for initial conditions (ic), boundary conditions (bc), and the mean squared error for the PDE residual (f). The detailed breakdown of results aids in the assessment and fine-tuning of the deep learning PINN model, showing its effectiveness in solving the Pennes bioheat equation.

The Mean squared error (MSE) is a metric used in deep learning and other machine learning tasks to evaluate model performance, particularly in regression issues. It calculates the average squared difference between expected and actual values. During model training, MSE is frequently employed as a loss function.

8.5.1 Simulation Results for the Kidney

Figure 8.3 shows the simulation result of the 1D-Biothermal profile in the kidney tissue.

Figure 8.4 shows the 2-D representation of the PINN predicted biothermal kidney profile.

```
...  500, Loss : 1.05208e+04, ic : 3.753e+02, bc : 4.003e+02, f : 9.745e+03
     1000, Loss : 2.55595e+03, ic : 7.862e+02, bc : 8.112e+02, f : 9.585e+02
     1500, Loss : 2.01790e+03, ic : 9.359e+02, bc : 9.609e+02, f : 1.212e+02
     2000, Loss : 2.00814e+03, ic : 9.587e+02, bc : 9.837e+02, f : 6.578e+01
     2500, Loss : 2.00810e+03, ic : 9.601e+02, bc : 9.851e+02, f : 6.290e+01
     3000, Loss : 2.00810e+03, ic : 9.601e+02, bc : 9.851e+02, f : 6.285e+01
     3500, Loss : 2.00810e+03, ic : 9.601e+02, bc : 9.851e+02, f : 6.285e+01
     4000, Loss : 2.00810e+03, ic : 9.601e+02, bc : 9.851e+02, f : 6.284e+01
     4500, Loss : 2.00810e+03, ic : 9.601e+02, bc : 9.851e+02, f : 6.284e+01
     5000, Loss : 2.00810e+03, ic : 9.601e+02, bc : 9.851e+02, f : 6.284e+01
     5500, Loss : 1.65318e+03, ic : 9.521e+02, bc : 5.606e+02, f : 1.405e+02
     6000, Loss : 1.62452e+03, ic : 9.511e+02, bc : 5.269e+02, f : 1.465e+02
     6500, Loss : 1.61868e+03, ic : 9.460e+02, bc : 5.257e+02, f : 1.470e+02
     7000, Loss : 1.61133e+03, ic : 9.407e+02, bc : 5.196e+02, f : 1.510e+02
     7500, Loss : 1.59657e+03, ic : 9.224e+02, bc : 5.057e+02, f : 1.684e+02
     8000, Loss : 1.57949e+03, ic : 9.223e+02, bc : 4.738e+02, f : 1.834e+02
     8500, Loss : 1.57235e+03, ic : 9.211e+02, bc : 4.654e+02, f : 1.859e+02
     8998, Loss : 1.56723e+03, ic : 9.246e+02, bc : 4.531e+02, f : 1.895e+02

...  tensor(2008.1041, device='cuda:0', grad_fn=<AddBackward0>)
```

Fig. 8.2 PINN model training for the kidney

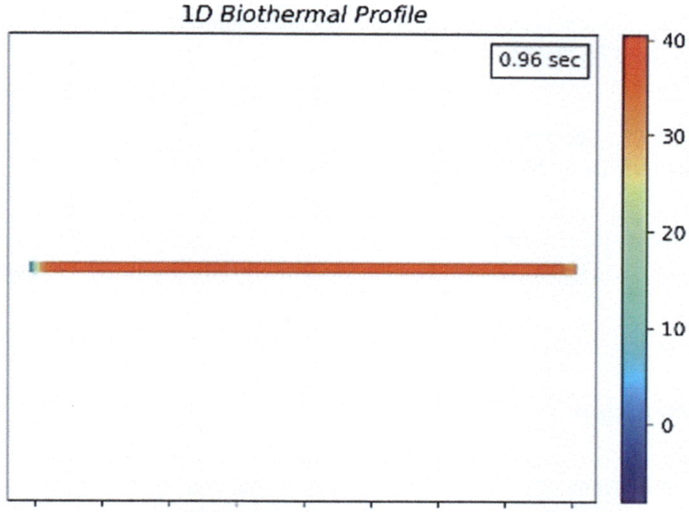

Fig. 8.3 Simulation of the 1D-biothermal profile (kidney)

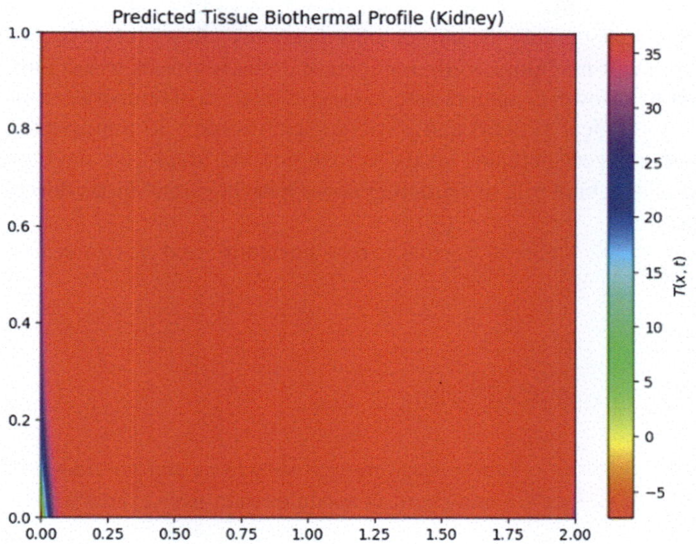

Fig. 8.4 2-D representation of the biothermal PINN profile (kidney)

8.5.2 PINN Training for the Skin Tissue

Figure 8.6 indicates that the training process stops at iteration 5793 with a loss of 6.14457e+02, mean squared error for the initial condition (IC): 2.806e+00, mean square for the boundary condition (BC): 4.630e+02, and the mean squared error for the PDE residual (f): 1.486e+02.

8.5.3 Simulation Results for the Skin

Figure 8.7 shows the simulation result of the 1D-Biothermal profile in the skin tissue.

Figure 8.9 shows the 2-D representation of the PINN predicted biothermal skin profile.

8.6 Conclusion

Accurate simulations of temperature distribution are critical for effective treatment planning. The models shown in Tables 8.1 and 8.2 show how the approximated properties for the kidney and skin tissues as simulated in Figs. 8.2, 8.3, 8.4, 8.5, 8.6, 8.7, 8.8, 8.9 and 8.10 can allow health care providers to estimate how heat will be transported into tissues after various interventions such as thermal therapy or surgical procedures. Understanding the motion of temperature allows for more accurate and targeted treatment planning, while reducing the possibility of tissue injury during thermal therapy. However, a more detailed representation of temperature distribution in complex biological systems and physiological diversity to improve the applicability in personalized medicine needs be incorporated. Real-time monitoring techniques may play a vital role in validating and improving simulations during medical procedures.

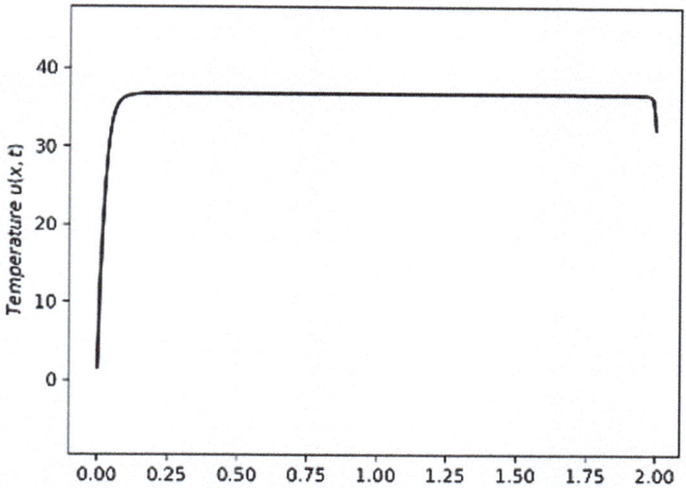

Fig. 8.5 Temperature distribution curve in the kidney

```
500, Loss : 1.18696e+04, ic : 3.758e+02, bc : 4.008e+02, f : 1.109e+04
1000, Loss : 2.66206e+03, ic : 7.899e+02, bc : 8.149e+02, f : 1.057e+03
1500, Loss : 2.02778e+03, ic : 9.425e+02, bc : 9.675e+02, f : 1.178e+02
2000, Loss : 2.01591e+03, ic : 9.661e+02, bc : 9.911e+02, f : 5.866e+01
2500, Loss : 2.01586e+03, ic : 9.676e+02, bc : 9.926e+02, f : 5.560e+01
3000, Loss : 2.01586e+03, ic : 9.677e+02, bc : 9.927e+02, f : 5.554e+01
3500, Loss : 2.01586e+03, ic : 9.677e+02, bc : 9.927e+02, f : 5.554e+01
4000, Loss : 2.01586e+03, ic : 9.677e+02, bc : 9.927e+02, f : 5.554e+01
4500, Loss : 2.01585e+03, ic : 9.676e+02, bc : 9.927e+02, f : 5.553e+01
5000, Loss : 2.01582e+03, ic : 9.676e+02, bc : 9.927e+02, f : 5.553e+01
5500, Loss : 6.37892e+02, ic : 5.513e+00, bc : 4.891e+02, f : 1.433e+02
5793, Loss : 6.14457e+02, ic : 2.806e+00, bc : 4.630e+02, f : 1.486e+02

tensor(2015.8236, device='cuda:0', grad_fn=<AddBackward0>)
```

Fig. 8.6 PINN model training for the skin

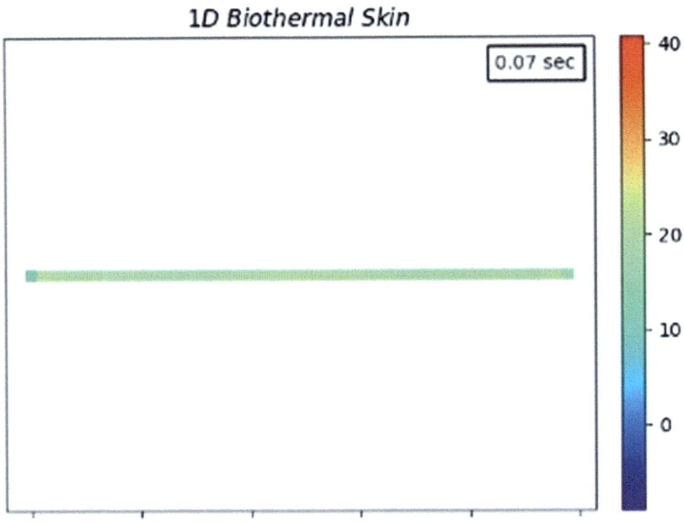

Fig. 8.7 Simulation after 0.07 s of the skin tissue in 1-D

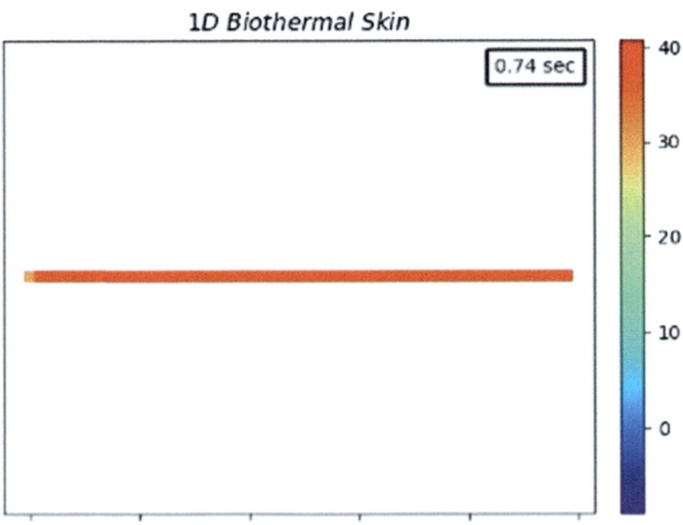

Fig. 8.8 Simulation after 0.74 s of the skin tissue in 1-D

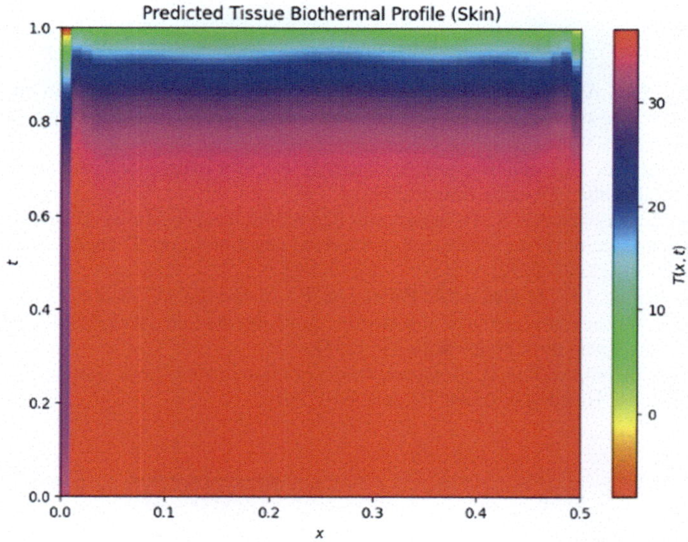

Fig. 8.9 Simulation of the biothermal profile for the skin in 2D

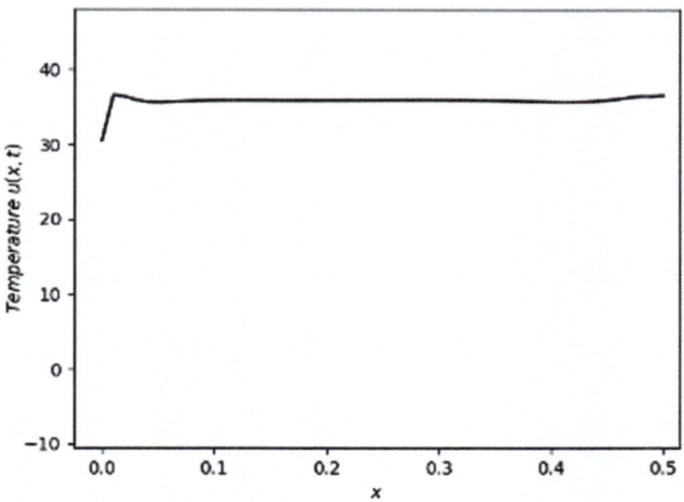

Fig. 8.10 2D curve of the temperature distribution in the skin

Acknowledgements The authors acknowledge the motivated research efforts of Orifa Timothy (Department of Physics, Federal University of Technology, Minna, Nigeria) as a promising young researcher.

References

Aijaz, M., Dar, J.G., Almanjahie, I.M., Alshahrani, F.: Temperature distribution in tumour tissue during targeted destruction by heat: a hyperbolic bioheat equation approach. Case Stud. Therm. Eng. **50**, 103491 (2023)

Osilla, E.V., Marsidi, J.L., Shumway, K.R., Sharma, S.: Physiology, Temperature Regulation. StatPearls Publishing, Treasure Island (2023)

Ostadhossein, R., Hoseinzadeh, S.: The solution of Pennes' bio-heat equation with a convection term and nonlinear specific heat capacity using Adomian decomposition. J. Therm. Anal. Calorim. **147**(22), 12739–12747 (2022)

Pioch, F., Harmening, J.H., Müller, A.M., Peitzmann, F.J., Schramm, D., Moctar, O.E.: Turbulence modeling for physics-informed neural networks: comparison of different RANS models for the backward-facing step flow. Fluids **8**(2), 43 (2023)

Sarkar, D., Haji-Sheikh, A., Jain, A.: Temperature distribution in multi-layer skin tissue in presence of a tumor. Int. J. Heat Mass Transf. **91**, 602–610 (2015)

Chapter 9
Performance Evaluation of Machine Learning Classification of Brain Tumors with WEKA and Python Programming

Abstract The manual interpretation of brain tumors MRI scans by medical personnel can be time-consuming and require a significant amount of manpower. This highlights the need for more efficient and accurate methods of interpreting MRI scans, such as machine learning implementations. By developing and implementing machine learning models capable of interpreting MRI scans, we can improve the speed and accuracy of diagnosis for brain tumors, and provide better patient outcomes. Therefore, the performance evaluation of both WEKA and Python as it correlates to image classification with the aid of appropriate filters and classifiers is necessary for the advancement of medical imaging. The objectives of the chapter are to: (i) evaluate the performance of machine learning classification of brain tumors with WEKA and PYTHON programming using MRI scan images (ii) deploy the best model for the brain tumor classification in WEKA and PYTHON programming.

Keywords Brain tumors · MRI scans · WEKA programming · PYTHON libraries · Machine learning implementations

9.1 Introduction

The Python library is a group of modules with classes and methods that other programs can utilize to carry out various tasks. The python libraries that will be used for this chapter includes.

1. OpenCv: Real-world computer vision is the goal of the Open Source Computer Vision programming library. The cross-platform library is available for free usage under the open source BSD license. Deep learning frameworks like TensorFlow, Torch, Caffe, and others are supported by OpenCV. Its implemented as Cv2 in python code. It was created using C and C++ as the programming languages (Deepthi et al. 2019). The applications of OpenCV include;

    a. 2D and 3D feature toolkits
    b. Facial recognition system

B. O. Awojoyogbe and M. O. Dada, *Digital Molecular Magnetic Resonance Imaging*, Series in BioEngineering, https://doi.org/10.1007/978-981-97-6370-2_9

   c. Mobile robotics
   d. Object detection
   e. Segmentation and recognition, etc.

2. NumPy: NumPy is an open-source programming language that specializes in numerical processing using multi-dimensional ndarrays, where the arrays permit element-by-element operations, also known as broadcasting. It is possible to apply the linear algebra formalism without first altering the NumPy arrays if necessary. Additionally, the size of the arrays can be changed dynamically. This eliminates the concerns that slow down efficient programming in other languages. When you want to remove specific components from an arrangement, you can apply a mask to it rather than generating a new arrangement (Bressert 2012).
3. Pandas: A Python module called pandas offers quick, adaptable, and expressive data structures that are intended to make working with "relational" or "labeled" data simple and natural. It seeks to serve as the essential, high-level building block for using Python for actual, useful data analysis. Series (1-dimensional) and DataFrame (2-dimensional) are the two main panda's data structures, and they can handle the bulk of common use cases in finance, statistics, social science, and many branches of engineering (McKinney and Team 2015).
4. Scikit-image: A Python programming language image processing package called scikit-image (formerly known as scikits.image) is free and open-source.it is called as Skimage in python language environment It contains algorithms for feature identification, analysis, filtering, morphology, segmentation, geometric transformations, color space manipulation, and more. It is made to work with Python's NumPy and SciPy scientific and mathematical libraries. Stéfan van der Walt's scikits.image project served as the inspiration for scikit-image. Its name refers to the idea that it is a third-party modification to SciPy called a "SciKit" (SciPy Toolkit), which was independently created and distributed. Later, other developers completely rewrote the original codebase. In November 2012, scikit-image and scikit-learn were among the scikits that were deemed "well-maintained and popular" (Wikipedia 2022c).
5. Matplotlib: Matplotlib is a low-level Python implementation of MATLAB-like plots that offers a wide range of customizability options. Almost any graph may be made quickly and easily (Stančin and Jović 2019). This Python package makes use of the scatterplot library to show multidimensional samples. Each one- and two-dimensional projection of the sample is plotted in these visuals to show covariance. For the Python programming language and its NumPy numerical mathematics extension, Matplotlib is a graphing library. John D. Hunter is credited for creating Matplotlib (Wikipedia 2022b).
6. SciPy: SciPy is a set of modules based on NumPy that can be used to tackle a variety of computational issues, including those related to statistics, linear algebra, sparse matrices, linear programming, image processing, differential equations, and symbolic mathematics, among others. The SciPy data structures and algorithms have broad domain compatibility. Because of its high-level

syntax, SciPy is usable and effective by programmers of all backgrounds and experience levels (Danial 2022).

7. Seaborn: Python's seaborn package allows you to create statistical visuals. It strongly interacts with panda's data structures and offers a high-level interface to matplotlib. The seaborn library's functions present a declarative, dataset-oriented API that makes it simple to convert data-related issues into visuals that can resolve them. When given a dataset and a plot specification, seaborn automatically converts data values into visual characteristics like color, size, or style, computes statistical adjustments internally, and embellishes the plot with helpful axis labels and a caption. In order to facilitate comparisons between conditional subsets of data or across various variable pairs in a dataset, many seaborn algorithms can produce figures with numerous panels (Waskom 2021).
8. Scikit-learn: Python library Scikit-learn offers a standardized interface for putting machine learning algorithms into practice. As part of the machine learning workflow, it also comprises various auxiliary tasks like data preprocessing operations, data resampling methods, assessment criteria, and search interfaces for tuning and performance optimization of algorithms. Various input variables are routinely subjected to common transformations, such as the whitening transformation, differencing, and log-odds ratios. Scikit-learn offers a number of useful functions to carry out these techniques for data modification and preprocessing (Bisong 2019).
9. tqdm: The progress bar library tqdm was created with speed and extensibility in mind. Although there are ports in other languages, it is written in Python. The acronym tqdm stands for "progress" in Arabic (taqadum,), and it also stands for "I love you so much" in Spanish (tequierodemasiado) (Costa-Luis, n.d.). The following programming issues are addressed by tqdm;

    a. preventing unnecessary printing, such as just showing every nth iteration
    b. displaying iteration rate
    c. displaying elapsed and predicted completion times, and
    d. providing all of the aforementioned information on a single continuously updated line.

10. PIL: One of the essential libraries for manipulating images in the Python programming language is the Python Imaging Library, or PIL for short. The PIL library, designed for Python beginners, can be used to do many image processing tasks. Examples include image inversion, binary conversion, cropping, adding text to photos, adjusting brightness and contrast, image filtering, blurring, contouring, and smoothing, among other things (Gujar et al. 2013).
11. Os: Python's OS module offers tools for communicating with the operating system. OS is included in the basic utility modules for Python. This module offers a portable method of using functionality that is dependent on the operating system. It can be used to read or write files, change paths, and aid in command-line reading of all lines in all files. Along with managing high-level file and

directory operations, it is also used to create temporary files and directories (GeeksForGeeks 2022).

12. Streamlit: Streamlit is an open source, Python-based platform for creating and deploying interactive data science dashboards and machine learning models. it offers a platform that converts data scripts into shareable web applications.

9.2 Image Filters

The bulk of images have some sort of unwanted noise that degrades the resolution and quality of the images. Following the removal of these noises from the image data, the data are typically processed for analysis. Most engineering and scientific challenges need filtering in one of its many forms. In comparison to other accessible methodologies, it offers researchers excellent findings. Filtering is being used to solve issues by researchers in the fields of electrical and electronic engineering, mechanical engineering, chemistry, and physics. In image processing, one of the most popular uses of filtering is to improve resolution or properly accentuate edges. Based on the duality of the intended frame for the analysis of the project, filters will be selected in both python code and WEKA software for image preprocessing.

9.3 WEKA Software

The WEKA program comes with a variety of interfaces for data processing and analysis, such as the explorer, experimenter, Knowledge Flow, Workbench, and Simple CLI. The simplest way to utilize WEKA is through the Explorer graphical user interface. By selecting items from a menu and filling up forms, users can access all of its amenities. The WEKA software only accepts data in two formats namely; the ARFF format which stands for Attribute-Relation File Format (It is an ASCII text file that lists instances that share a certain set of characteristics) and the CSV format which stands for Comma Separated Values.

9.4 Spyder

Spyder is a free and open source scientific environment created by and for scientists, engineers, and data analysts in Python, for Python. The functionality of a comprehensive development tool's powerful editing, analysis, debugging, and profiling skills are combined with a scientific package's data exploration, interactive execution, deep inspection, and beautiful visualization features to create a singular product (Wikipedia 2022a).

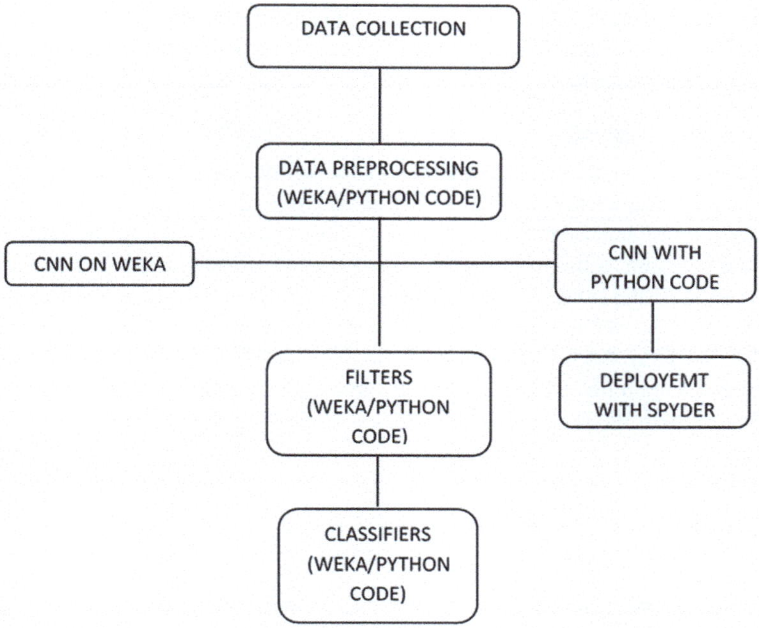

Fig. 9.1 Flowchart describing the method used in this study

9.5 Data Collection

The Brain tumor data used in this chapter was downloaded from kaggle.com. It is a data of brain MRI images for brain tumor detection. The dataset is divided into two unequal parts, with one folder containing MRI images with no brain tumor while the other contains MRI images of previously detected presence of brain tumor. The first folder is named no while the second folder is named yes. The folder named no, contains ninety-eight healthy MRI images of the brain while the second folder contains one hundred and fifty five MRI images of brains diagnosed with the presence of brain tumor. The Flowchart describing the method used in this chapter is shown in Fig. 9.1.

9.6 Data Preprocessing Both on WEKA and Python

Preprocessing of the data is done both on WEKA and with the use of python code in jupyter notebook. This step involves the input of the data meant for analysis into the WEKA interface and the python code frame of reference i.e. it involves calling the folder that has the data with the use of python code in jupyter notebook. Since WEKA only accept ARFF format or CSV format, the brain tumor dataset utilized in this study needs to be converted into an ARFF format file. A program called weka-imagefilter-arff-builder is either used to perform the transformation or the transformation can be triggered directly from within WEKA. For the former, the application must be in a folder containing all the images of the downloaded brain tumor dataset in order to build an arff formatted dataset with accurate information about the images. To maintain the distinct classes of the images as indicated by their respective folder name, the names of the respective images were all manually changed to provide a more general difference for each class (no and yes). By clicking on it to launch the program, a file in the arff format was created that contained an accurate representation of the images. Afterwards, data analysis is then done by opening the explorer tab in the WEKA program. While for the latter (that is, generation of arff file within WEKA), the folder containing the two subfolders is accessed through the preprocessor's open file button, and the preprocessor output image (Fig. 9.2) will be displayed and will the image dataset that can be used for data training.

9.7 Filters

The arff dataset generated for the image is a nominal data type, hence subsequent data training with the use of algorithms save for CNN, is seemingly impossible. There is a need for image filters which will then extract features and convert them to numeric data type. The dataset is subsequently filtered both on WEKA and with python codes. The filters used in WEKA are;

1. Binary pattern pyramid filter
2. Gabor filter
3. Simple color histogram filter and
4. Edge histogram filter.

9.8 Classifiers

The classifiers used are available in the WEKA software and also in python code, therefore after the application of the filters on the data, the net step involves training of the data with classifiers. The classifiers used are;

1. Logistic regression
2. K-nearest neighbor
3. Decision tree
4. Support Vector Machine.

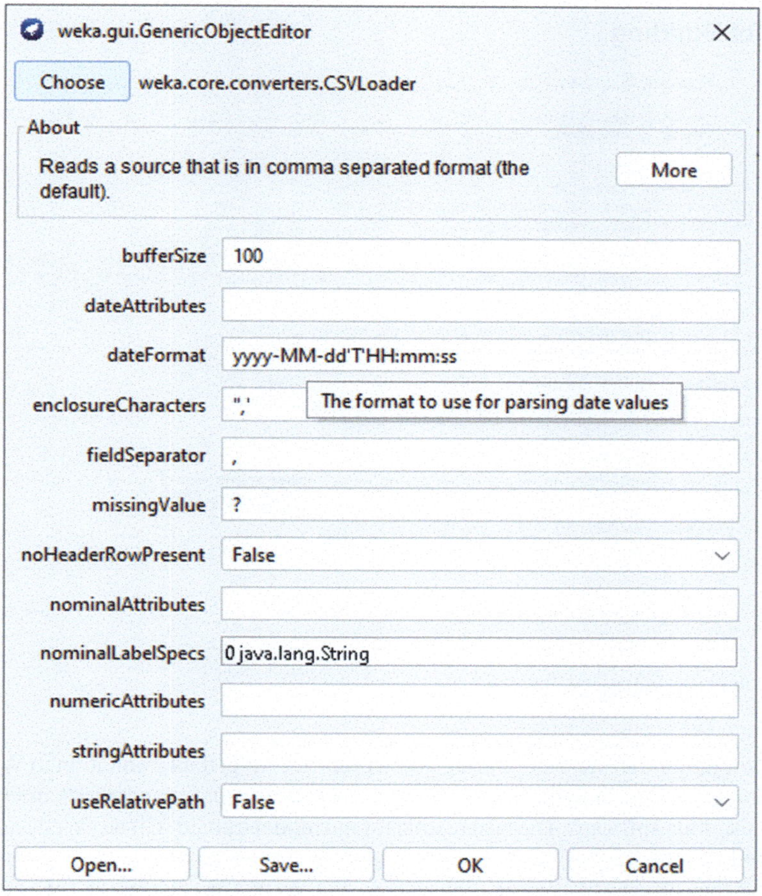

Fig. 9.2 Preprocessor output image

9.9 Data Classification with CNN on WEKA and Python

Convolutional neural network (CNN) is implemented on WEKA as Dl4jMlpClassifier. It is not accessible on the regular WEKA software but subsequent download from the package manager. For application CNN on images, filters are not needed to convert the nominal data to numeric rather the classifier is configured so as to provide more layers for the training. The python code used for the Convolutional neural network training of the data was downloaded from kaggle repository, the image shown in Fig. 9.3, captures the code used in building the model;

After the data training with CNN in python code, the result is then deployed with the aid of spider.

Model Building

```
model = tf.keras.models.Sequential()

model.add(tf.keras.layers.Conv2D(32, kernel_size=(2, 2), input_shape=(128, 128, 3), padding = 'Same'))
model.add(tf.keras.layers.Conv2D(32, kernel_size=(2, 2),  activation ='relu', padding = 'Same'))


model.add(BatchNormalization())
model.add(tf.keras.layers.MaxPooling2D(pool_size=(2, 2)))
model.add(tf.keras.layers.Dropout(0.25))

model.add(tf.keras.layers.Conv2D(64, kernel_size = (2,2), activation ='relu', padding = 'Same'))
model.add(tf.keras.layers.Conv2D(64, kernel_size = (2,2), activation ='relu', padding = 'Same'))

model.add(BatchNormalization())
model.add(tf.keras.layers.MaxPool2D(pool_size=(2,2), strides=(2,2)))
model.add(tf.keras.layers.Dropout(0.25))

model.add(tf.keras.layers.Flatten())

model.add(tf.keras.layers.Dense(512, activation='relu'))
model.add(tf.keras.layers.Dropout(0.5))
model.add(tf.keras.layers.Dense(2, activation='softmax'))

model.compile(loss=tf.keras.losses.BinaryCrossentropy(),
              optimizer=tf.keras.optimizers.Adam(),
              metrics=["accuracy"])

print(model.summary())
```

Fig. 9.3 Code for the model

9.10 Discussion

The brain tumor dataset is put through data preprocessing, data evaluation in WEKA software, as well as in python codes with the help of the jupyter notebook contained in the anaconda software. The best results for the model trained will be saved and then deployed with the help of Spyder, which is also integrated into the anaconda program, based on the comparative nature of this study. For WEKA the percentage of correctly classified instances and incorrectly classified instances were used in gauging how well a classifier performed, while in python code, accuracy percentage on trained data and test score were used to check the performance of the mathematical method used.

9.11 Data Filtering and Training in WEKA with Classifiers

The computer ran into problems leading to the shutting down of WEKA based on the size of the data meant for training after filtering. Once a filter is added to the data, there is an increase in the number of attributes and this prove to be an insurmountable task for the program to run based on the limited specifications of the system used. Initially the dataset contains 98 images of "no" tumor and 155 images for "yes"

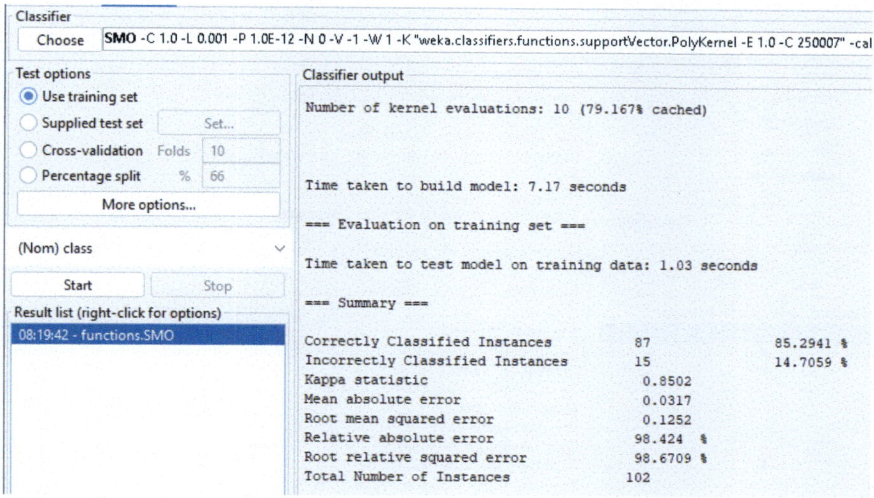

Fig. 9.4 Result of the support vector machine after training

tumor, but upon running into the aforementioned problem, data reduction to about 65 images for both "no" and "yes" is considered for the data training and testing.

The filters were then applied to the data and it generated 961 attributes with the sum of weights and instances equals to 102, thereafter classifiers are used to train the data and the results are of the following;

1. SMO which is WEKA's implementation of Support vector machine is used in training the data and the result is displayed in Fig. 9.4.
2. IBK which is WEKA's implementation for k-nearest neighbor is used in training the data and the result is shown in Fig. 9.5;
3. REPTree which is WEKA's implementation of decision tree is used in training the data and the result is shown in Fig. 9.6.
4. Logistics which is WEKA's implementation of logistic regression is used in training the data and Fig. 9.7 show the results.

The result shown for the data trained by logistics implied that the training could not be completed due to the limitations of the computer's allocated memory. The java heap size limit is closely tied to the memory of the computer and enlarging the heap size to 12 gigabyte is not recommended since the RAM of the computer is just 6 gigabyte.

9.12 Data Training with CNN Implementation on WEKA

Dl4jMlpClassifier which is WEKA's implementation of CNN is used in training the data and the result obtained is shown in Fig. 9.8.

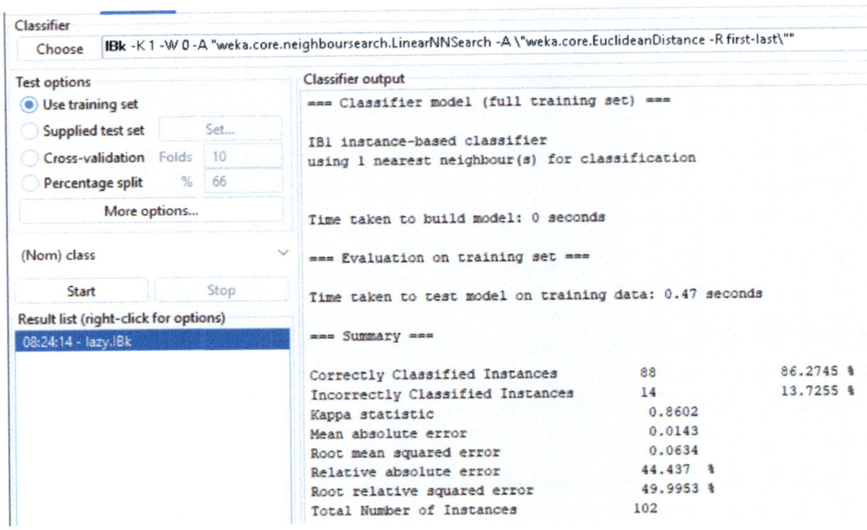

Fig. 9.5 Result of k-nearest neighbor after training

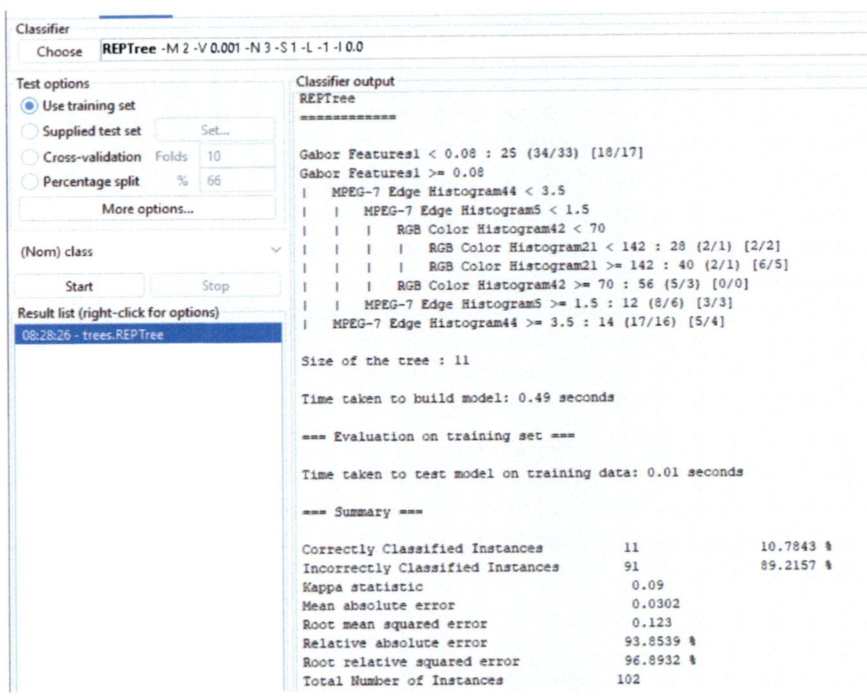

Fig. 9.6 Result of decision tree after training

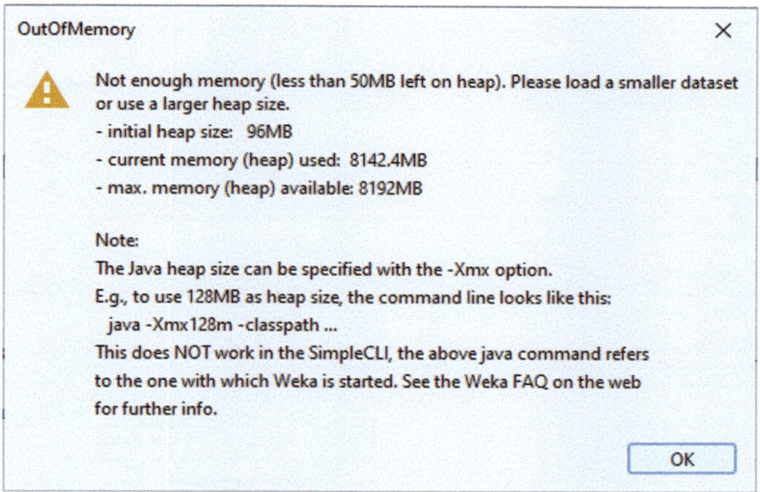

Fig. 9.7 Error test shown upon training with logistic regression

9.13 Data Training and Filtering with Python Code

Data filtering and subsequent training is done with the use of a python code downloaded from kaggle repository. The downloaded python script contains different filter which is used to extract features from the image contained in the data. The filters itemized, were used for image processing before the application of mathematical models for training the data. The following are the algorithms (mathematical models) used in training the data and their respective results;

1. Logistic Regression (Fig. 9.9)
2. K-Nearest Neighbor (Fig. 9.10)
3. Decision Tree (Fig. 9.11)
4. Support Vector Machine (SVM) (Fig. 9.12)

9.14 Data Training with CNN in Python Code

The data trained with the CNN in python code, images were then tested on the model to ensure the model accuracy and precision, based on the classes of the brain tumor. Figures 9.13 and 9.14 are the images that show the result of the test;

Classifier output

```
=== Run information ===

Scheme:       weka.classifiers.functions.Dl4jMlpClassifier -S 1 -cache-mode MEMORY -early-stopping "weka.
Relation:     image_dataset__brain
Instances:    253
Attributes:   2
              image_path
              class
Test mode:    evaluate on training data

=== Classifier model (full training set) ===

Network Configuration:
NeuralNetConfiguration(weightInit=XAVIER, biasInit=0.0, dist=weka.dl4j.distribution.Disabled@66, l1=NaN,
Model Summary:

==========================================================================================
VertexName (VertexType)       nIn,nOut   TotalParams   ParamsShape            Vertex Inputs
==========================================================================================
input (InputVertex)           -,-        -             -                      -
Dense layer (DenseLayer)      784,32     25,120        W:{784,32}, b:{1,32}   [input]
Output layer (OutputLayer)    32,2       66            W:{32,2}, b:{1,2}      [Dense layer]
------------------------------------------------------------------------------------------
            Total Parameters:  25,186
        Trainable Parameters:  25,186
           Frozen Parameters:  0
==========================================================================================


Time taken to build model: 19.68 seconds
```

a Convolutional neural network on WEKA.

```
=== Evaluation on training set ===

Time taken to test model on training data: 1.18 seconds

=== Summary ===

Correctly Classified Instances         215               84.9802 %
Incorrectly Classified Instances        38               15.0198 %
Kappa statistic                          0.6775
Mean absolute error                      0.277
Root mean squared error                  0.339
Relative absolute error                 58.3382 %
Root relative squared error             69.5812 %
Total Number of Instances              253

=== Detailed Accuracy By Class ===

                 TP Rate  FP Rate  Precision  Recall   F-Measure  MCC      ROC Area  PRC Area  Class
                 0.755    0.090    0.841      0.755    0.796      0.680    0.923     0.913     no
                 0.910    0.245    0.855      0.910    0.881      0.680    0.923     0.942     yes
Weighted Avg.    0.850    0.185    0.849      0.850    0.848      0.680    0.923     0.931

=== Confusion Matrix ===

   a   b   <-- classified as
  74  24 |   a = no
  14 141 |   b = yes
```

b Result after training with CNN on WEKA

Fig. 9.8 **a** Convolutional neural network on WEKA. **b** Result after training with CNN on WEKA

Logistic Regression

```
parameters ={"C":[0.01,0.1,1],'penalty':['l2'], 'solver':['lbfgs','saga'], 'max_iter':[50000]}# l1 lasso l2 ridge
lr=LogisticRegression()
logreg_cv = GridSearchCV(lr, parameters, cv=10)
logreg_cv.fit(X_train, y_train)

print("tuned hpyerparameters :(best parameters) ",logreg_cv.best_params_)
print("accuracy :",logreg_cv.best_score_)
```

```
tuned hpyerparameters :(best parameters)  {'C': 0.1, 'max_iter': 50000, 'penalty': 'l2', 'solver': 'lbfgs'}
accuracy : 0.7102564102564102
```

```
model=logreg_cv.predict(X_test)
#check accuracy of our model on the test data
logreg_cv.score(X_test, y_test)
```

```
0.7419354838709677
```

Fig. 9.9 Logistic regression code and output

K - Nearest Neighbours

```
parameters = {'n_neighbors': [1, 2, 3, 4, 5, 6, 7, 8, 9, 10],
              'algorithm': ['auto', 'ball_tree', 'kd_tree', 'brute'],
              'p': [1,2]}

KNN = KNeighborsClassifier()
```

```
knn_cv = GridSearchCV(KNN, parameters, cv=10)
knn_cv.fit(X_train,y_train)
print("tuned hpyerparameters :(best parameters) ",knn_cv.best_params_)
print("accuracy :",knn_cv.best_score_)
```

```
tuned hpyerparameters :(best parameters)  {'algorithm': 'auto', 'n_neighbors': 7, 'p': 1}
accuracy : 0.6519230769230769
```

```
#check accuracy of our model on the test data
knn_cv.score(X_test, y_test)
```

```
0.8064516129032258
```

Fig. 9.10 K-nearest neighbour code and output

9.15 Model Deployment on Spyder

After the model testing, the model developed was then deployed with the aid of spyder and Figs. 9.15 and 9.16 are the interface and the result generated.

Decision Tree

```
parameters = {'criterion': ['gini', 'entropy'],
      'splitter': ['best', 'random'],
      'max_depth': [2*n for n in range(1,10)],
      'max_features': ['sqrt'],
      'min_samples_leaf': [1, 2, 4],
      'min_samples_split': [2, 5, 10]}

tree = DecisionTreeClassifier()
```

```
tree_cv = GridSearchCV(tree, parameters, cv=10)
tree_cv.fit(X_test,y_test)
print("tuned hpyerparameters :(best parameters) ",tree_cv.best_params_)
print("accuracy :",tree_cv.best_score_)
```

```
C:\Users\OLUMIDE\anaconda3\envs\wahala\lib\site-packages\sklearn\model_selection\_split.py:700: UserWarning: The least populate
d class in y has only 9 members, which is less than n_splits=10.
  warnings.warn(
```

```
tuned hpyerparameters :(best parameters)  {'criterion': 'gini', 'max_depth': 18, 'max_features': 'sqrt', 'min_samples_leaf': 2,
'min_samples_split': 5, 'splitter': 'best'}
accuracy : 0.9333333333333332
```

```
#check accuracy of our model on the test data
tree_cv.score(X_test,y_test)
```

```
0.9354838709677419
```

Fig. 9.11 Decision tree code and output

```
X_train, X_test, y_train, y_test =  train_test_split(X,y,test_size=0.2,random_state=2)
parameters = {'kernel':('linear',"sigmoid"),
              'C': [1000,100,300]}
svm = SVC()
svm_cv = GridSearchCV(svm, parameters, cv=5)
svm_cv.fit(X_train, y_train)

print("tuned hpyerparameters :(best parameters) ",svm_cv.best_params_)
print("accuracy :",svm_cv.best_score_)
```

```
tuned hpyerparameters :(best parameters)  {'C': 100, 'kernel': 'linear'}
accuracy : 0.653
```

Fig. 9.12 Support vector machine code and output

```
classification = np.where(res == np.amax(res))[1][0]
imshow(img)
print(str(res[0][classification]*100) + '% Confidence This Is ' + names(classification))
100.0% Confidence This Is No, Its not a tumor
```

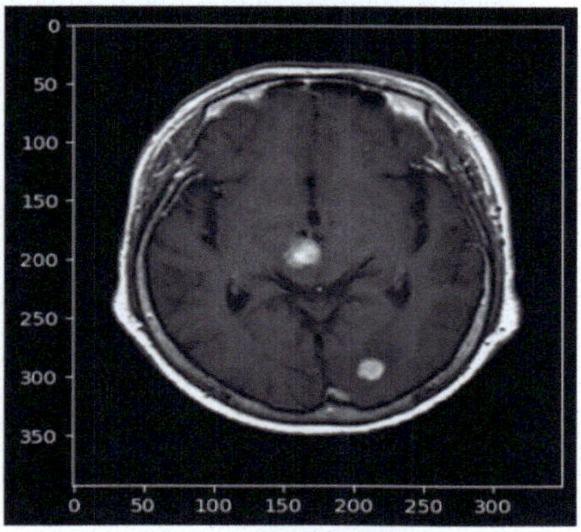

Fig. 9.13 CNN implementation depicting the absence of brain tumor

```
classification = np.where(res == np.amax(res))[1][0]
imshow(img)
print(str(res[0][classification]*100) + '% Confidence This Is yes, ' + names(classification))
99.9994158744812% Confidence This Is yes, Its a Tumor
```

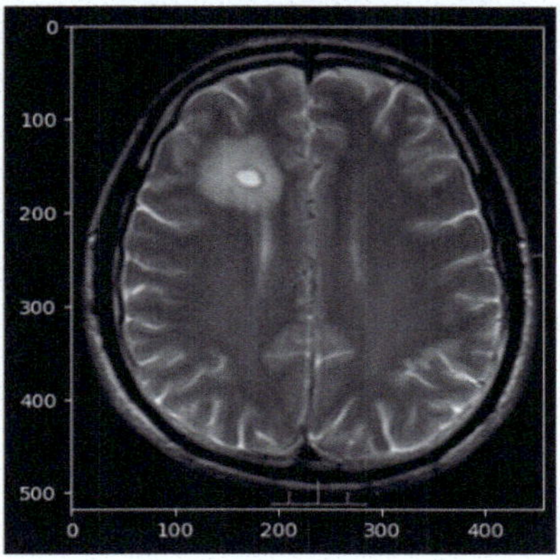

Fig. 9.14 CNN implementation depicting the presence of brain tumor

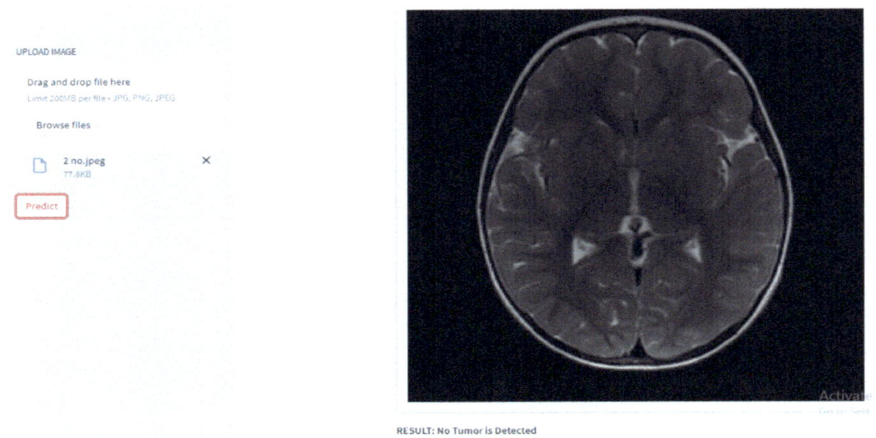

Fig. 9.15 Model deployment with Spyder

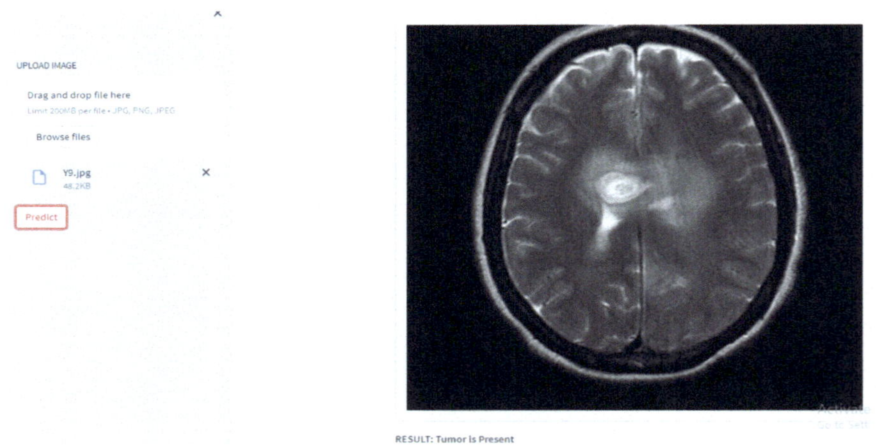

Fig. 9.16 Result after model deployment

9.16 Conclusion

The need for a machine learning model that will alleviate the workload and assist medical personnel in diagnosis of brain tumor cannot be under-emphasized. This chapter has shown that with sufficient data a model that can detect brain tumors with adequate accuracy is possible.

Reduction of data after filtering using filters in WEKA was necessary due to the capability of the specifications of the system used. This chapter has shown that WEKA can be used to classify image data based on the image filters available and the configuration of the computer used in training the machine learning model. The result obtained from the data training on WEKA based on correctly classified instances includes;

1. SMO—85.2941%
2. IBK—86.2745%
3. RepTree—10.7843%
4. CNN—84.9802%.

IBK which is WEKA's implementation for K-nearest neighbor performed better than the other algorithms, while logistic regression could not be used in training the data due to the limitations of the computer specification. The data was subsequently trained using python code and the model results generated based on accuracy are as follows;

1. logistic regression—0.71%
2. K-Nearest Neighbor—0.61%
3. Decision tree—0.93%
4. SVM—0.65%
5. CNN—0.99.

As shown by the model result generated, the classifier with the best result is CNN. Comparing the model generated from the model trained on WEKA and with the use of python code, we have shown that based on accuracy of correctly classified instances, the best model trained is of the CNN algorithm implemented with python code. After which the model is then deployed with the use of Spyder, and further testing was carried out for accurate detection of the presence of brain tumor. The result has proven satisfactory based on the scope of this study. It is therefore recommended that a higher computational resource especially making use of graphic processing unit (GPU) in the system specification be available for further study of much larger dataset. Application of machine learning on larger dataset will yield a much better accuracy rate, so for future studies it is recommended that a much larger data on brain tumor be made available so as to improve and increase the accuracy and precision rate of the model developed after data training.

Acknowledgements The authors acknowledge the motivated research efforts of Olasehinde Emmanuel Olumide (Department of Physics, Federal University of Technology, Minna, Nigeria) as a promising young researcher.

References

Bisong, E.: Building Machine Learning and Deep Learning Models on Google Cloud Platform, pp. 59–64. Apress, Berkeley, CA (2019)

Bressert, E.: SciPy and NumPy: An Overview for Developers. O'Reilly Media, Inc: Sebastopol, United States (2012)

Danial, A.: NumPy and SciPy. In: Python for MATLAB Development, pp. 335–492. Apress, Berkeley, CA (2022)
Deepthi, T.H., Gaayathri, R., Shanthosh, S., Gebin, A.S., Nithya, R.A.: Firearm recognition using convolutional neural network. In: 2018 First International Conference on Secure Cyber Computing and Communication, ICSCCC
GeeksforGeeks: OS module in python with examples. https://www.geeksforgeeks.org/os-module-python-examples/ (2022)
Gujar, M.H., Mhatre, M.P., Ghanate, M.S., Kadam, M.S., Kurle, M.D., Shitole, S.: Python Based Image Processing (2013)
McKinney, W., Team, P.D.: Pandas-powerful python data analysis toolkit. In: Pandas—Powerful Python Data Analysis Toolkit, p. 1625 (2015)
Stančin, I., Jović, A.: An overview and comparison of free Python libraries for data mining and big data analysis. In: 2019 42nd International Convention on Information and Communication Technology, Electronics and Microelectronics (MIPRO), pp. 977–982, IEEE, May 2019
Waskom, M.L.: Seaborn: statistical data visualization. J. Open Source Softw. **6**(60), 3021 (2021)
Wikipedia Contributors: Anaconda (Python distribution). Wikipedia. https://en.wikipedia.org/wiki/Anaconda_(Python_distribution) (19 October 2022a)
Wikipedia Contributors: Matplotlib. Wikipedia. https://en.wikipedia.org/wiki/Matplotlib (6 October 2022b)
Wikipedia Contributors: Scikit-image. https://en.wikipedia.org/wiki/Scikit-image (12 December 2022c)

Chapter 10
Neural Radiance Fields (NeRFs) Technique to Render 3D Reconstruction of Magnetic Resonance Images

Abstract In medical diagnostics, magnetic resonance imaging (MRI) is a vital tool that offers comprehensive understanding of anatomical structures. Nevertheless, the thorough investigation of the intrinsic three-dimensional complexity of MRI data is restricted by conventional 2D visualization techniques. This restriction makes it difficult to interpret and comprehend complex anatomical details accurately, which reduces the possibility of making better clinical decisions. Advanced 3D rendering methods are clearly needed in the field of medical imaging. Current methodologies, although beneficial, display certain limitations. To overcome these constraints, a new approach is needed that can work in unison with MRI data to provide a more engaging and educational visualization. The task at hand involves utilizing Neural Radiance Fields (NeRFs) to improve the rendering of Magnetic Resonance Images. NeRFs have shown great promise in the fields of computer graphics and vision, but there is still much to learn about how to use them in the context of medical imaging, especially MRI data. The main goal of this chapter is to apply NeRFs to establish a strong framework for the three-dimensional rendering of MRI data, which will enhance the precision and depth of medical image interpretation. This will advance medical imaging technology and make a significant contribution to the visualization of anatomical structures with never-before-seen precision and detail. The outcomes would not only hold significance for the field of medical imaging but also have broader implications for enhancing diagnostic capabilities and patient care.

Keywords Neural Radiance Fields · 3D reconstruction · Magnetic resonance imaging · Computer graphics and vision

10.1 Introduction

While the creation of 3D volumetric models can be challenging, MRI provides good 2D anatomical details (Pujol et al. 2016). Due to physical constraints, individual thin slices take several minutes, which may compromise subject comfort and motion artifacts. Surface clarity is diminished by noise, and low resolutions make it difficult

B. O. Awojoyogbe and M. O. Dada, *Digital Molecular Magnetic Resonance Imaging*, Series in BioEngineering, https://doi.org/10.1007/978-981-97-6370-2_10

to visualize small objects. Adjustments are necessary because of gradient distortions. Metal inserts have the ability to shift fields, resulting in voids. Real-time 3D is hampered by long computations (Sengupta et al. 2022). More progress is needed to fully realize the potential of MRI through photorealistic spatial modeling, as sturdy frameworks still fall short of the 3D utilities for CT scans. In computer graphics and computer vision, Neural Radiance Fields (NeRFs) have become a ground-breaking method for representing and rendering three-dimensional scenes. NeRFs employ neural networks to learn a continuous function that directly maps 3D coordinates to scene radiance values, in contrast to conventional methods that rely on explicit geometric models or point-based representations. Because of this, complex scenes can be represented in a more flexible and data-driven manner, which makes NeRFs particularly appealing for applications such as 3D reconstruction and medical image rendering (Dursun 2023). NeRFs' fundamental concept is to represent a scene as a volumetric function in which a radiance value is mapped to each point in three dimensions. A neural network is trained to approximate this volumetric function, taking 3D coordinates as input and outputting the corresponding radiance. The network is trained using a set of images, each with known camera poses, allowing it to capture the intricate details of the scene (Tewari et al. 2022).

With the development of three-dimensional (3D) volumetric visualization, medical imaging modalities such as CT, MRI, and ultrasound have been revolutionized. This is because doctors can now plan treatments and diagnose patients with greater accuracy, thanks to the ability to see the patient's anatomy in unprecedented details (Myszczynska et al. 2020). In contrast to traditional planar images that are 2D and only capture isolated anatomical cross-sections, 3D rendering creates interpretable spatial perspectives that can be explored along any orientation, revealing complex morphological relationships between different structures (Liu et al. 2022). This allows clinicians to accurately pinpoint the spread of pathology and the subtleties that are invisible along individual sequential views (Ahmad et al. 2021). Particularly for magnetic resonance imaging (MRI), 3D perspectives generated via segmentation or shading from multiple coregistered scans can enhance the appreciation of subtle characteristics. MRI is an excellent tool for studying complex structures such as the brain, spinal cord, cartilage, and visceral organs because it does not have any ionizing side effects and has a unique sensitivity to changes in soft tissue (Westen et al. 2019). When compared to conventional radiological readings, quantitative morphometric analysis shows that 3D allows for the early detection of CREDIT (Weinhardt et al. 2018). Enhancements in surgical planning and diagnosis confidence are qualitative and cover a wide range of conditions, including trauma, inflammation, degeneration, and cancer. When compared to CT or X-rays, dynamic 3D visualizations, using methods such as interpolated voxel shifting, allow for non-invasive, radiation-free pinpoint studies of intricate joint movements and muscle activations during motions, going beyond static diagnosis. While maintaining underlying signal intensities for flexible post-hoc analysis, photorealistic depth perception outperforms individual stacks (Cho and Roh 2019). However multiple challenges continue to hinder the ubiquitous adoption of 3D. However, a number of obstacles still stand in the way of the widespread adoption of 3D MRI perspectives (Ching et al. 2018). First

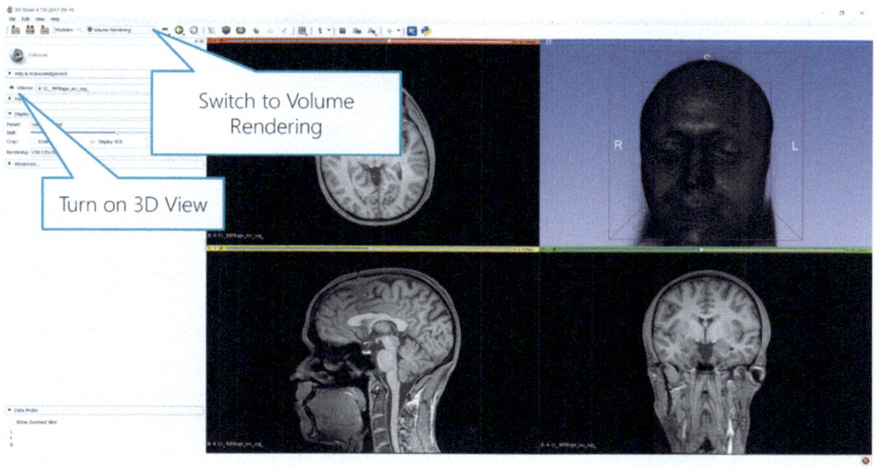

Fig. 10.1 3D rendering of a brain MRI

off, time-constrained staffing is discouraged by related acquisition overhead related to specialized protocols, positional calibration, respiratory compensation, sequence repetition, and post-processing workload. Current algorithms face challenges from signal heterogeneities, motion distortions leading to missing data, metal inserts, and tune-ups customized for scanner specifics (Zhang et al. 2021). Finally, in order to ensure reliability, visual analytics must strike a balance between efficiency and the potential for perceptual gaps and cognitive anchoring biases. This calls for quantitative validations using histological ground truth whenever feasible. Recent advances in deep learning show remarkable success in computer vision for generative 3D modeling based only on 2D supervision, such as images. Specifically, neural radiance calculated using coordinates (Fig. 10.1).

10.2 Magnetic Resonance Imaging (MRI)

Magnetic resonance imaging (MRI) is a non-invasive medical imaging modality that uses non-ionizing radiation to produce functional mappings and detailed 3D anatomical models of the human body (Lalitha et al. 2022). It generates images based on the distribution of protons (hydrogen atoms) in tissues and organs by using strong magnetic fields, magnetic field gradients, and radio waves. The spin of atomic nuclei, such as the single proton of hydrogen, is a quantum mechanical property that is used in MRI. The proton spins align either parallel or anti-parallel in a strong external magnetic field, creating a net tissue magnetization vector. Then, just enough energy is added using carefully timed radiofrequency (RF) pulses to tilt this vector away from equilibrium. During relaxation, protons release characteristic radiofrequency signals that are used to encode the spatial origin.

10.2.1 Application of MRI

Compared to CT scans, magnetic resonance imaging (MRI) offers the superior soft tissue contrast required for neurological conditions, musculoskeletal disorders, analysis of the circulatory system, and oncological studies (Shahzad and Mati 2020). It is widely used in brain imaging, tumors, blood vessel mapping, and other fields for diagnosis, early interventions, and treatment planning. Examining non-bony or soft tissues such as the abdomen, brain, muscles, ligaments, tendons, pelvis, spinal cord, or nerves is one area where MRI is especially helpful (Panzer et al. 2015). By distinguishing between the brain's gray and white matters, magnetic resonance imaging (MRI) enables medical professionals to identify tumors and aneurysms. Furthermore, the medical professional can determine which parts of the brain use more oxygen during various mental functions by performing a functional MRI (Fig. 10.2).

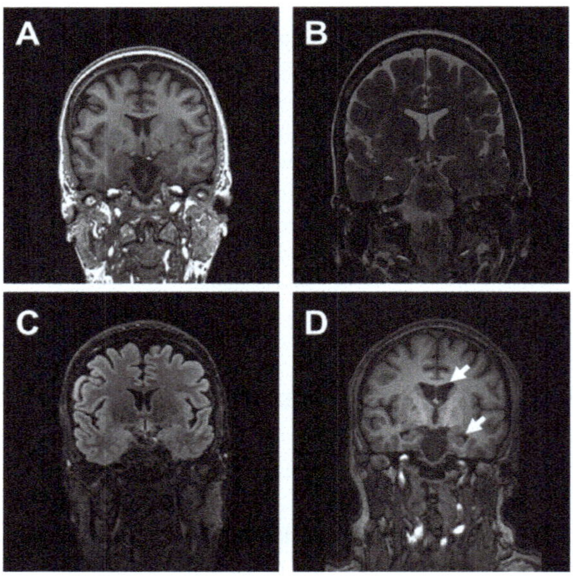

Fig. 10.2 Functional magnetic resonance images

10.3 Virtual Studio Code

One of the reasons why VS Code is considered one of the best development environment tools is its versatility. VS Code supports a wide range of programming languages, including popular languages like Python, Java, and C++ . Additionally, the editor provides a range of features, including syntax highlighting, code completion, and debugging tools, which significantly enhance the development experience. Another reason for its popularity is its active and supportive community, which provides a wealth of extensions and plugins that further extend the editor's functionality. Finally, the fact that VS Code is free and open-source makes it accessible to developers worldwide, making it an ideal choice for individuals and teams of all sizes.

10.3.1 Python Programming Language

Python is a widely used general-purpose, high level programming language. It was created by Guido van Rossum in 1991 and further developed by the Python Software Foundation. It was designed with an emphasis on code readability, and its syntax allows programmers to express their concepts in fewer lines of code. Because of its ease of reading and ease of use, Python is a popular choice among both novices and experts. Because Python's code is simple to read and comprehend, developers can work together and maintain codebases more easily. Furthermore, Python's syntax places a strong emphasis on code clarity, which lowers the possibility of mistakes and facilitates debugging. Python's popularity can also be attributed to its versatility. Numerous fields, such as web development, data analysis, machine learning, and scientific computing, can benefit from the use of Python. Its vast module and package library makes it simple to complete complicated tasks without writing a lot of code.

Another factor in Python's appeal is that it is open-source. The Python community offers a multitude of resources, tutorials, and documentation; it is vibrant and helpful. Further increasing Python's functionality is the plethora of third-party libraries and frameworks that have been created as a result of its open-source nature.

The usage of Python in cutting-edge fields such as artificial intelligence and machine learning has also contributed to its rise in popularity. Python's extensive library of machine learning frameworks and tools has made it the preferred language for data scientists and machine learning engineers. Its simplicity and versatility make it an ideal language for developing machine learning models.

10.3.2 PyTorch

PyTorch is an optimized Deep Learning tensor library based on Python and Torch and is mainly used for applications using GPUs and CPUs. PyTorch stands out as a robust and versatile deep learning framework, widely employed in applications such as image recognition and natural language processing. As a Python-based framework, PyTorch offers a user-friendly experience, attracting machine learning developers due to its readability and ease of use. Its accessibility facilitates a quicker learning curve for practitioners entering the field. One of PyTorch's distinguishing features is its exceptional GPU support, empowering developers to harness the computational power of graphics processing units for accelerated model training and inference. This capability is instrumental in handling the complex computations inherent in deep learning tasks, enhancing efficiency and reducing training times. PyTorch's use of reverse-mode auto-differentiation is a key innovation. This technique enables dynamic computation graphs, allowing models to adapt and modify structures on-the-fly. This dynamic approach is particularly beneficial in scenarios where model architectures may change during runtime, providing flexibility and facilitating experimentation with different network configurations.

In essence, PyTorch's blend of simplicity, GPU acceleration, and dynamic computation graph capabilities positions it as a preferred choice in the deep learning landscape. Its user-centric design and support for rapid experimentation make it an invaluable tool for researchers and practitioners pushing the boundaries of machine learning applications.

10.3.3 How NeRF Works

Although Neural Radiance Fields (NeRF) constitute a recently developed technique for creating new perspectives of a scene from a set of 2D images taken from various angles. This technique creates photorealistic 3D visualizations by using a neural network to represent the radiance, or color and intensity of light, at each point in a 3D space within a scene. Learning a continuous 3D function that maps a scene point's spatial coordinates and viewing direction to its corresponding radiance value is the fundamental concept behind NeRF. The network is able to infer the radiance at each point from various perspectives by using this function, which it learns from a set of training images taken from various angles. NeRF has proven to be more effective than more conventional techniques such as multi-view stereo and voxel-based methods in producing photorealistic 3D visualizations from real-world data. NeRF shows promise, but it has limitations as well. One of these is that it requires a dense set of input views that cover the entire scene from multiple viewpoints. This can be difficult to obtain in the medical domain, where privacy regulations may make data collection more difficult. Furthermore, the high dimensionality of MRI data, low contrast or missing information in the images, and computational costs associated

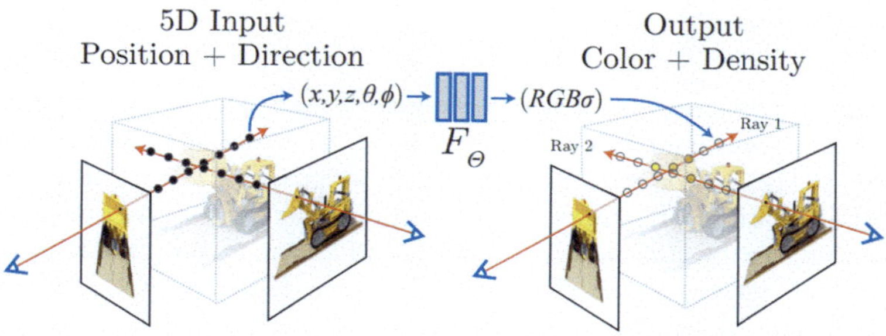

Fig. 10.3 NeRFs process (Mildenhall et al. 2021)

with processing these data present additional challenges for NeRF in the context of MRI scans. In an effort to address NeRF's shortcomings, numerous adaptations have been made to apply it to various applications (Fig. 10.3).

10.3.4 Training NeRFs

The difficulty in training these architectures stems from the fact that the target density and color are unknown. As a result, we require a (differentiable) way to translate them into 2D pictures. After that, these pictures are contrasted with the ground truth photos to create a rendering loss that we can use to optimize the network (Fig. 10.4).

Volume rendering is used to map the neural field output back to a 2D image, as seen in the image above. The input image or pixel can be used to compute the standard L2 loss in an autoencoder manner. It should be noted that volume rendering is a widely used technique in computer graphics. Let's quickly review how it functions.

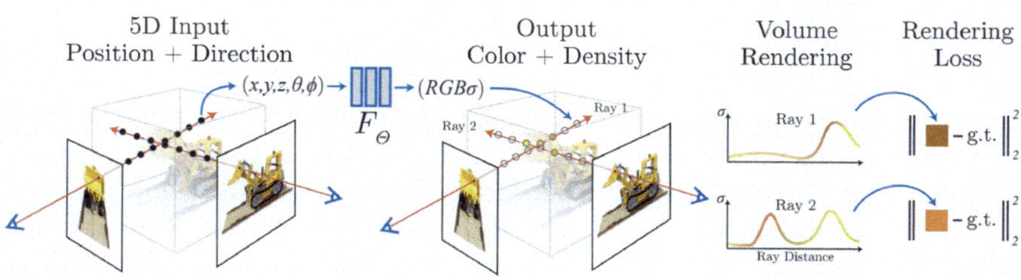

Fig. 10.4 NeRFs training process (Mildenhall et al. 2021)

10.3.5 Volume Rendering

Ray marching is the process by which we emit rays at each pixel and sample at various times teps when obtaining coordinates from the original images. Every sample point has a color, a volume density, and a spatial location. These are the neural field's inputs. The ray is a function of origin *o*, direction *d*, and its samples at time *t*. This can be formulated by $\mathrm{r(t)} = \mathrm{o} + \mathrm{dt}$ both the volume and color are dependent on the ray and are donated according to Mildenhall et al. (2021) as $\sigma(r(t))$ and $\mathrm{c}(r(t))$, respectively. To map them back to the image, all we have to do is integrate these rays and acquire the color of each pixel.

$$\mathrm{C(r)} = \int_{\mathrm{t_n}}^{\mathrm{t_f}} \mathrm{T(t)}\sigma(\mathrm{r(t)})\mathrm{c(r(t), d)dt} \tag{10.1}$$

Given that t_n and t_f are the bound of ray and T(t) its transmittance. The transmittance is a measure of how much the ray can penetrate the 3D space to a certain point and it is defined as

$$\mathrm{T(t)} = \exp\left(-\int_{\mathrm{t_n}}^{\mathrm{t_f}} \sigma(\mathrm{r(s)})\mathrm{ds}\right) \tag{10.2}$$

The aforementioned technique, wherein you generate images or video by tracing a ray and integrating along it is also referred as neural rendering or differentiable rendering in bibliography.

10.3.6 Medical Neural Radiance Fields (MedNeRF)

MedNeRF is an adaptation of the generative radiance fields (GRAF) model for rendering CT projections given a few or even a single-view X-ray in the medical domain. The approach not only synthesizes realistic images but also captures the data manifold and provides a continuous representation of how the attenuation and volumetric depth of anatomical structures vary with the viewpoint without 3D supervision (Iddrisu et al. 2023).

This approach employs GRAF as a backbone architecture to generate reconstructions from slices of MRI data that have undergone slice interpolation and feature extraction via NeRF. GRAF is a generative model that uses a continuous representation of 3D space to generate high-quality images of scenes from any viewpoint. One of the key contributions of GRAF is its hierarchical sampling approach, which enables it to generate high-resolution images efficiently. The model also incorporates

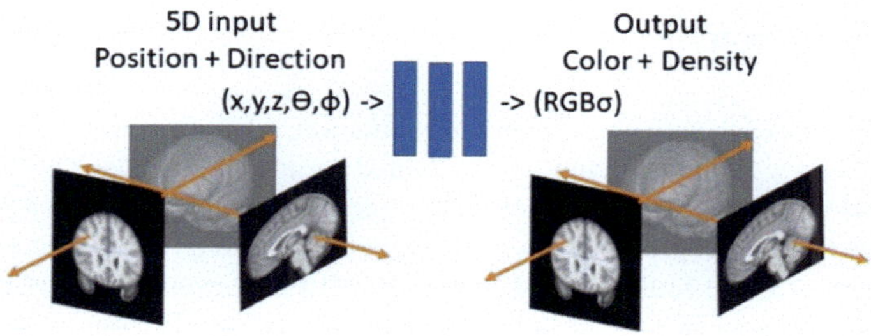

Fig. 10.5 Overview of MedNeRF

a novel attention mechanism that allows it to focus on relevant parts of the scene when generating images (Corona-Figueroa et al. 2022) (Fig. 10.5).

10.3.7 Data Preprocessing

There are a few key reasons why MRI images are often preprocessed before fitting them to Medical Neural Radiance Fields (MedNeRF) models:

i. Registration: Individual MRI scan slices need to be spatially aligned to a consistent coordinate system for proper 3D reconstruction by MedNeRF. Registration corrects for any patient motion between scans using affine/deformable algorithms.
ii. Intensity normalization: MRI intensities are arbitrary depending on scan parameters and can vary across patients or scan sessions. Intensity normalization maps intensities to a standard range (e.g. 0–1) for MedNeRF stability.
iii. Noise reduction: MRI images contain noise from scanner electronics and patient physiology. Reducing noise with filters improves signal quality for MedNeRF training.
iv. Artifact mitigation: Common MRI artifacts such as bias fields, metal-induced distortions negatively impact MedNeRF modeling. Detecting and correcting helps MedNeRF convergence.

Images used for this study are already processed; hence, they can be directly passed into the MedNeRF model.

Comparisons were made between the model's outcomes, two other benchmarks, and a study was conducted on the impact of specific variables. Both qualitative and quantitative assessments are presented, and all models are trained for 100,000 iterations with a batch size of 8. To uniformly sample points on a sphere's surface, we select projection parameters (u, v) that elevate slightly horizontally between 70–85 degrees and have umin = 0, umax = 1 for a complete 360-degree vertical rotation. During training, we only offer a fifth of the views (72-views each at five degrees) and allow the model to render the remaining ones.

10.4 Conclusion

NeRFs offer a flexible and data-driven representation, adapting to the complexity of real-world scenes without relying on predefined geometric models. This flexibility allows NeRFs to generate novel views of complex 3D scenes, based on a partial set of 2D images. NeRFs represent a 3D scene in the weights of a fully connected neural network, known as the Multi-Layer Perception (MLP). They use a rendering loss to reproduce input views of a scene and are trained to map directly from viewing direction and spatial location to opacity and color. NeRFs are computationally intensive and processing of complex scenes can take hours or days. However, new algorithms are available that dramatically improve performance.

NeRFs excel at capturing fine details, making them suitable for applications where high-resolution reconstructions are essential, such as medical imaging.

Neural Radiance Fields (NeRFs) show a stunning potential to improve realism in 3D renders by capturing minute details and complex scene properties. Neural network-based models (NeRFs) are able to simulate intricate lighting conditions, non-Lambertian surfaces, and subtle view-dependent effects because of the continuous volumetric representation that they have learnt (Gao et al. 2022). One reason for the network's superiority in achieving photorealism is its ability to synthesize high-fidelity images with accurate global context awareness (Mildenhall et al. 2021). This realism is extremely helpful in medical imaging applications because it makes anatomical structures accurate and realistic, which improves diagnostic interpretation and analysis.

One of the main drawbacks of Neural Radiance Fields (NeRFs) is how much computing power they require to render. The ray-marching method used to create synthetic images requires a lot of processing power, which makes real-time applications difficult (Mildenhall et al. 2021). This computational overhead is caused by the complexity of the volumetric representation and the requirement to query the neural network along rays for every pixel (Tewari et al. 2022). NeRFs are excellent at capturing the intricacies of scenes, but because of their computational requirements, they are not as suitable for situations that call for quick and dynamic rendering, which restricts their usefulness in time-sensitive fields such as virtual environments or interactive simulations (Mildenhall et al. 2021).

Another drawback of Neural Radiance Fields (NeRFs) is their high data requirements for training and generalization. To reach peak performance, you need a large dataset with a variety of scenes and viewpoints (Mildenhall et al. 2021). Overfitting caused by insufficient data can prevent NeRFs from generalizing to new situations. Furthermore, the richness of the training data has a significant impact on the neural network's ability to model complex scene details (Tewari et al. 2022). Acquiring such extensive datasets in practice can be difficult, particularly in fields with little or specialized data. This could prevent NeRFs from being widely used in situations where a sufficient amount of training data is not easily accessible (Martin-Brualla et al. 2021).

Acknowledgements The authors acknowledge the motivated research efforts of Abel Temitayo Adejorin (Department of Physics, Federal University of Technology, Minna, Nigeria) as a promising young researcher.

References

Ahmad, Z., Rahim, S., Zubair, M., Abdul-Ghafar, J.: Artificial intelligence (AI) in medicine, current applications and future role with special emphasis on its potential and promise in pathology: present and future impact, obstacles including costs and acceptance among pathologists, practical and philosophical considerations. A comprehensive review. Diagn. Pathol. **16**, 1–16 (2021)

Ching, T., Himmelstein, D.S., Beaulieu-Jones, B.K., Kalinin, A.A., Do, B.T., Way, G.P., Ferrero, E., Agapow, P.M., Zietz, M., Hoffman, M.M., Greene, C.S.: Opportunities and obstacles for deep learning in biology and medicine. J. R. Soc. Interface **15**(141), 20170387 (2018)

Cho, S.Y., Roh, H.T.: Taekwondo enhances cognitive function as a result of increased neurotrophic growth factors in elderly women. Int. J. Environ. Res. Public Health **16**(6), 962 (2019)

Corona-Figueroa, A., Frawley, J., Bond-Taylor, S., Bethapudi, S., Shum, H. P., Willcocks, C.G.: Mednerf: medical neural radiance fields for reconstructing 3d-aware CT-projections from a single x-ray. In: 2022 44th Annual International Conference of the IEEE Engineering in Medicine & Biology Society (EMBC), pp. 3843–3848. IEEE, July 2022

Dursun, İ.: Creating digital twins from computer drawings and laser scans: a comparison. In: Proceedings of the Institution of Civil Engineers-Civil Engineering, vol. 176, no. 4, pp. 164–176. Emerald Publishing Limited (2023)

Gao, K., Gao, Y., He, H., Lu, D., Xu, L., Li, J.: Nerf: Neural radiance field in 3d vision, a comprehensive review (2022). arXiv preprint arXiv:2210.00379

Iddrisu, K., Malec, S., Crimi, A.: 3D reconstructions of brain from MRI scans using neural radiance fields. In: International Conference on Artificial Intelligence and Soft Computing, pp. 207–218. Springer Nature Switzerland, Cham (2023)

Lalitha, S., Sanjana, T., Bhavana, H.T., Bhan, I., Harshith, G.: Medical imaging modalities and different image processing techniques: state of the art review. Disruptive Dev. Biomed. Appl., 17–36 (2022)

Liu, W., Liu, Z., Paull, L., Weller, A., Schölkopf, B.: Structural causal 3d reconstruction. In: European Conference on Computer Vision, pp. 140–159. Springer Nature Switzerland, Cham (2022)

Martin-Brualla, R., Radwan, N., Sajjadi, M.S., Barron, J.T., Dosovitskiy, A., Duckworth, D.: Nerf in the wild: neural radiance fields for unconstrained photo collections. In: Proceedings of the IEEE/CVF Conference on Computer Vision and Pattern Recognition, pp. 7210–7219 (2021)

Mildenhall, B., Srinivasan, P.P., Tancik, M., Barron, J.T., Ramamoorthi, R., Ng, R.: Nerf: representing scenes as neural radiance fields for view synthesis. Commun. ACM **65**(1), 99–106 (2021)

Myszczynska, M.A., Ojamies, P.N., Lacoste, A.M., Neil, D., Saffari, A., Mead, R., Hautbergue, G.M., Holbrook, J.D., Ferraiuolo, L.: Applications of machine learning to diagnosis and treatment of neurodegenerative diseases. Nat. Rev. Neurol. **16**(8), 440–456 (2020)

Panzer, S., Mc Coy, M.R., Hitzl, W., Piombino-Mascali, D., Jankauskas, R., Zink, A.R., Augat, P.: Checklist and scoring system for the assessment of soft tissue preservation in CT examinations of human mummies. PLoS One **10**(8), e0133364 (2015)

Pujol, S., Baldwin, M., Nassiri, J., Kikinis, R., Shaffer, K.: Using 3D modeling techniques to enhance teaching of difficult anatomical concepts. Acad. Radiol. **23**(4), 507–516 (2016)

Sengupta, S., Yan, X., Hoyt, T.L., Drake, G., Gunderman, A., Chen, Y.: Minimal artifact actively shimmed metallic needles in MRI. Magn. Reson. Med. **87**(1), 541–550 (2022)

Shahzad, K., Mati, W.: Advances in magnetic resonance imaging (MRI). In: Advances in Medical and Surgical Engineering, pp. 121–142. Academic Press (2020)
Tewari, A., Thies, J., Mildenhall, B., Srinivasan, P., Tretschk, E., Yifan, W., Lassner, C., Sitzmann, V., Martin-Brualla, R., Lombardi, S., Golyanik, V.: Advances in neural rendering. In: Computer Graphics Forum, vol. 41, no. 2, pp. 703–735 (2022)
Weinhardt, V., Shkarin, R., Wernet, T., Wittbrodt, J., Baumbach, T., Loosli, F.: Quantitative morphometric analysis of adult teleost fish by X-ray computed tomography. Sci. Rep. **8**(1), 16531 (2018)
Westen, S.C., Warnick, J.L., Albanese-O'Neill, A., Schatz, D.A., Haller, M.J., Entessari, M., Janicke, D.M.: Objectively measured adherence in adolescents with type 1 diabetes on multiple daily injections and insulin pump therapy. J. Pediatr. Psychol. **44**(1), 21–31 (2019)
Zhang, X., Wang, S., Li, W., Lu, X.: Heterogeneous sensors-based feature optimisation and deep learning for tool wear prediction. Int. J. Adv. Manuf. Technol. **114**, 2651–2675 (2021)

Chapter 11
Computational Molecular Magnetic Resonance Fingerprinting in Digital Health

Abstract In this study, we present mathematical and computational concepts (generally applicable to the analysis of biological and non-biological systems) at the molecular level with special application to digital health. Digital health design can be likened to complex systems and are often dominated by large numbers of processes. For example, when deviations occur in normal human system, human disease conditions are produced. Understanding these processes is important not just in unravelling the causes of diseases, but also the manner of disease propagation and the best plan for treatment. The inadequate understanding of the molecular dynamics of diseases is one reason why many diseases remain incurable and become life-threatening. Computational molecular magnetic resonance imaging now provides new ways of visualizing molecular dynamics and the management of human diseases. Currently, the available imaging equipment no longer matches the increasing number of patients requiring healthcare. The few available imaging machines are costly to maintain while financial difficulties are making acquisition of new ones near impossible. Obviously, experimental methods alone are no longer enough for efficient diagnosis, therapy and prognosis. These challenges may now require the development of appropriate mathematical tools and sophisticated computer simulations based on the Bloch NMR flow equations to complement laboratory and clinical observations. Such mathematical techniques have the potential to provide insight into the imaging of molecular interactions through the analysis of relaxation processes as observed in computational molecular magnetic resonance. The aim of this study is to develop computational magnetic resonance fingerprinting (MRF) methods based on Bloch NMR flow equations and artificial intelligence for differentiating intra-axial brain tumors. We calculate simultaneous measurement of multiple tissue properties in terms of T1 and T2 relaxation times for differentiating intra-axial brain tumors. Machine learning models for contralateral white matter, peritumoral white matter and solid tumors dataset are developed with graphic user interface (GUI) and shiny package for model implementation.

Keywords Computational magnetic resonance fingerprinting · Bloch NMR flow equations · Artificial intelligence · Molecular dynamics · Relaxation parameters

B. O. Awojoyogbe and M. O. Dada, *Digital Molecular Magnetic Resonance Imaging*, Series in BioEngineering, https://doi.org/10.1007/978-981-97-6370-2_11

11.1 Introduction

Magnetic resonance fingerprinting (MRF) is a relatively new but increasingly used concept for fast and multi parametric quantitative magnetic resonance imaging (MRI). Acquisitions of MRF produce unique magnetization evolutions per voxel, called fingerprints, due to temporal varying magnetic resonance (MR) sequence schedules by comparing each fingerprint to a dictionary of simulated fingerprints with known MR parameters (such as relaxation times).

In order to get beyond the limits of dictionary matching, neural networks (NNs) have been developed. It is possible to construct the dictionary matching issue as a regression from the fingerprints to the MR parameters. Several techniques have been used on MRF, with impressive outcomes in terms of reconstruction speed and accuracy (Christen et al. 2014; Chen et al. 2016, 2019; Boux et al. 2021; Buonincontri and Sawiak 2016; Badve et al. 2015, 2017). By considering NN-based MRF reconstruction as an inverse problem, we may learn both the forward and backward processes from MR parameters to fingerprints. The forward process's knowledge that is already available may improve in the separation of MR physical processes and subsequently enhance the estimation of MR parameters for the backward process. In order to achieve this, we use invertible neural networks (INNs) (Dinh et al. 2017). T1, T2, B0, and proton density mapping were the major MRF options in the beginning. New research, however, has demonstrated that it is possible to measure associated B1+, T2*, perfusion, and hemodynamic features. There are numerous alternatives available for each step in the acquisition and reconstruction process due to the rather open MRF framework (Ropella-Panagis et al. 2019; Cohen et al. 2018; Cruz et al. 2019; Cloos et al. 2019).

This study provides and creates machine learning models in which diagnosis can be done locally without undergoing full MRI scans. These processes are based on short time acquisition. This helps to check patient's treatment performance and in follow-up studies. A new opportunity to speed up image collection through the rejection of spatial under sampling errors is presented by the combination of temporal and spatial incoherence that is possible in MRF.

Machine learning algorithms were designed and trained using dataset of 900 MRF sample points comprising of T1 and T2 relaxation times, T_0 relaxation rates and transverse magnetisation (M_y). The parameters are based on MRF data processing performed at MR scanner based on fingerprinting computation in another study. The time independent NMR Bloch flow equation is used to simulate the acquisition parameters from the used sequence and all possible T1 and T2 are combined. We calculate simultaneous measurement of multiple tissue parameters of T1 and T2 relaxation times and assess the utility of MR fingerprinting in differentiating common types of adult intra-axial brain tumors. This enabled us to evaluate the diagnostic yield of MRI dataset computation of transverse magnetization (M_y) from the solutions of NMR flow equation for tumors in the brain regions of interest.

Bloch equations are a set of macroscopic equations that are used to calculate the nuclear magnetization $M = (M_x, M_y, M_z)$ as a function of time when relaxation times T_1 and T_2 are present. This presentation provides the mathematical model using Time-independent Bloch NMR Flow Equation as derived by Awojoyogbe (2002, 2003, 2004):

$$v^2\frac{\partial^2 M_y}{\partial x^2} + 2v\frac{\partial^2 M_y}{\partial x \partial t} + T_o\frac{\partial M_y}{\partial t} + vT_o\frac{\partial M_y}{\partial x} + \frac{\partial^2 M_y}{\partial t^2} + (T_g + \gamma^2 B_1^2)M_y = F_o\gamma B_1, \tag{11.1}$$

where $T_o = \frac{1}{T_1} + \frac{1}{T_2}$, $T_g = \frac{1}{T_1 T_2}$ and $F_o = \frac{M_o}{T_1}$.

Equation (11.1) is a general second order differential equation with constant coefficient derived from the Bloch NMR equation and can be applied to any fluid flow problem. At any given time t, we can obtain information about the system, provided that appropriate boundary conditions are applied. From Eq. (11.1), we have obtained the wave equation and solved them in terms of NMR parameters by the application of appropriate initial or boundary conditions. Hence, we could get very important information about the dynamics of any system. It should be noted, however, that the term $F_o\gamma B_1(x, t)$ is the forcing function or disturbance $\left(F_o = \frac{M_o}{T_1}\right)$ where M_o is the equilibrium magnetization. If the function is zero, there is no disturbance and we have a freely vibrating system; else, the system is undergoing a forced vibration.

Equation (11.1) is applicable under the following conditions: (i) the selected slice must be homogeneous with one compartment (either of tissue or of fluid compartment), (ii) rate of fluid influx and efflux should be very small and equal (such as in some slow blood flow junctions) or static fluid (e.g. edema), (iii) the size of the chemical substance to be investigated must be significantly very small compared to the tissue compartment in which they are found, (iv) the geometry of the compartment should be defined and could be approximated to common geometrical system. Assuming that the excitation coil is large enough to cover the slice under investigation, the amount of radiofrequency (RF) power received by any of the fluid particles should be the same, irrespective of the position of the particle spin. From the Bloch NMR flow equations, the equation of motion for spins moving with a variable velocity v(x) is given as (Awojoyogbe et al. 2011):

$$v^2\frac{d^2 M_y}{dx^2} + vT_0\frac{dM_y}{dx} + \left(T_g + \gamma^2 B_1^2(x)\right)M_y = \frac{M_0}{T_1}\gamma B_1(x). \tag{11.2}$$

Assuming we apply RF pulse with maximum magnitude, $M_0 = 0$, Eq. (11.2) becomes:

$$v^2\frac{d^2 M_y}{dx^2} + v\left(T_0 + \frac{dv}{dx}\right)\frac{dM_y}{dx} + \left(\gamma^2 B_1^2 + T_g\right)M_y = 0 \tag{11.3}$$

Let $v(x) = \frac{x}{\delta}$

$$\gamma B_1 = \alpha\gamma G x$$

$$\frac{dv}{dx} = \frac{1}{\delta}$$

Then, Eq. (11.3) becomes:

$$\frac{x^2}{\delta^2}\frac{d^2M_y}{dx^2} + \frac{x}{\delta}\left(T_0 + \frac{1}{\delta}\right)\frac{dM_y}{dx} + \left(\alpha^2\gamma^2B_1^2 + T_g\right)M_y = 0 \tag{11.4}$$

Multiply Eq. (11.4) by δ^2 gives:

$$x^2\frac{d^2M_y}{dx^2} + x(\delta T_0 + 1)\frac{dM_y}{dx} + \left(\alpha^2\gamma^2G^2x^2\delta^2 + T_g\delta^2\right)M_y = 0 \tag{11.5}$$

Equation (11.5) is similar to Bessel differential equation with the solution (Dada and Awojoyogbe 2021)

$$M_y(x) = x^{-\frac{\delta T_0}{2}}\left[C_1J_n(\alpha\gamma G\delta x) + C_2Y_n(\alpha\gamma G\delta x)\right] \tag{11.6}$$

where J_n is the Bessel function of the first kind while Y_n is the Bessel function of the second kind (Kreyszig 2011), C_1 and C_2 are constants. Since the MRI signals have to be measurable at all values of x, we need to set $C_2 = 0$ in order to get rid of the influence of Y_n at the origin. Hence, Eq. (11.6) becomes:

$$M_y(x) = C_1x^{-\frac{\delta T_0}{2}}J_n(\alpha\gamma G\delta x) \tag{11.7}$$

where M_y = transverse magnetization, M_0 = equilibrium magnetization, T_1 = spin lattice relaxation time, T_2 = spin spin relaxation time, T_0 = relaxation rate, $B_{1(x)}$ = radio frequency magnetic field, γ = gyromagnetic ratio, v = fluid velocity, G = gradient magnetic field, τ = length of time, α = gradient pulse duration.

$$n = \frac{\delta}{2}\sqrt{T_o^2 - 4T_g} \tag{11.8}$$

11.2 Computation Results of Magnetic Resonance Fingerprinting Dataset

R program was used to simulate and generate randomly consistent clinical magnetic resonance fingerprinting data. The Bloch NMR flow equation was used to prepare the data in Excel, where the value of transverse magnetization (M_y) was calculated and T1 and T2 relaxation times are presented. The Exploratory Data Analysis was performed on the data and different visualisation and processing were also performed.

Tables 11.1, 11.2 and 11.3 show the summary of the first 15 sample data out of 300 for each data points of Contralateral, Peritumoral, White Matter, and Solid Tumor, respectively.

All datasets are saved as a CSV (Comma Separated Value) files. This allows data to be saved in a table structure format. Saving files as CSV offers many advantages such as:

I. CSV files are easy to create and read using text editor.
II. CSV provides a straightforward information schema.
III. CSV is simple to implement and parse. The CSV parser is much simpler compared to other parser due to the inherent simplicity of this format.
IV. CSV is compact. Compared to other formats you do not need to care about tags. One needs to start and end tags separately for each column and row which can be unwieldy.
V. CSV is faster to handle; it is highly readable. It does not require data encoding or conversion to binary before storing. Therefore, it makes it easy to be manipulated and edited.

11.3 Garson Features Result Analysis

Figures 11.1, 11.3 and 11.5 show the plot of linear regression information. The relaxation rate T0, transverse or spin-spin lattice relaxation time T2, and longitudinal or spin lattice relaxation time T1 (I3, I2, I1), respectively, are the input which is connected to the weights (Anitha and Kuldeep 2015) through the bias (B1). The weight is the hidden layer with 5 neurons (H1, H2, H3, H4, H5). The output transverse magnetization My (O1) is obtained which gives the squared of the input values (Figs. 11.2 and 11.4).

The Garson function evaluates relative variable importance using the Garson algorithm. By deconstructing the model weights, this function determines the relative weights of explanatory variables for a single response variable. Finding all weighted connections between the network's layers can help to determine each variable's importance. In other words, all weights identifying the individual input node that passes through the hidden layer and connect to the response variable are identified. This is repeated for all other explanatory variables until a list of all weights that are specific to each input variable is obtained. The connections are calculated for every input node and scaled compared to all other inputs. Each explanatory variable that describes the connection to the response variable in the model is given a single value.

From the chart obtained in Figs. 11.2, 11.4 and 11.6 (from the application of the Garson algorithm to CW, PW and ST data), it is possible to note that, in the decision to give the classification, the relaxation rate T0 of the spins which may be changing from pixel to pixel received by the tranverse magnetisation M_y, has the greater influence as observe in Figs. 11.2 and 11.4, while T1s have greater influence in Fig. 11.6. The results indicate relative importance as the absolute magnitude from 0 to 25, 0 to 2 and 0 to 1, respectively.

Table 11.1 Computed data for contralateral white matter

No	T1 (s)	T2 (s)	T0 (s^{-1})	M_y	Tumor class
1	1.090115	0.060671	17.39961	37.6103	Glioblastoma_ multiforme
2	0.995724	0.066021	16.15108	39.70191	Glioblastoma_ multiforme
3	0.93088	0.063033	16.93901	43.27516	Glioblastoma_ multiforme
4	1.111814	0.080025	13.3956	− 11.9779	Glioblastoma_ multiforme
5	0.973093	0.05965	17.79217	− 27.604	Glioblastoma_ multiforme
6	1.066266	0.07519	14.23746	− 38.6182	Glioblastoma_ multiforme
7	0.700494	0.070096	15.69366	− 37.2333	Glioblastoma_ multiforme
8	0.847611	0.065639	16.41454	14.48856	Glioblastoma_ multiforme
9	0.819698	0.078055	14.03146	42.73802	Glioblastoma_ multiforme
10	0.938144	0.064611	16.54325	− 41.356	Glioblastoma_ multiforme
11	0.977743	0.056146	18.83345	38.08396	Glioblastoma_ multiforme
12	0.924039	0.082271	13.23721	− 38.4999	Glioblastoma_ multiforme
13	1.103389	0.050681	20.63757	5.826529	Glioblastoma_ multiforme
14	0.748177	0.073213	14.99544	− 40.1155	Glioblastoma_ multiforme
15	0.915147	0.06849	15.69335	43.2255	Glioblastoma_ multiforme
16	0.865134	0.060213	17.76363	35.70711	Metastasis
17	0.913623	0.059347	17.94458	38.8516	Metastasis
18	0.945617	0.080242	13.51982	37.55123	Metastasis
19	0.86848	0.068839	15.67808	35.32274	Metastasis
20	0.95536	0.077835	13.89445	− 41.6957	Metastasis
21	0.891882	0.069299	15.55145	− 40.3238	Metastasis
22	0.916804	0.081676	13.33417	31.09605	Metastasis
23	0.935011	0.067971	15.7816	− 2.50618	Metastasis
24	0.896585	0.060525	17.63739	28.11907	Metastasis
25	0.879794	0.077402	14.05615	16.64153	Metastasis
26	0.960299	0.071048	15.11632	22.36072	Metastasis
27	0.945884	0.06593	16.2247	39.67453	Metastasis
28	0.937052	0.079402	13.66126	− 32.4317	Metastasis

(continued)

Table 11.1 (continued)

No	T1 (s)	T2 (s)	T0 (s^{-1})	M_y	Tumor class
29	0.969161	0.075336	14.30577	− 32.7861	Metastasis
30	0.924105	0.078985	13.7427	− 38.0248	Metastasis
31	0.873465	0.048395	21.80823	− 39.0263	Low_grade_glioma
32	0.889918	0.086462	12.68947	− 41.1369	Low_grade_glioma
33	0.875503	0.056746	18.76458	− 43.1375	Low_grade_glioma
34	0.876974	0.093329	11.85505	− 29.1037	Low_grade_glioma
35	0.871754	0.095647	11.60226	− 16.1072	Low_grade_glioma
36	0.878695	0.099906	11.1475	− 2.00081	Low_grade_glioma
37	0.874201	0.059637	17.91215	− 43.4893	Low_grade_glioma
38	0.867827	0.064267	16.71229	41.20721	Low_grade_glioma
39	0.851548	0.071886	15.08515	43.06365	Low_grade_glioma
40	0.86634	0.071525	15.1355	43.96398	Low_grade_glioma
41	0.854813	0.075623	14.39326	− 0.42516	Low_grade_glioma
42	0.870441	0.098768	11.27358	6.266845	Low_grade_glioma
43	0.885119	0.069894	15.43721	27.35311	Low_grade_glioma
44	0.891844	0.068096	15.8064	− 40.6098	Low_grade_glioma
45	0.879305	0.041016	25.51816	8.738514	Low_grade_glioma

11.4 SNNs Plots for CW, PW and ST Features

The plot in Figs. 11.7a, 11.9a and 11.11a show the iterative fit error as a black line and the iterative test error as a red line. Both lines have a strongly decreasing trend, demonstrating that the algorithm quickly converges. Also, Figs. 11.7b, 11.9b and 11.11b show that the optimal fit would yield a line through zero with gradient one. The optimal line is shown in black and a linear fit to the actual data is shown in red. To evaluate the network performance, we have plotted the receiver's operating characteristic, ROC lines approach the upper-left corner; this is considered a better network performance. Figures 11.8, 11.10 and 11.12 show the ROC curves for both phases (training to the left and test to the right).

LHS shows a plot of iterative fit error and iterative test error; RHS shows a plot of regression error for the trained model with CW data.

Figure 11.10b shows a plot of iterative fit error and iterative test error and Fig. 11.10a shows a plot of regression error for the trained model with PW data.

Figure 11.11a shows a plot of iterative fit error and iterative test error; Fig. 11.11b shows a plot of regression error for the trained model with ST data.

Table 11.2 Computed data for peritumoral white matter

No	T1 (s)	T2 (s)	T0 (s^{-1})	M_y	Tumor class
1	1.95782	0.157328	6.866907	20.53131	Glioblastoma_ multiforme
2	1.087621	0.139069	8.110109	9.447232	Glioblastoma_ multiforme
3	1.105886	0.130183	8.585729	35.20434	Glioblastoma_ multiforme
4	1.065562	0.14148	8.006611	− 33.6681	Glioblastoma_ multiforme
5	1.441253	0.091398	11.63498	23.57986	Glioblastoma_ multiforme
6	0.975353	0.16579	7.057002	7.327258	Glioblastoma_ multiforme
7	1.832432	0.125794	8.495246	− 8.70711	Glioblastoma_ multiforme
8	1.647664	0.136184	7.949922	1.8157	Glioblastoma_ multiforme
9	1.597968	0.167485	6.59649	− 18.0093	Glioblastoma_ multiforme
10	1.558549	0.107622	9.933442	11.25942	Glioblastoma_ multiforme
11	1.618673	0.126867	8.500038	30.58406	Glioblastoma_ multiforme
12	1.399708	0.106825	10.07556	− 5.68275	Glioblastoma_ multiforme
13	1.951498	0.131718	8.104384	− 3.45338	Glioblastoma_ multiforme
14	1.175946	0.158724	7.150605	− 12.5154	Glioblastoma_ multiforme
15	1.686708	0.188679	5.892873	10.25356	Glioblastoma_ multiforme
16	1.283651	0.096142	11.18031	35.79543	Metastasis
17	1.525791	0.120333	8.965647	19.70048	Metastasis
18	1.325377	0.108316	9.986763	− 23.6887	Metastasis
19	1.170001	0.123092	8.978694	34.60645	Metastasis
20	1.292351	0.155625	7.19949	− 32.3164	Metastasis
21	1.616199	0.105033	10.13957	8.898739	Metastasis
22	1.657067	0.122168	8.788906	− 18.1174	Metastasis
23	1.232929	0.11184	9.752423	23.96806	Metastasis
24	1.251974	0.163658	6.909037	34.65444	Metastasis
25	1.461003	0.129248	8.421502	26.99807	Metastasis
26	1.557645	0.126875	8.523786	21.26019	Metastasis
27	1.610664	0.146212	7.460232	− 30.5865	Metastasis
28	1.408323	0.093968	11.35197	13.61693	Metastasis

(continued)

Table 11.2 (continued)

No	T1 (s)	T2 (s)	T0 (s^{-1})	M_y	Tumor class
29	1.265277	0.073646	14.36882	27.21015	Metastasis
30	0.982193	0.153055	7.551721	11.41576	Metastasis
31	0.726442	0.095161	11.88511	4.336828	Low_grade_glioma
32	1.284825	0.126101	8.708444	28.94428	Low_grade_glioma
33	1.050974	0.149162	7.655637	− 39.9356	Low_grade_glioma
34	1.110903	0.043387	23.94837	1.916923	Low_grade_glioma
35	1.162287	0.10658	10.24297	− 36.3826	Low_grade_glioma
36	1.122094	0.135992	8.244553	36.08806	Low_grade_glioma
37	1.357948	0.110661	9.77299	− 17.2974	Low_grade_glioma
38	1.097812	0.190992	6.146716	− 34.7017	Low_grade_glioma
39	1.248678	0.054775	19.05746	21.18025	Low_grade_glioma
40	1.0955	0.077173	13.87066	− 7.56953	Low_grade_glioma
41	1.092442	0.121524	9.144241	33.22468	Low_grade_glioma
42	1.138355	0.097604	11.12393	37.22559	Low_grade_glioma
43	1.247518	0.05831	17.95131	5.38942	Low_grade_glioma
44	1.259275	0.109215	9.950374	− 36.625	Low_grade_glioma
45	1.095336	0.141466	7.981815	− 4.98961	Low_grade_glioma

11.5 Perceptron Profiles for CW, PW and ST Features

Here, Figs. 11.13, 11.15 and 11.17 show the distribution of the data. That is, T0 relaxation rate versus T1 relaxation time. This shows how they are distributed according to their tumor classes. In Fig. 11.13, there is a density at 0.85 s on T1 relaxation time and at 0.95 s. In Fig. 11.15, there is a density at 1.23 s on T1 relaxation time and at 1.30 s also in Fig. 11.17, the density lies at 1.75 s on T1 relaxation time. We have picked density and a lot of data points are seen. This means that the majority of diagnosis for each tumor class is expected in this range. On the axis of T_0, from 25 s^{-1}, 30 s^{-1} and 35 s^{-1} upward in Fig. 11.13, 11.15 and 11.17, respectively which show that only low grade glioma has the highest distribution and no other data point (Figs. 11.14 and 11.16).

GBMs and LGGs when T1 and T0 parameters of contralateral white matter and solid tumor were compared with differences in mean T1 remain significant. Also, several PW and ST region parameters were significantly different between LGGs and METs with difference in T0 of PW remaining significant. This shows that none of the other parameters significantly improved the ability of the model to discriminate LGG (Fig. 11.18).

Table 11.3 Computed data for solid tumor

No	T1 (s)	T2 (s)	T0 (s^{-1})	M_y	Tumor class
1	1.496597	0.13091	8.307039	− 15.4174	Glioblastoma_ multiforme
2	1.057555	0.15993	7.198332	33.74429	Glioblastoma_ multiforme
3	2.245018	0.125206	8.432238	25.49128	Glioblastoma_ multiforme
4	1.586293	0.115505	9.288023	29.24119	Glioblastoma_ multiforme
5	1.641383	0.115741	9.24923	− 21.4811	Glioblastoma_ multiforme
6	1.594551	0.100999	10.52821	− 30.517	Glioblastoma_ multiforme
7	1.937502	0.151272	7.126719	18.32855	Glioblastoma_ multiforme
8	0.771935	0.16546	7.33922	41.30292	Glioblastoma_ multiforme
9	2.105779	0.131262	8.093258	− 18.6563	Glioblastoma_ multiforme
10	1.40452	0.115856	9.343428	− 30.7372	Glioblastoma_ multiforme
11	1.699501	0.136062	7.938014	31.41325	Glioblastoma_ multiforme
12	1.745231	0.125432	8.545452	18.46543	Glioblastoma_ multiforme
13	2.113448	0.1419	7.520379	15.64315	Glioblastoma_ multiforme
14	1.565824	0.149414	7.331455	− 19.4247	Glioblastoma_ multiforme
15	1.277051	0.092044	11.64743	− 27.089	Glioblastoma_ multiforme
16	0.62501	0.093029	12.34937	42.70233	Metastasis
17	1.394901	0.091846	11.6047	34.55763	Metastasis
18	1.582887	0.088813	11.89132	− 7.32311	Metastasis
19	1.66952	0.101219	10.47856	22.70536	Metastasis
20	1.465029	0.15052	7.32621	− 20.701	Metastasis
21	1.183856	0.156449	7.236565	11.11296	Metastasis
22	1.227435	0.054991	18.99943	− 37.5846	Metastasis
23	1.147171	0.113844	9.655628	− 38.3488	Metastasis
24	1.243805	0.057582	18.1705	− 36.2899	Metastasis
25	1.121837	0.082024	13.08301	34.85516	Metastasis
26	1.611793	0.109809	9.727184	− 32.1552	Metastasis
27	1.633954	0.10818	9.855852	26.3822	Metastasis
28	0.918282	0.149809	7.764154	9.413875	Metastasis

(continued)

Table 11.3 (continued)

No	T1 (s)	T2 (s)	T0 (s^{-1})	M_y	Tumor class
29	1.610656	0.146411	7.45095	– 30.5528	Metastasis
30	1.521438	0.107596	9.951328	23.90628	Metastasis
31	1.641777	0.171464	6.441228	– 24.7446	Low_grade_glioma
32	1.456629	0.114082	9.452128	16.47751	Low_grade_glioma
33	1.245325	0.169128	6.715699	– 24.0361	Low_grade_glioma
34	1.615103	0.181658	6.124017	– 4.84635	Low_grade_glioma
35	1.495996	0.165061	6.726816	– 8.01755	Low_grade_glioma
36	1.751916	0.16415	6.662781	– 3.63433	Low_grade_glioma
37	1.849041	0.150584	7.181614	– 15.0665	Low_grade_glioma
38	1.629751	0.122325	8.788556	– 20.4054	Low_grade_glioma
39	1.394749	0.252729	4.673781	33.68358	Low_grade_glioma
40	1.659123	0.166665	6.602805	– 31.3111	Low_grade_glioma
41	1.681182	0.222154	5.096194	18.78053	Low_grade_glioma
42	1.995279	0.183737	5.943734	– 28.182	Low_grade_glioma
43	1.503111	0.171376	6.500395	– 3.5888	Low_grade_glioma
44	1.398082	0.140826	7.816233	15.81323	Low_grade_glioma
45	1.729973	0.180484	6.118712	24.94394	Low_grade_glioma

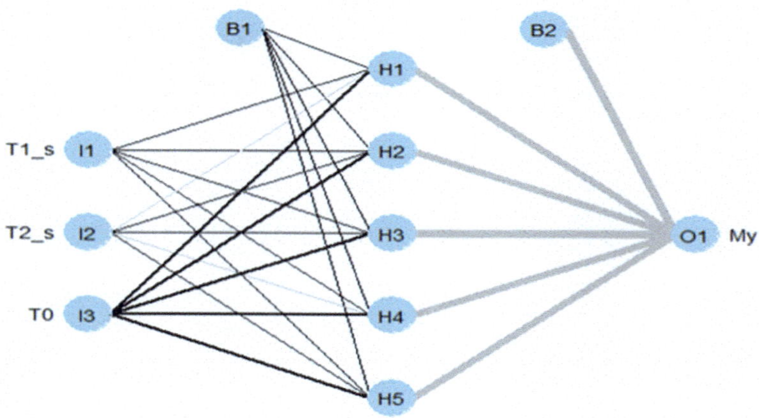

Fig. 11.1 Plot of linear regression for contralateral white matter (CW) features

11.6 Machine Learning Classification

The box and whisker plots in Figs. 11.19, 11.26 and 11.33 show the shape of the box plot with the data being distributed in T1 and T2 relaxation times, T_0 relaxation rate and transverse magnetisation and it also shows the outliers which is used to compare the dataset. The five horizontal lines in the box plot represent the minimum, first

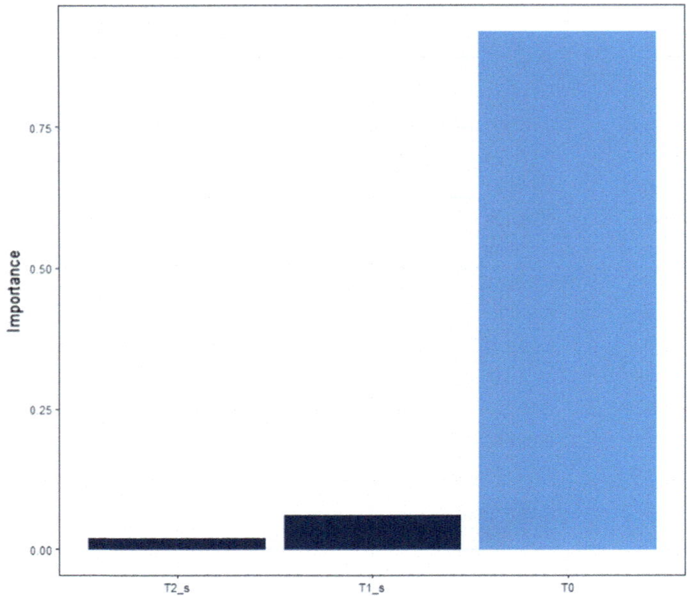

Fig. 11.2 CW chart with Garson model

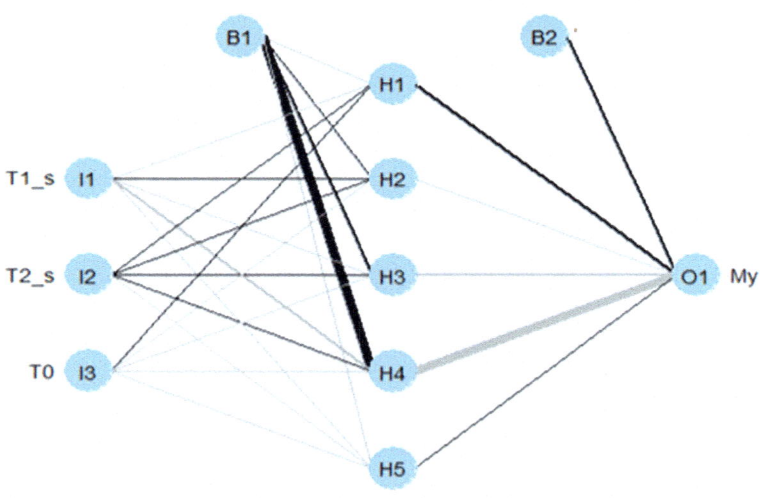

Fig. 11.3 Plot of linear regression for peritumoral white matter (PW) features

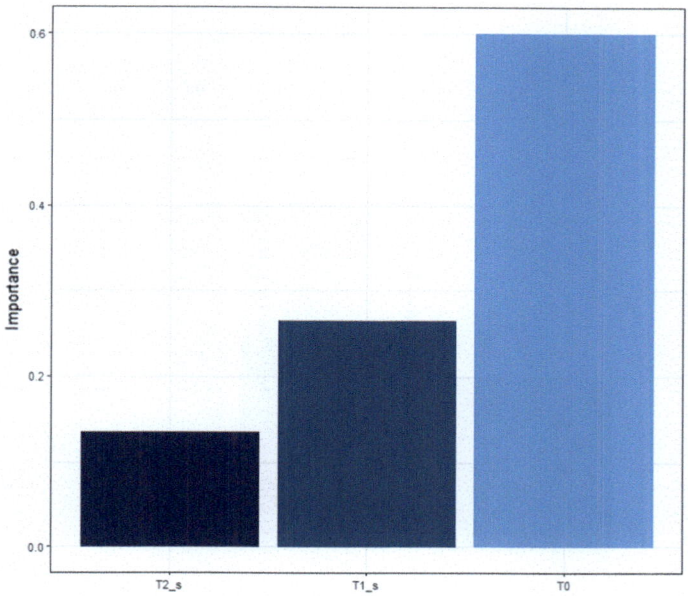

Fig. 11.4 PW chart with Garson model

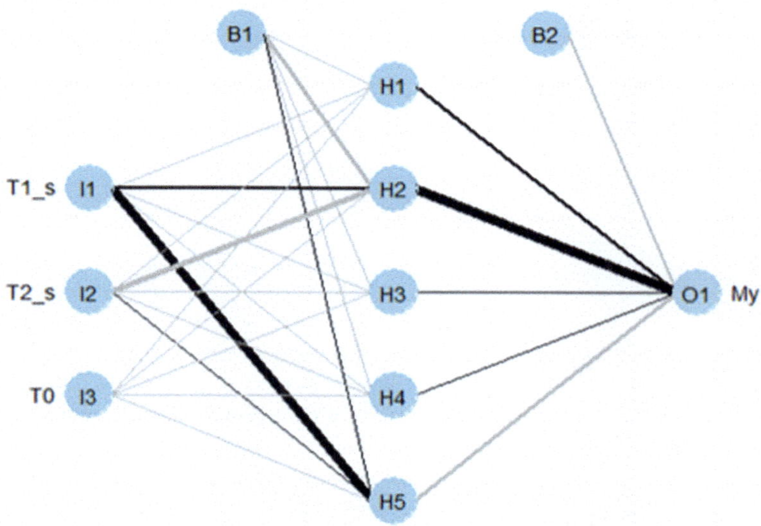

Fig. 11.5 Plot of linear regression with solid tumor (ST) features

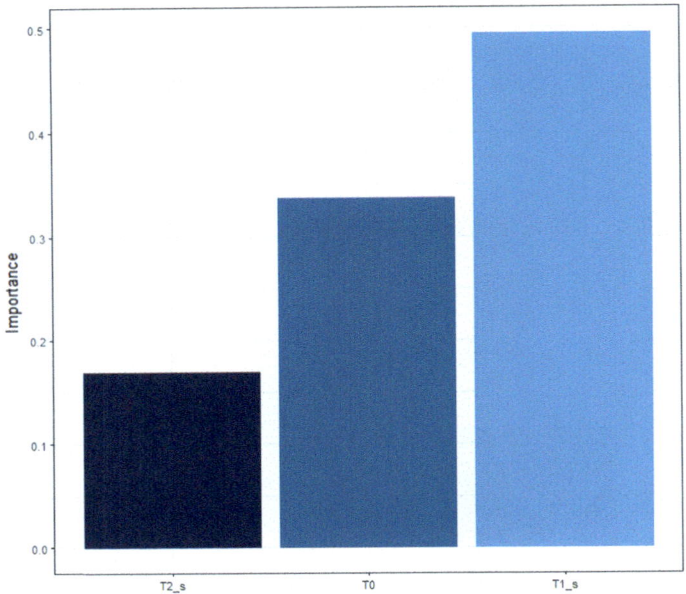

Fig. 11.6 ST chart with Garson model

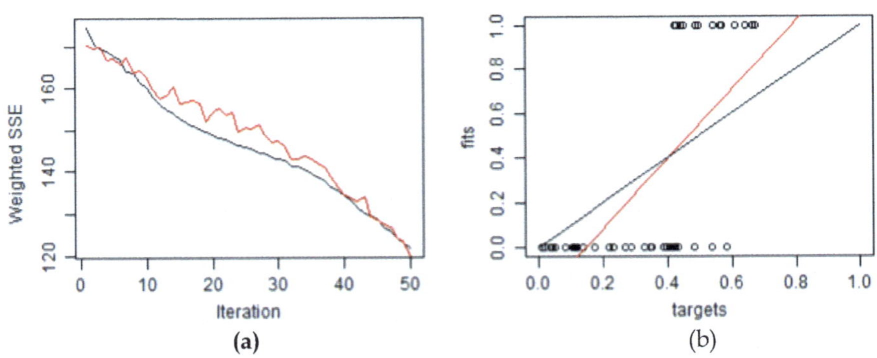

Fig. 11.7 SNNs plot for contralateral white matter (CW) features showing **a** iterative fit error and **b** iterative test error

quratile, median, thrid quartile, and maximum of the dataset. The minimum and maximum are found at the two opposite ends of the line that extends from both ends of the box. The range of the dataset is the length of this line. The box begins at the first quartile and ends at the third quartile, such that the interquartile range (IQR) of the dataset is the length of the box. The median is indicated by a thick line that divides the box in two (Figs. 11.20, 11.21, 11.22, 11.23, 11.24, 11.25, 11.27, 11.28, 11.29, 11.30, 11.31 and 11.32).

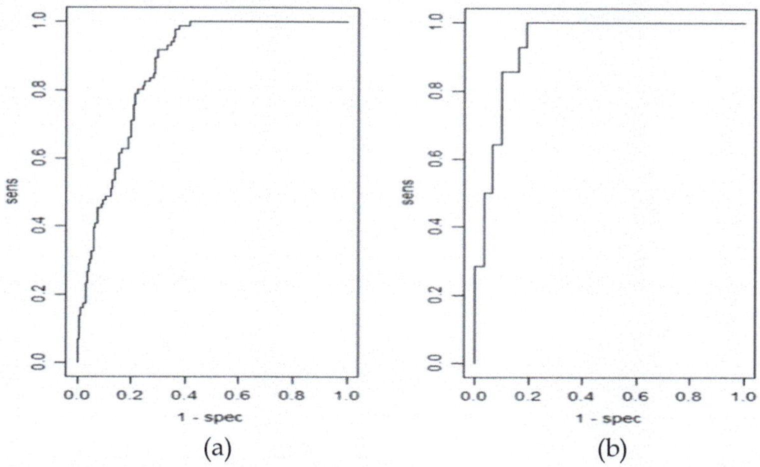

Fig. 11.8 Plot of receiver operating characteristics for **a** training and **b** testing CW data

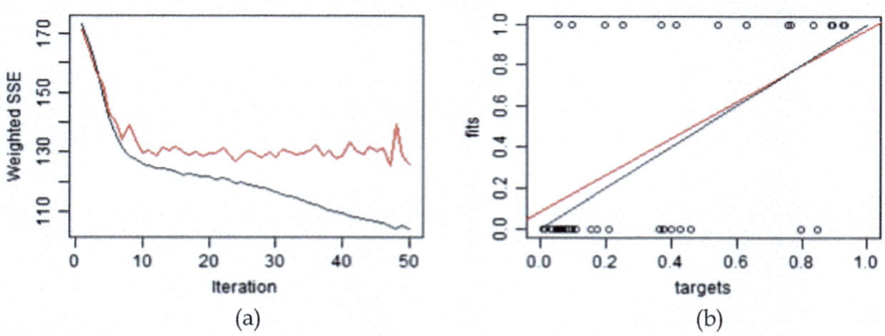

Fig. 11.9 SNNs plot for peritumoral white matter (PW) features showing **a** iterative fit error and **b** iterative test error

Evaluating the range and interquartile range (IQR) of transverse magnetization box plot for each attribute distribtions on one image for CW, PW and ST. CW had a range value of 93.459 and IQR value of 60.401, Pw had a range value of 93.3093 and IQR value of 50.1544 and ST had a range value 84.717 and IQR value of 44.864. The variations of the range and IQR shows that CW portrays the distribution of the dataset having the greatest value among the other dataset. Moreover, Figs. 11.20, 11.27 and 11.34 show the scattered box plots for the datapoint attribute features, highlighting the outliers of the dataset.

Figures 11.21, 11.28 and 11.35 have shown the shape of the box plot and how the data is distributed according to their tumor classes and it also shows the outliers. Considering CW and PW dataset, all T1 relaxation time features (Glioblastoma multiforme, low grade glioma and metastasis) have a constant mean value of 0.1 s. All T2 relaxation times features (Glioblastoma multiforme, low grade glioma and

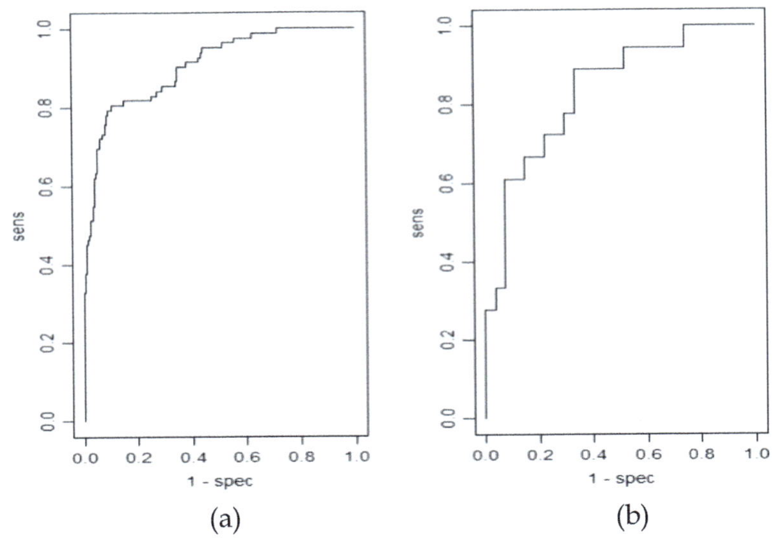

Fig. 11.10 Plot of receiver operating characteristics for **a** training and **b** testing PW data

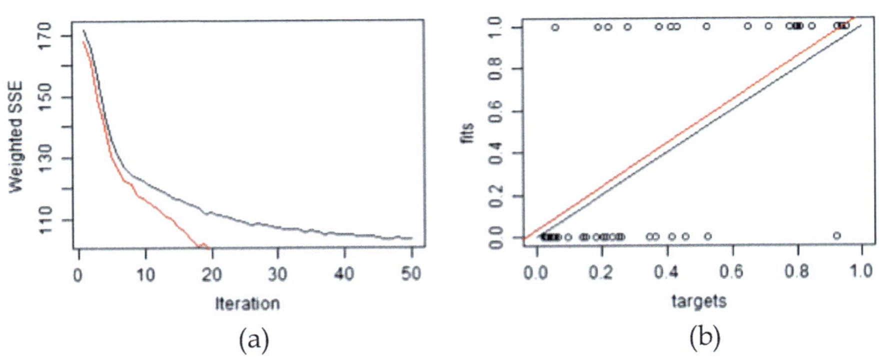

Fig. 11.11 SNNs plot for solid tumor (ST) features showing **a** iterative fit error and **b** iterative test error

metastasis) are symmetrically 0, whereas for T0 relaxation rate in CW and PW low grade, glioma is very distant from the rest of the tumors with outliers. Also, in ST, metastasis has more dispersed data. Comparing the range and interquartile range (IQR) for transverse magnetization in Figs. 11.21 and 11.28 shows that low grade glioma had the greatest of tumor, whereas Figs. 11.35 shows that only metastasis had a higher spread of tumor.

Figures 11.22, 11.29 and 11.36 show the peak of the densities plot and display where values are concentrated over the interval of the continuous variables. In T1 relaxation time, the mean value appears to be greater than the median value which

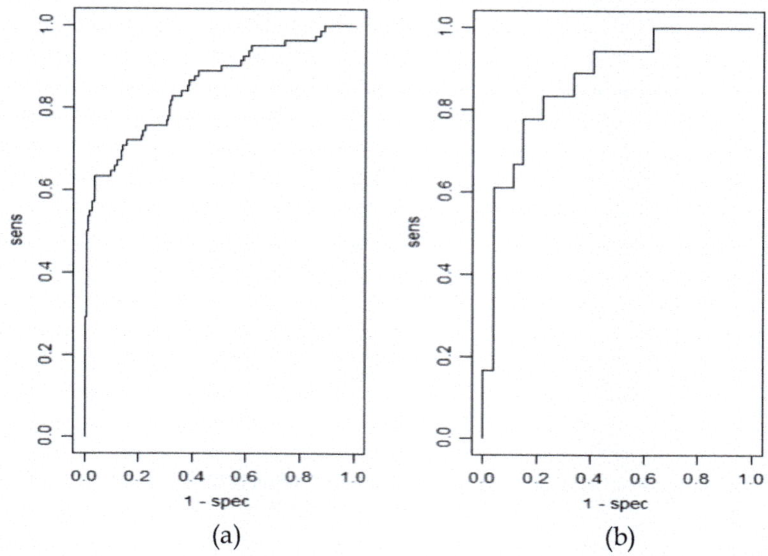

Fig. 11.12 Plots of receiver operating characteristics for **a** training and **b** testing ST data

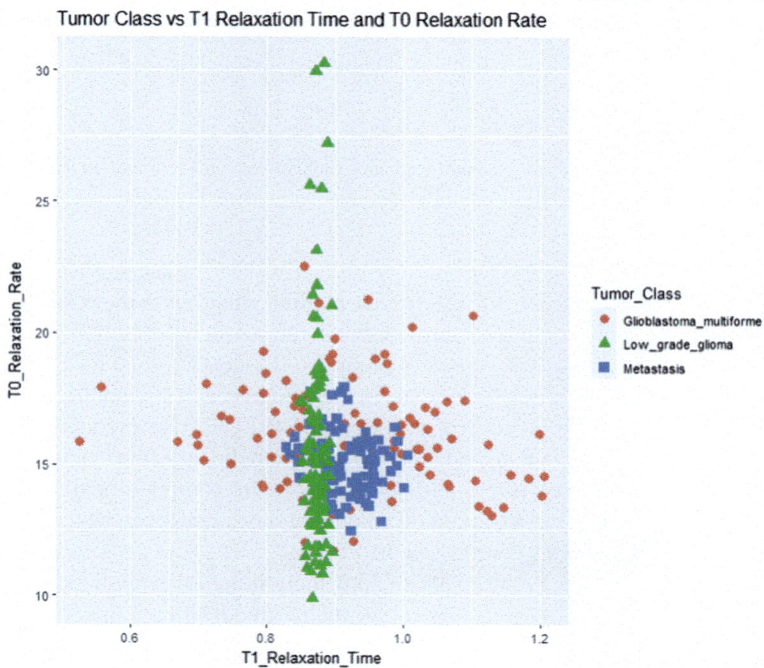

Fig. 11.13 Scatterplot of T1 relaxation time against T_0 relaxation rate for contralateral white matter (CW) according to the tumor class

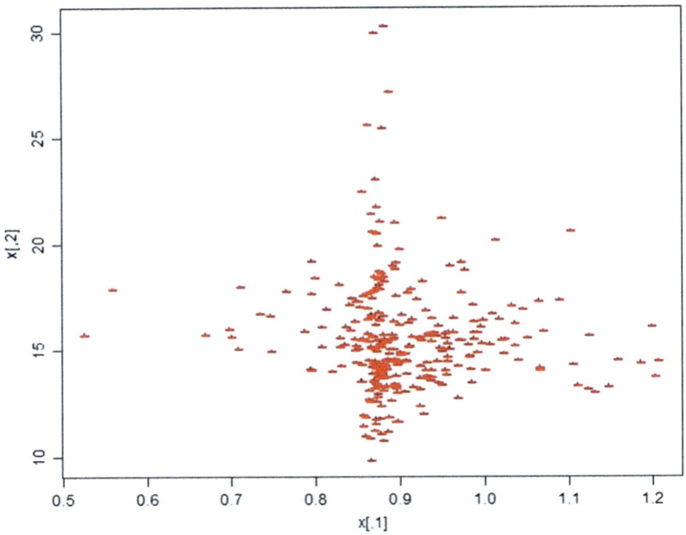

Fig. 11.14 Optimal weights for the perceptron training model for CW

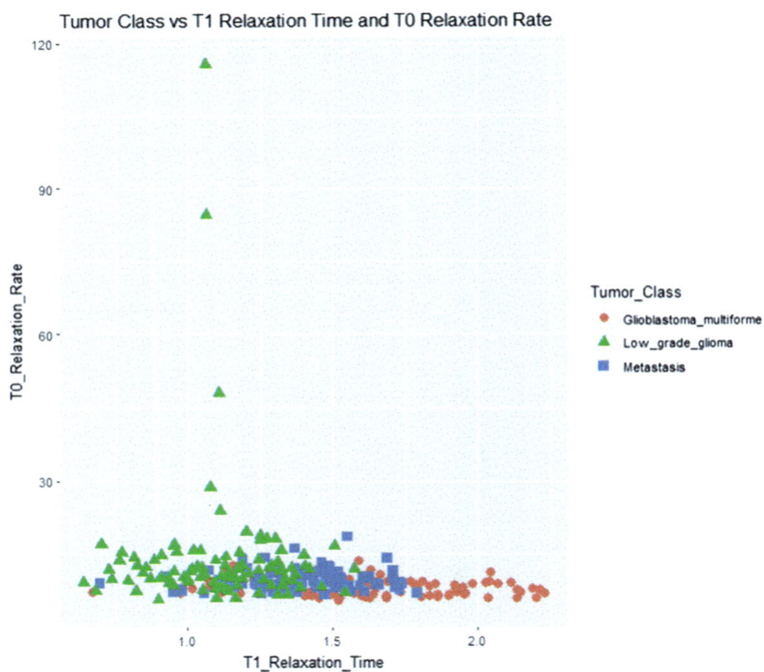

Fig. 11.15 Scatterplot of T1 relaxation time against T_0 relaxation rate for peritumoral white matter (PW) according to the tumor class

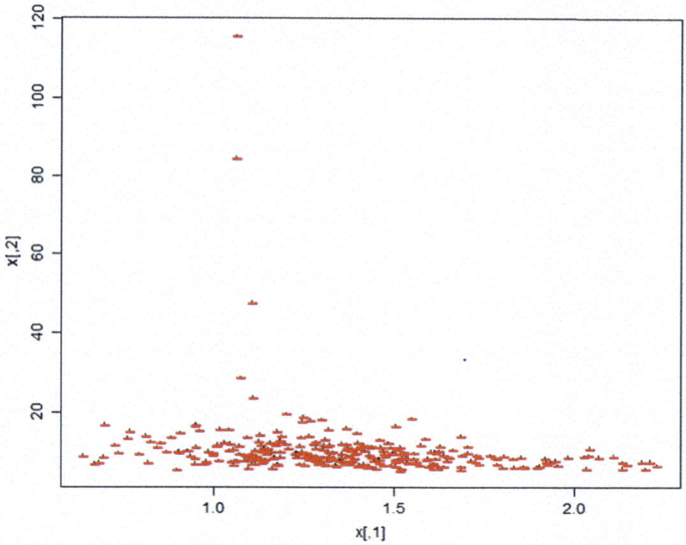

Fig. 11.16 Optimal weights for the perceptron training model for PW

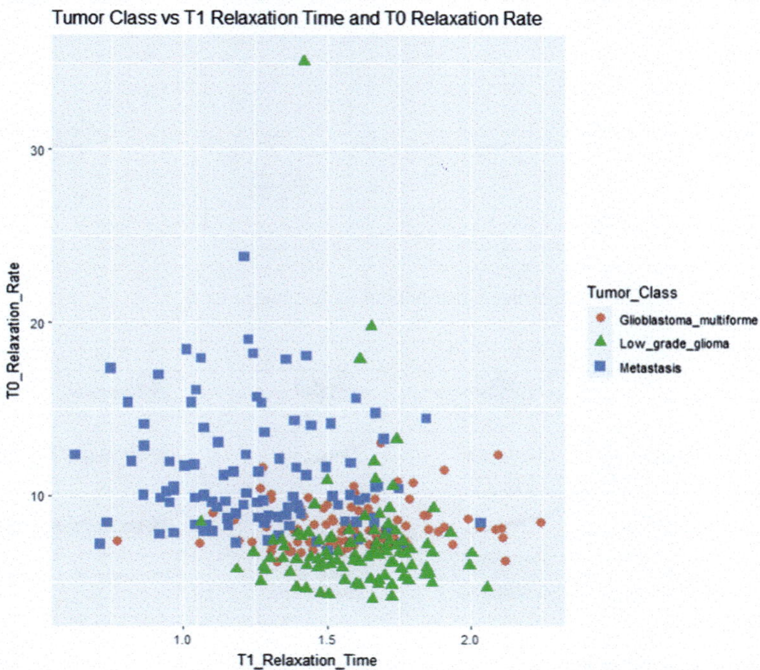

Fig. 11.17 Scatterplot of T1 relaxation time and T_0 relaxation rate for solid tumor (ST) according to tumor class

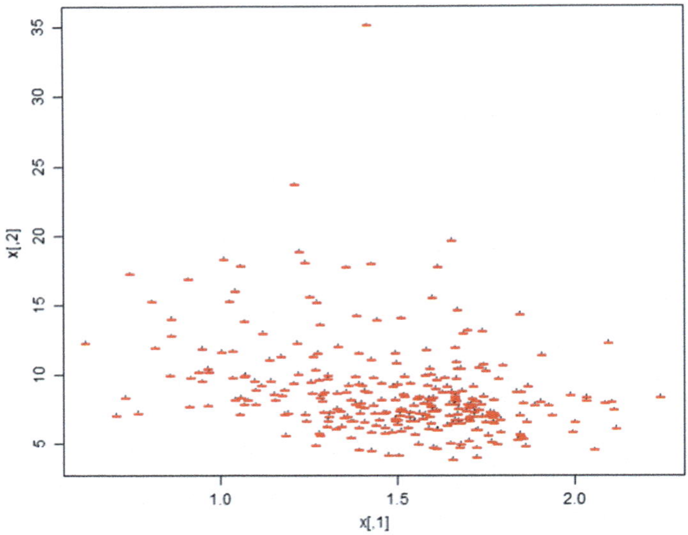

Fig. 11.18 Optimal weights for the perceptron training model for ST

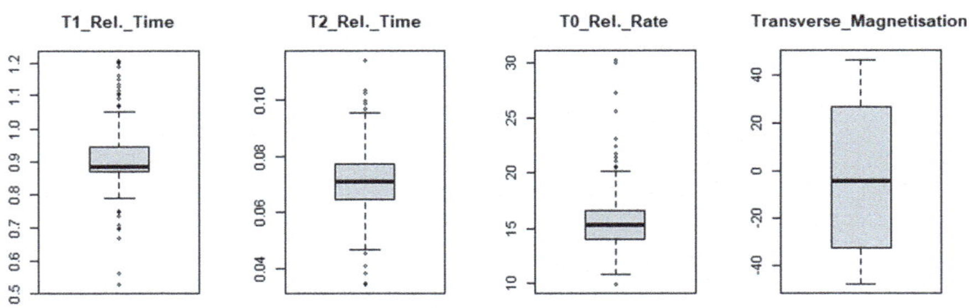

Fig. 11.19 Box plots for each attribute distributions on one image for CW

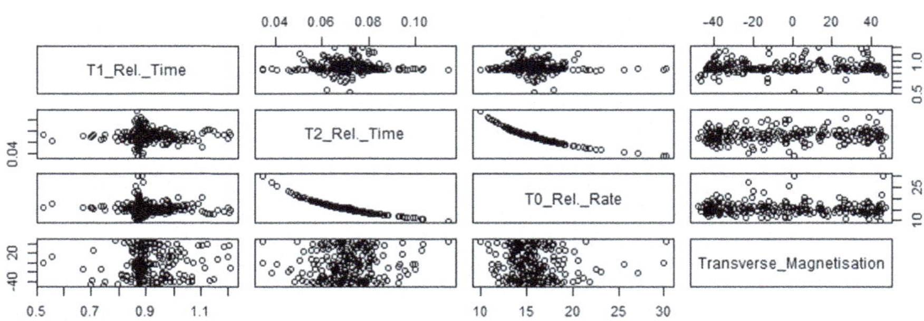

Fig. 11.20 Correlation plots for the datapoint attribute features for CW

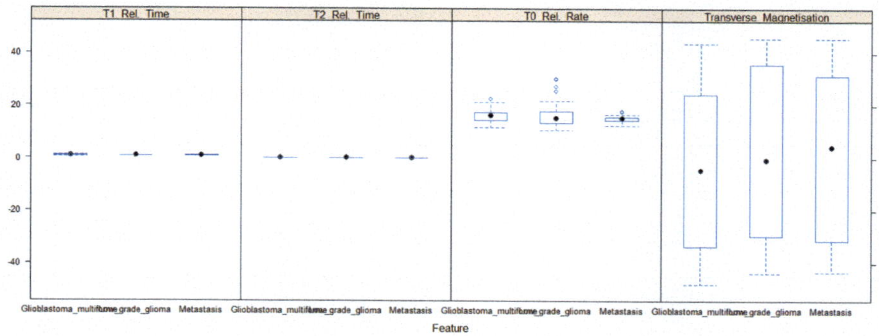

Fig. 11.21 Box and whisker plots for each attribute features for CW

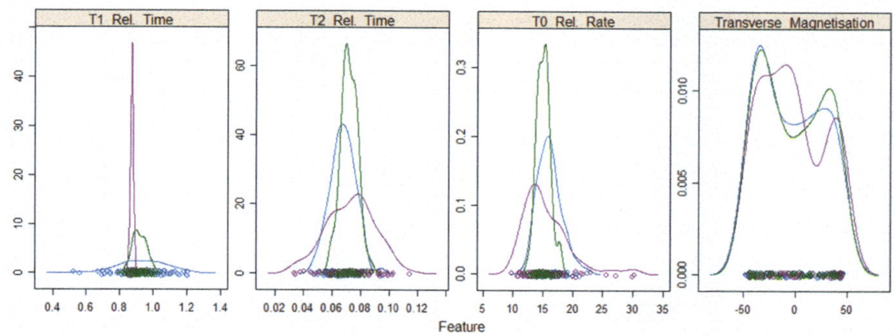

Fig. 11.22 Density plots for each attribute by class value for CW

shows to be right skewed for CW and PW, respectively. While for ST, the mean value is less than the median value which displays to be left skewed. In T2 relaxation time and T0 relaxation rate, the mean values show to be greater than the median value which means the densities plot are right skewed for the dataset (CW, PW and ST), respectively. In transverse magnetization, the mean values are less than the median value which means the densities are left skewed for CW, PW and ST, respectively.

Figures 11.23, 11.30 and 11.37 show nine (9) different techniques for comparing the estimated accuracy and kappa of the constructed models. This shows the creation of table with one algorithm for each row and evaluation matrices for each column having their accuracies and kappa for CW, PW and ST, respectively. Moreover, Figs. 11.24, 11.31 and 11.38 show the dot plot comparison that displays the model's accuracy and kappa (Tables 11.4, 11.5, and 11.6).

Figure 11.23 shows that negative binomial regression (nb) model for contralateral white matter datapoint has the highest and best performance with the accuracy of 79% and that of kappa 68%. In Fig. 11.38, it has shown that support vector machine (svm) model for peritumoral white matter datapoint has the highest and best performance with accuracy of 70% and that for kappa is 55%. Moreover, in Fig. 11.37, it can be

```
> results <- resamples(list(cart=fit.cart, knn=fit.knn, svm=fit.svm, rf=fit.rf, glmnet=fi
t.glmnet, svmR=fit.svmRadial, nb=fit.nb, c50=fit.c50, tree=fit.treebag))
> summary(results)

Call:
summary.resamples(object = results)

Models: cart, knn, svm, rf, glmnet, svmR, nb, c50, tree
Number of resamples: 10

Accuracy
            Min.   1st Qu.    Median      Mean   3rd Qu.      Max. NA's
cart   0.6250000 0.6666667 0.6875000 0.7000000 0.7083333 0.7916667    0
knn    0.2500000 0.3437500 0.4166667 0.4166667 0.4583333 0.5833333    0
svm    0.5000000 0.6250000 0.6875000 0.6791667 0.7395833 0.9166667    0
rf     0.5833333 0.7500000 0.7916667 0.7666667 0.8229167 0.8750000    0
glmnet 0.4583333 0.5416667 0.5833333 0.6041667 0.6562500 0.8333333    0
svmR   0.5000000 0.6250000 0.6875000 0.6791667 0.7395833 0.9166667    0
nb     0.6666667 0.7291667 0.7916667 0.7916667 0.8333333 0.9166667    0
c50    0.6250000 0.7083333 0.7708333 0.7541667 0.7916667 0.8750000    0
tree   0.6250000 0.6770833 0.7083333 0.7333333 0.8125000 0.8750000    0

Kappa
          Min.  1st Qu.  Median    Mean  3rd Qu.   Max. NA's
cart    0.4375 0.500000 0.53125 0.55000 0.562500 0.6875    0
knn    -0.1250 0.015625 0.12500 0.12500 0.187500 0.3750    0
svm     0.2500 0.437500 0.53125 0.51875 0.609375 0.8750    0
rf      0.3750 0.625000 0.68750 0.65000 0.734375 0.8125    0
glmnet  0.1875 0.312500 0.37500 0.40625 0.484375 0.7500    0
svmR    0.2500 0.437500 0.53125 0.51875 0.609375 0.8750    0
nb      0.5000 0.593750 0.68750 0.68750 0.750000 0.8750    0
c50     0.4375 0.562500 0.65625 0.63125 0.687500 0.8125    0
tree    0.4375 0.515625 0.56250 0.60000 0.718750 0.8125    0
```

Fig. 11.23 Machine learning models tested on CW dataset

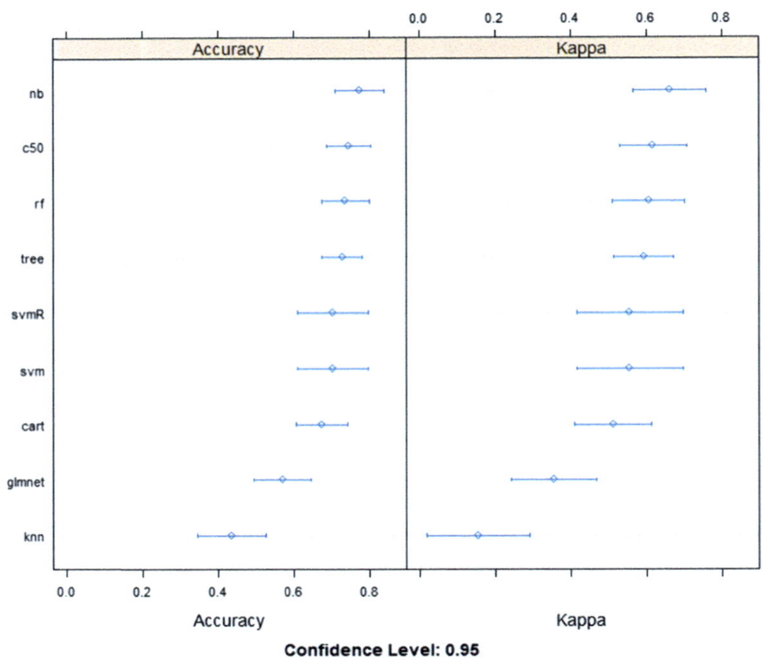

Fig. 11.24 Dot plot comparison of accuracy of models tested on CW dataset

```
> confusionMatrix(predictions, validation$Tumor_Class)
Confusion Matrix and Statistics

                          Reference
Prediction                 Glioblastoma_multiforme Low_grade_glioma Metastasis
  Glioblastoma_multiforme                       15                1          2
  Low_grade_glioma                               1               17          3
  Metastasis                                     4                2         15

Overall Statistics

               Accuracy : 0.7833
                 95% CI : (0.658, 0.8793)
    No Information Rate : 0.3333
    P-Value [Acc > NIR] : 1.152e-12

                  Kappa : 0.675

 Mcnemar's Test P-Value : 0.8335

Statistics by Class:

                     Class: Glioblastoma_multiforme Class: Low_grade_glioma
Sensitivity                                  0.7500                  0.8500
Specificity                                  0.9250                  0.9000
Pos Pred Value                               0.8333                  0.8095
Neg Pred Value                               0.8810                  0.9231
Prevalence                                   0.3333                  0.3333
Detection Rate                               0.2500                  0.2833
Detection Prevalence                         0.3000                  0.3500
Balanced Accuracy                            0.8375                  0.8750
                     Class: Metastasis
Sensitivity                     0.7500
Specificity                     0.8500
Pos Pred Value                  0.7143
Neg Pred Value                  0.8718
Prevalence                      0.3333
Detection Rate                  0.2500
```

Fig. 11.25 Naive Bayes (NB) models performance with CW dataset

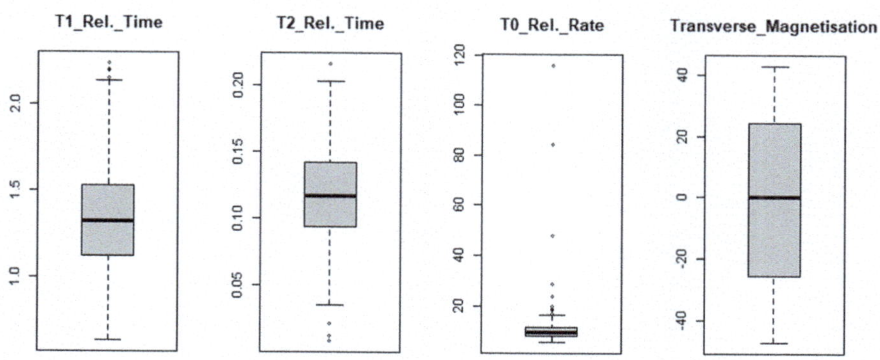

Fig. 11.26 Shows box plot for each attribute distribtions on one image for PW

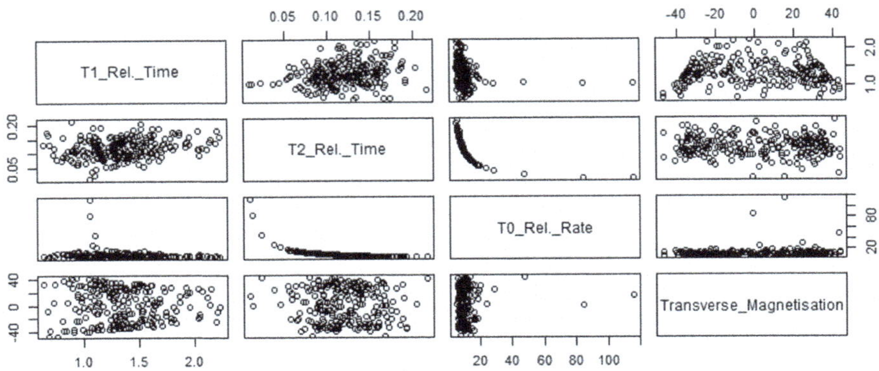

Fig. 11.27 Correlation plots for the datapoint attribute features for PW

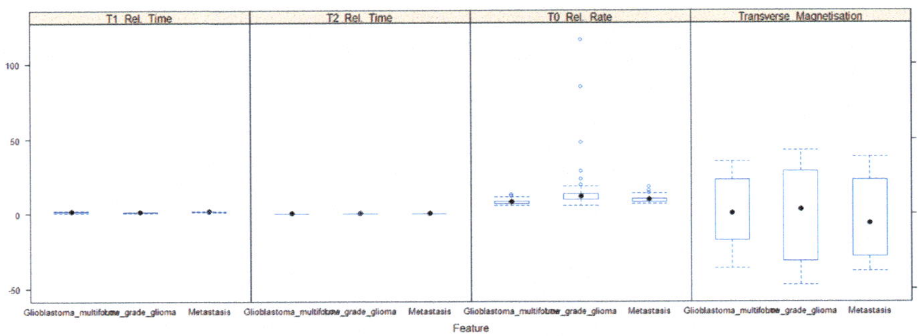

Fig. 11.28 Box and whisker plots for each attribute features for PW

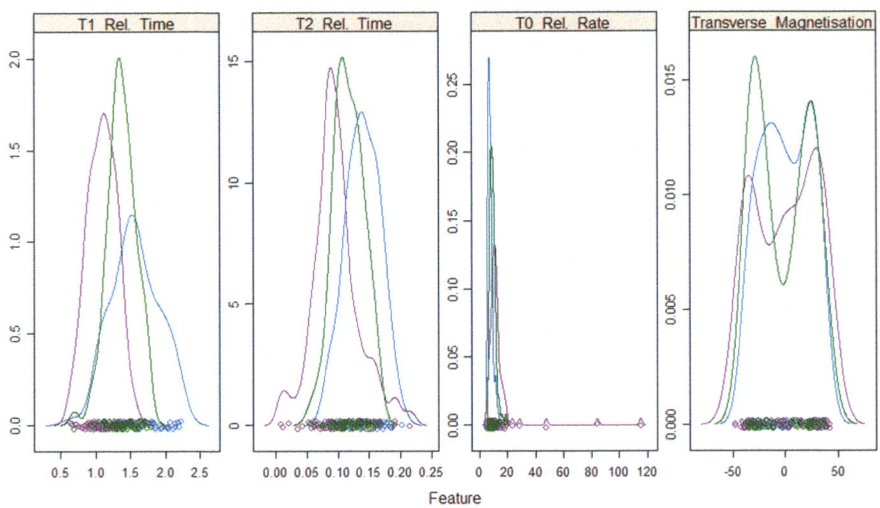

Fig. 11.29 Density plots for each attribute by class value for PW

```
> results <- resamples(list(cart=fit.cart, knn=fit.knn, svm=fit.svm, rf=fit.rf, glmnet=fi
t.glmnet, svmR=fit.svmRadial, nb=fit.nb, c50=fit.c50, tree=fit.treebag))
> summary(results)

Call:
summary.resamples(object = results)

Models: cart, knn, svm, rf, glmnet, svmR, nb, c50, tree
Number of resamples: 10

Accuracy
            Min.   1st Qu.    Median      Mean   3rd Qu.      Max. NA's
cart   0.2916667 0.5416667 0.5416667 0.5458333 0.5833333 0.7083333    0
knn    0.3750000 0.4583333 0.5833333 0.5625000 0.6562500 0.7083333    0
svm    0.5416667 0.6458333 0.7291667 0.7041667 0.7500000 0.8333333    0
rf     0.4166667 0.6041667 0.6875000 0.6583333 0.7395833 0.8333333    0
glmnet 0.5833333 0.6250000 0.6875000 0.6791667 0.7083333 0.7916667    0
svmR   0.5416667 0.6458333 0.7291667 0.7041667 0.7500000 0.8333333    0
nb     0.4583333 0.6354167 0.6666667 0.6541667 0.7291667 0.7916667    0
c50    0.5416667 0.6354167 0.6875000 0.6791667 0.7083333 0.8750000    0
tree   0.5000000 0.5625000 0.6250000 0.6291667 0.6979167 0.7916667    0

Kappa
          Min.  1st Qu.  Median    Mean  3rd Qu.   Max. NA's
cart   -0.0625 0.312500 0.31250 0.31875 0.375000 0.5625    0
knn     0.0625 0.187500 0.37500 0.34375 0.484375 0.5625    0
svm     0.3125 0.468750 0.59375 0.55625 0.625000 0.7500    0
rf      0.1250 0.406250 0.53125 0.48750 0.609375 0.7500    0
glmnet  0.3750 0.437500 0.53125 0.51875 0.562500 0.6875    0
svmR    0.3125 0.468750 0.59375 0.55625 0.625000 0.7500    0
nb      0.1875 0.453125 0.50000 0.48125 0.593750 0.6875    0
c50     0.3125 0.453125 0.53125 0.51875 0.562500 0.8125    0
tree    0.2500 0.343750 0.43750 0.44375 0.546875 0.6875    0
```

Fig. 11.30 Machine learning models tested on PW dataset

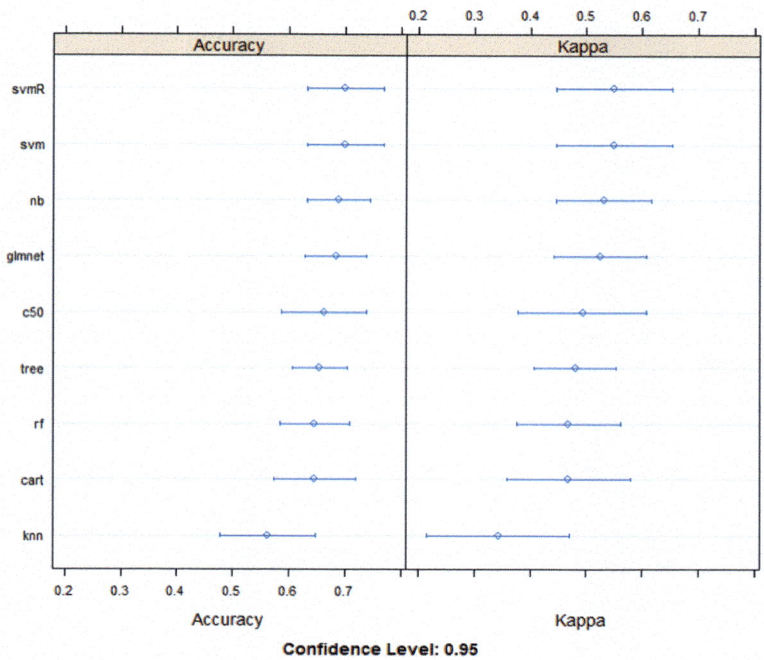

Fig. 11.31 Dot plot comparison of accuracy of models tested on PW dataset

```
> confusionMatrix(predictions, validation$Tumor_Class)
Confusion Matrix and Statistics

                        Reference
Prediction               Glioblastoma_multiforme Low_grade_glioma Metastasis
  Glioblastoma_multiforme                     12                1          4
  Low_grade_glioma                             2               14          3
  Metastasis                                   6                5         13

Overall Statistics

               Accuracy : 0.65
                 95% CI : (0.516, 0.7687)
    No Information Rate : 0.3333
    P-Value [Acc > NIR] : 5.312e-07

                  Kappa : 0.475

 Mcnemar's Test P-Value : 0.745

Statistics by Class:

                     Class: Glioblastoma_multiforme Class: Low_grade_glioma
Sensitivity                                  0.6000                  0.7000
Specificity                                  0.8750                  0.8750
Pos Pred Value                               0.7059                  0.7368
Neg Pred Value                               0.8140                  0.8537
Prevalence                                   0.3333                  0.3333
Detection Rate                               0.2000                  0.2333
Detection Prevalence                         0.2833                  0.3167
Balanced Accuracy                            0.7375                  0.7875
                     Class: Metastasis
Sensitivity                     0.6500
Specificity                     0.7250
Pos Pred Value                  0.5417
Neg Pred Value                  0.8056
Prevalence                      0.3333
Detection Rate                  0.2167
```

Fig. 11.32 SVM model performance with PW dataset

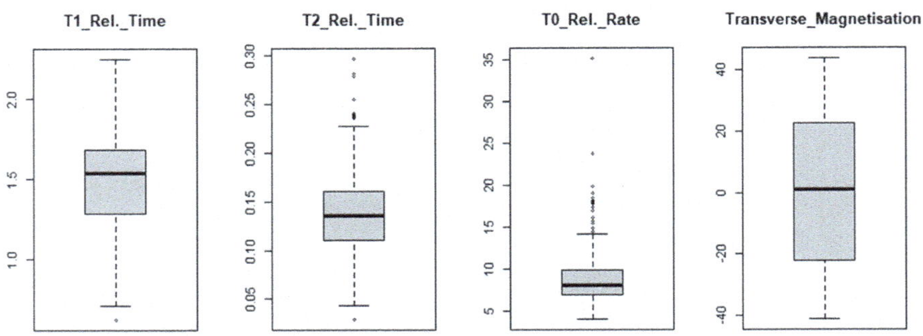

Fig. 11.33 Box plots for each attribute distributions on one image for ST

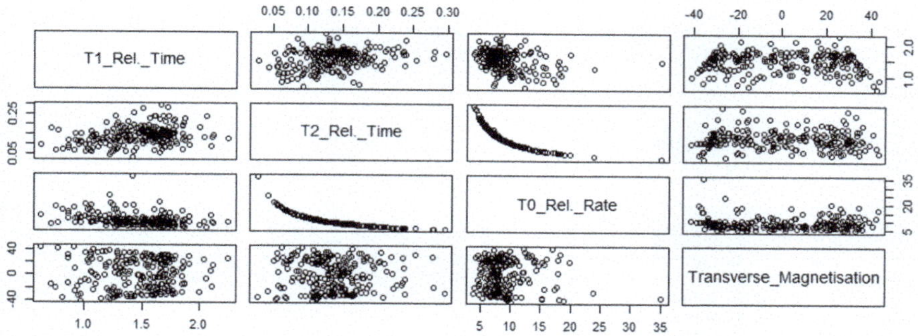

Fig. 11.34 Box plots for the datapoint attribute features for ST

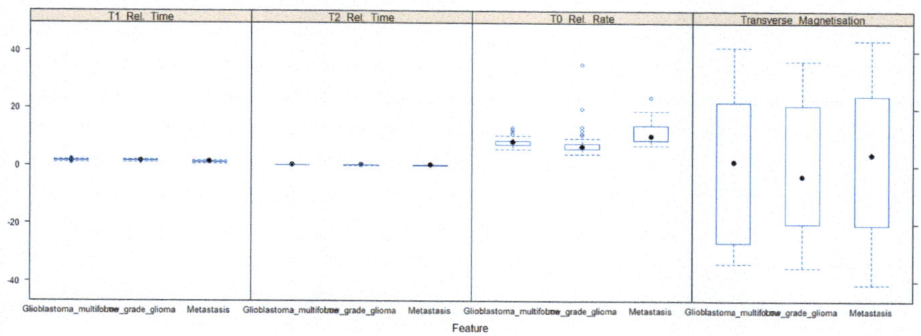

Fig. 11.35 Box and whisker plots for each attribute features for ST

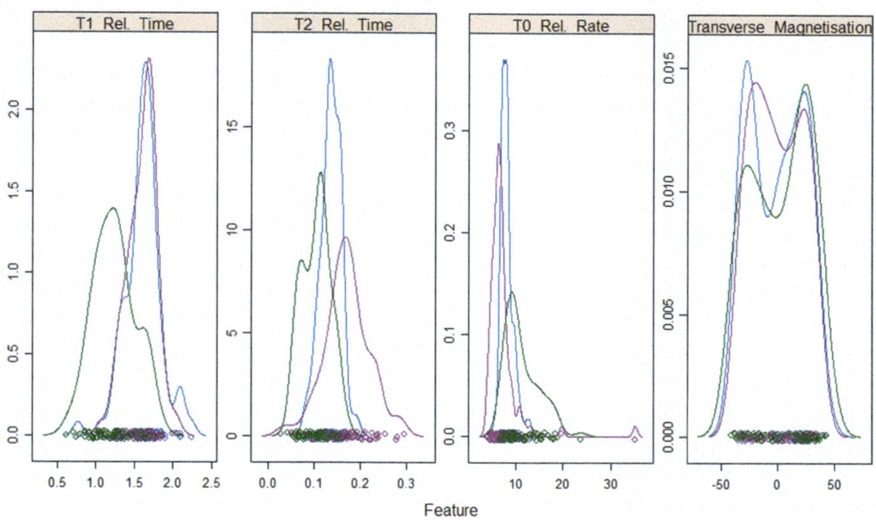

Fig. 11.36 Density plots for each attribute by class value for ST

```
> results <- resamples(list(cart=fit.cart, knn=fit.knn, svm=fit.svm, rf=fit.rf, glmnet=fi
t.glmnet, svmR=fit.svmRadial, nb=fit.nb, c50=fit.c50, tree=fit.treebag))
> summary(results)

Call:
summary.resamples(object = results)

Models: cart, knn, svm, rf, glmnet, svmR, nb, c50, tree
Number of resamples: 10

Accuracy
            Min.   1st Qu.    Median      Mean   3rd Qu.      Max. NA's
cart   0.6250000 0.6250000 0.6666667 0.6625000 0.6666667 0.7916667    0
knn    0.4583333 0.4687500 0.5833333 0.5500000 0.5833333 0.6666667    0
svm    0.5416667 0.6770833 0.7083333 0.7041667 0.7395833 0.8333333    0
rf     0.5416667 0.6666667 0.6875000 0.6708333 0.7083333 0.7500000    0
glmnet 0.6250000 0.6770833 0.7083333 0.7250000 0.7812500 0.8333333    0
svmR   0.5416667 0.6770833 0.7083333 0.7041667 0.7395833 0.8333333    0
nb     0.5833333 0.6666667 0.6875000 0.6916667 0.7395833 0.7916667    0
c50    0.6250000 0.6354167 0.7083333 0.7041667 0.7500000 0.8333333    0
tree   0.5000000 0.6250000 0.6458333 0.6333333 0.6666667 0.7083333    0

Kappa
         Min.  1st Qu.  Median    Mean  3rd Qu.   Max. NA's
cart   0.4375 0.437500 0.50000 0.49375 0.500000 0.6875    0
knn    0.1875 0.203125 0.37500 0.32500 0.375000 0.5000    0
svm    0.3125 0.515625 0.56250 0.55625 0.609375 0.7500    0
rf     0.3125 0.500000 0.53125 0.50625 0.562500 0.6250    0
glmnet 0.4375 0.515625 0.56250 0.58750 0.671875 0.7500    0
svmR   0.3125 0.515625 0.56250 0.55625 0.609375 0.7500    0
nb     0.3750 0.500000 0.53125 0.53750 0.609375 0.6875    0
c50    0.4375 0.453125 0.56250 0.55625 0.625000 0.7500    0
tree   0.2500 0.437500 0.46875 0.45000 0.500000 0.5625    0
```

Fig. 11.37 Several models tested on the ST dataset

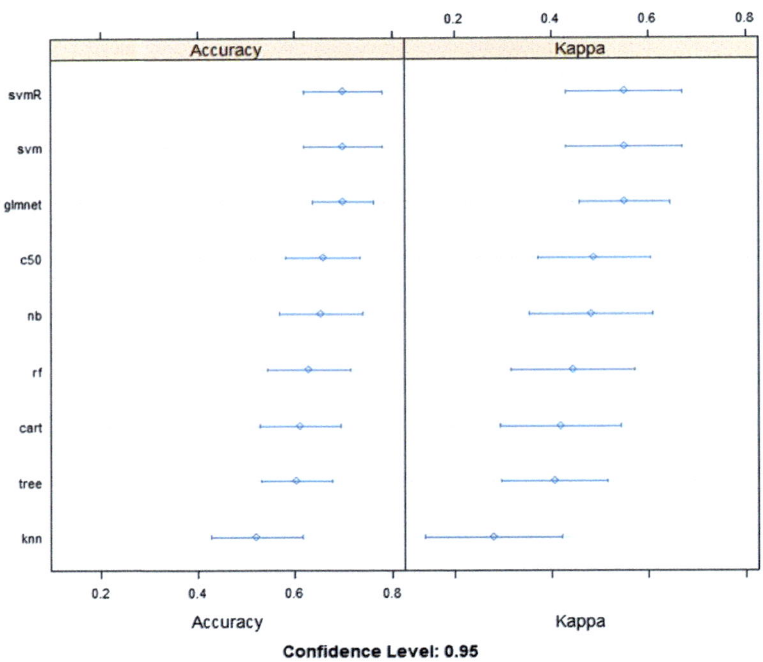

Fig. 11.38 Dot plots comparing accuracy of models for ST dataset

Table 11.4 Machine learning classification with contralateral white matter CW (summarised attribute distribution values for CW)

	T1 relaxation time (s)	T2 relaxation time (s)	T0 relaxation rate (s^{-1})	Transverse magnetization
Minimum	0.5265	0.03428	9.886	− 47.894
1st Qu	0.8682	0.06434	14.095	− 33.016
Median	0.8846	0.07035	15.323	− 8.173
Mean	0.9069	0.07085	15.649	− 3.119
3rd Qu	0.941	0.07722	16.705	27.385
Maximum	1.2079	0.11452	30.303	45.565

Table 11.5 Machine learning classification with peritumoral white matter PW (summarised attribute distribution values for PW)

	T1 relaxation time (s)	T2 relaxation time (s)	T0 relaxation rate (s^{-1})	Transverse magnetisation
Minimum	0.6403	0.008712	5.578	− 47.4191
1st Qu	1.1276	0.096031	7.837	− 25.3163
Median	1.3235	0.118333	9.169	0.2376
Mean	1.3584	0.119577	10.282	− 0.235
3rd Qu	1.5411	0.141329	11.193	24.8381
Maximum	2.2323	0.203272	115.731	45.8902

Table 11.6 Machine learning classification with solid tumor ST (summarised attribute distribution values for ST)

	T1 relaxation time (s)	T2 relaxation time (s)	T0 relaxation rate (s^{-1})	Transverse magnetization
Minimum	0.625	0.05213	3.981	− 41.064
1st Qu	1.334	0.11364	6.968	− 21.969
Median	1.544	0.13614	7.964	3.97
Mean	1.504	0.13868	8.647	1.384
3rd Qu	1.69	0.15923	9.695	22.895
Maximum	2.245	0.29606	19.786	43.653

observed that the glmnet model for solid tumor datapoint has the highest and best performance having accuracy of 72% and for kappa is 58% (Fig. 11.39).

```
> confusionMatrix(predictions, validation$Tumor_Class)
Confusion Matrix and Statistics

                          Reference
Prediction                 Glioblastoma_multiforme Low_grade_glioma Metastasis
  Glioblastoma_multiforme                       17                4          6
  Low_grade_glioma                               2               13          0
  Metastasis                                     1                3         14

Overall Statistics

               Accuracy : 0.7333
                 95% CI : (0.6034, 0.8393)
    No Information Rate : 0.3333
    P-Value [Acc > NIR] : 2.803e-10

                  Kappa : 0.6

 Mcnemar's Test P-Value : 0.06468

Statistics by Class:

                     Class: Glioblastoma_multiforme Class: Low_grade_glioma
Sensitivity                                  0.8500                  0.6500
Specificity                                  0.7500                  0.9500
Pos Pred Value                               0.6296                  0.8667
Neg Pred Value                               0.9091                  0.8444
Prevalence                                   0.3333                  0.3333
Detection Rate                               0.2833                  0.2167
Detection Prevalence                         0.4500                  0.2500
Balanced Accuracy                            0.8000                  0.8000
                     Class: Metastasis
Sensitivity                     0.7000
Specificity                     0.9000
Pos Pred Value                  0.7778
Neg Pred Value                  0.8571
Prevalence                      0.3333
Detection Rate                  0.2333
```

Fig. 11.39 Glmnet model performance with ST dataset

11.7 Shiny App Deployment with Contralateral, Peritumoral White Matter and Solid Tumor

Figures 11.40, 11.45 and 11.50 present the deployment processes with contralateral, peritumoral white matter and solid tumor datasets, respectively. The following steps are important for model deployment with negative binomial (nb), support vector machine (SVM) and glmnet models (Figs. 11.41, 11.42, 11.43, 11.44, 11.46, 11.47, 11.48 and 11.49).

Step I: Open R-studio and setting the directory. The code for shiny app is created. In Figs. 11.41, 11.46 and 11.51, the Red pointed arrow shows the running app by clicking on the Run App button to see the result.

Step II: The Blue pointing arrow shows to browse upload MR Relaxometry data in csv file. The location of the selected file is opened with the Black pointing arrow.

Step III: After successfully uploading the dataset in Step 2, the tumor was predicted. This is shown for the first 10 columns for each tumor classification.

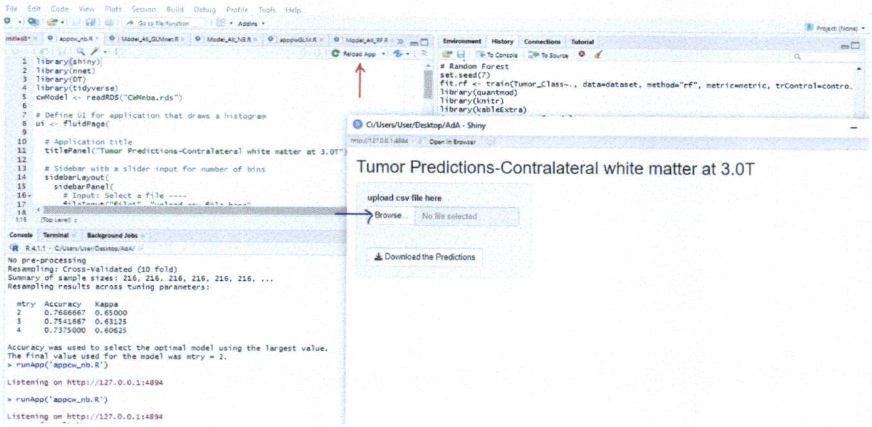

Fig. 11.40 App deployment with contralateral white matter

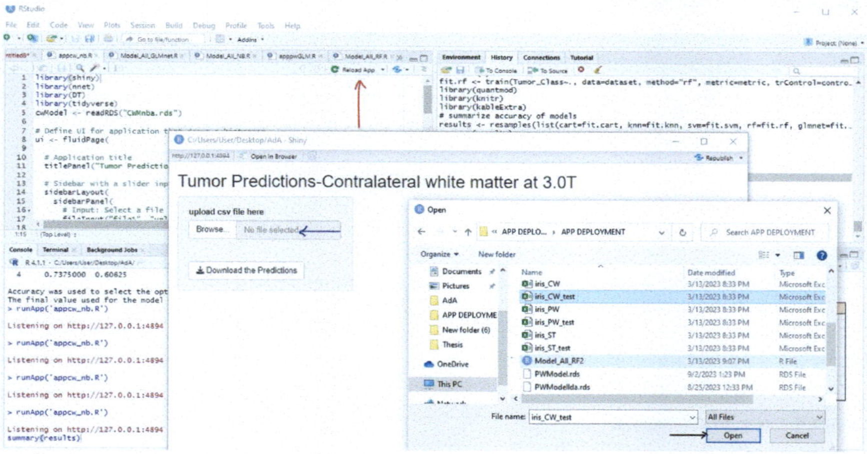

Fig. 11.41 Contralateral white matter dataset for app deployment

Figures 11.42, 11.43 and 11.44 for contralateral white matter, Figs. 11.47, 11.48, and 11.49 for peritumoral white matter, and Figs. 11.52, 11.53, and 11.54 for solid tumor.

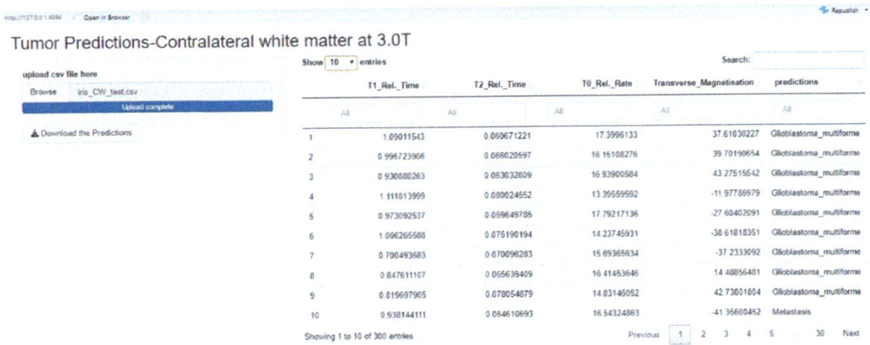

Fig. 11.42 Contralateral white matter tumor predictions column (1–10)

Fig. 11.43 Contralateral white matter tumor predictions column (101–110)

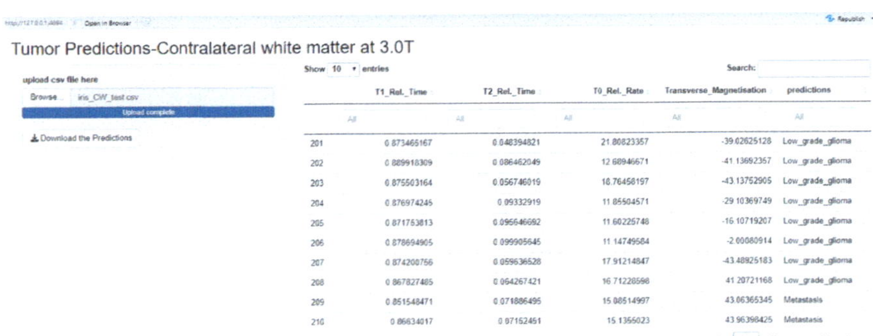

Fig. 11.44 Contralateral white matter tumor predictions column (201–210)

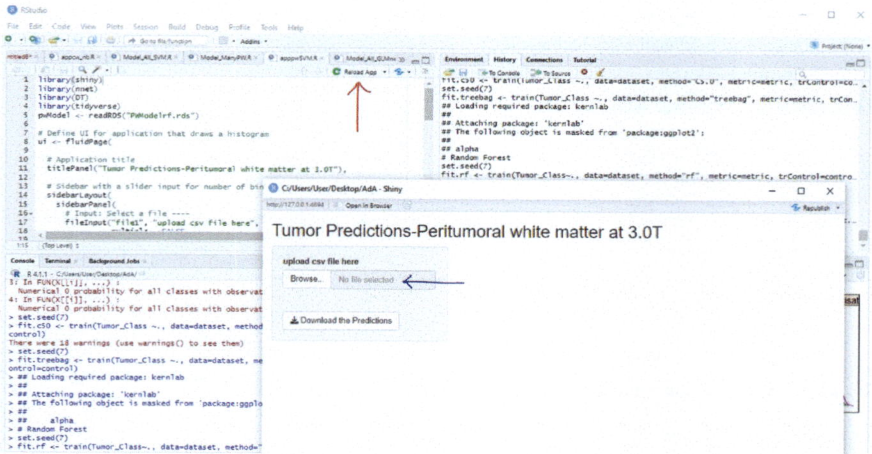

Fig. 11.45 App deployment with peritumoral white matter

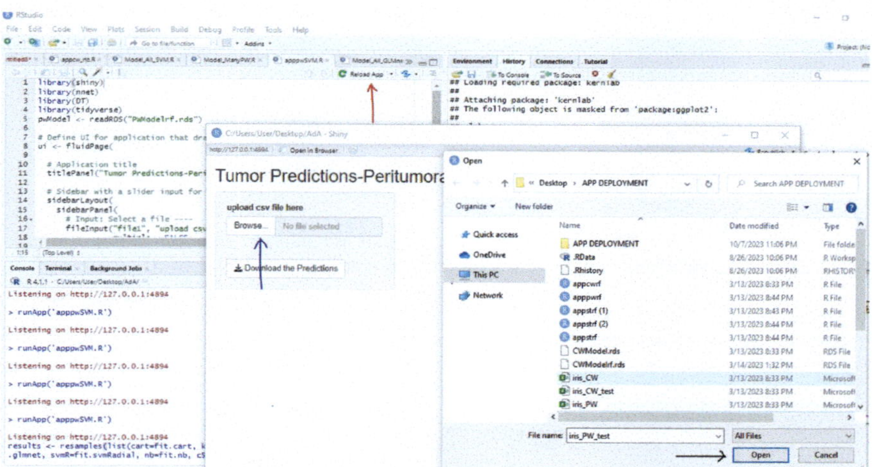

Fig. 11.46 Peritumoral white matter dataset for app deployment

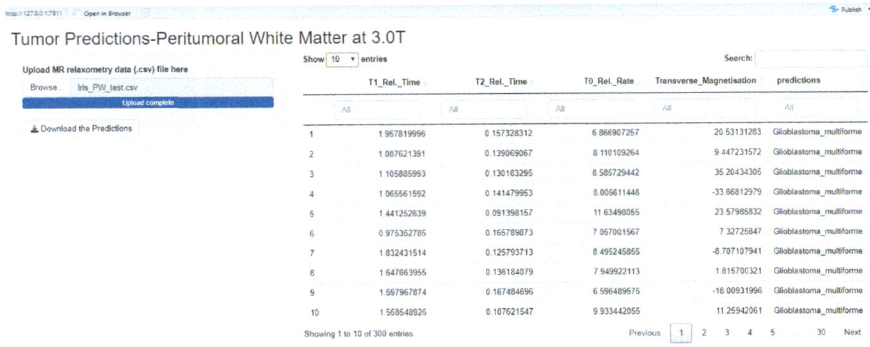

	T1_Rel._Time	T2_Rel._Time	T0_Rel._Rate	Transverse_Magnetisation	predictions
1	1.957819996	0.157328312	6.866907257	20.53131283	Glioblastoma_multiforme
2	1.087621391	0.139069067	8.110109264	9.447231572	Glioblastoma_multiforme
3	1.105885993	0.130183295	8.585729442	35.20434305	Glioblastoma_multiforme
4	1.065561592	0.141479953	8.006611448	-33.66812979	Glioblastoma_multiforme
5	1.441252639	0.091398157	11.63498055	23.57985832	Glioblastoma_multiforme
6	0.975362785	0.165789873	7.057001567	7.32725847	Glioblastoma_multiforme
7	1.832431514	0.125793713	8.495245855	-8.707107941	Glioblastoma_multiforme
8	1.647663955	0.136184079	7.949922113	1.815700321	Glioblastoma_multiforme
9	1.597967874	0.167484696	6.596489575	-18.00931996	Glioblastoma_multiforme
10	1.558548926	0.107621547	9.933442055	11.25942061	Glioblastoma_multiforme

Fig. 11.47 Peritumoral white matter tumor predictions column (1–10)

	T1_Rel._Time	T2_Rel._Time	T0_Rel._Rate	Transverse_Magnetisation	predictions
101	1.283650831	0.096142025	11.18030681	35.79542522	Metastasis
102	1.525791101	0.120333335	8.96564692	19.70047636	Metastasis
103	1.325376907	0.108315832	9.986763024	-23.68868631	Metastasis
104	1.170001132	0.123092159	8.978694339	34.6064523	Metastasis
105	1.292350638	0.15562493	7.199489511	-32.31639315	Metastasis
106	1.616199102	0.105032766	10.13957411	8.898739305	Metastasis
107	1.657066805	0.122168288	8.788906191	-18.11736191	Metastasis
108	1.23292931	0.111839979	9.752422934	23.96806305	Metastasis
109	1.251974415	0.163658124	6.909036944	34.65443606	Metastasis
110	1.461003497	0.129248385	8.421501663	26.99806628	Metastasis

Fig. 11.48 Peritumoral white matter tumor predictions column (101–110)

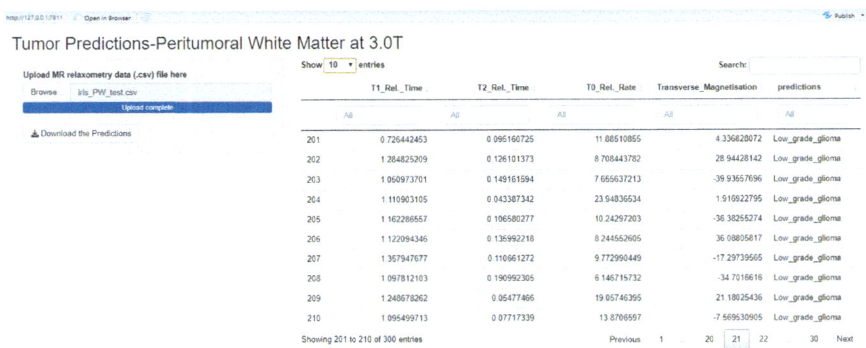

	T1_Rel._Time	T2_Rel._Time	T0_Rel._Rate	Transverse_Magnetisation	predictions
201	0.726442453	0.095160725	11.88510855	4.336828072	Low_grade_glioma
202	1.284825209	0.126101373	8.708443782	28.94428142	Low_grade_glioma
203	1.060973701	0.149161594	7.665637213	-39.93557696	Low_grade_glioma
204	1.110903105	0.043387342	23.94836534	1.916922795	Low_grade_glioma
205	1.162286557	0.106580277	10.24297203	-36.38255274	Low_grade_glioma
206	1.122094346	0.135992218	8.244552605	36.08805817	Low_grade_glioma
207	1.357947677	0.110661272	9.772990449	-17.29739565	Low_grade_glioma
208	1.097812103	0.190992305	6.146715732	-34.7016616	Low_grade_glioma
209	1.248678262	0.05477466	19.05746395	21.18025436	Low_grade_glioma
210	1.095499713	0.07717339	13.8706597	-7.569530905	Low_grade_glioma

Fig. 11.49 Peritumoral white matter tumor predictions column (201–210)

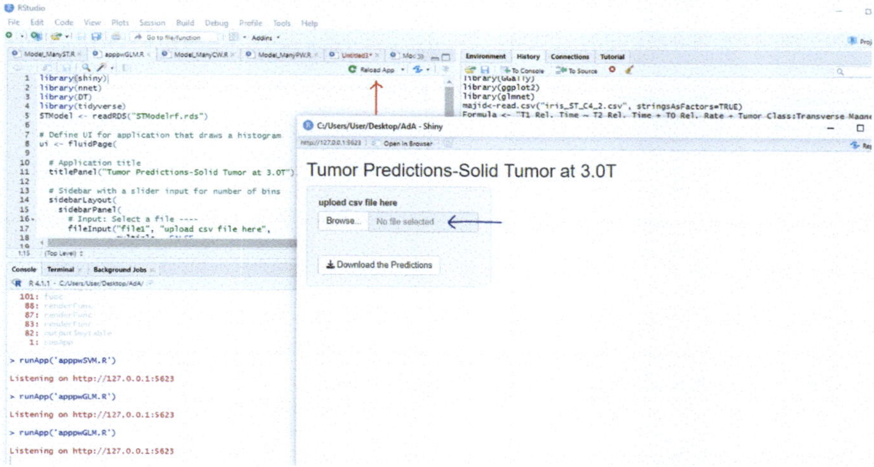

Fig. 11.50 App deployment with solid tumor

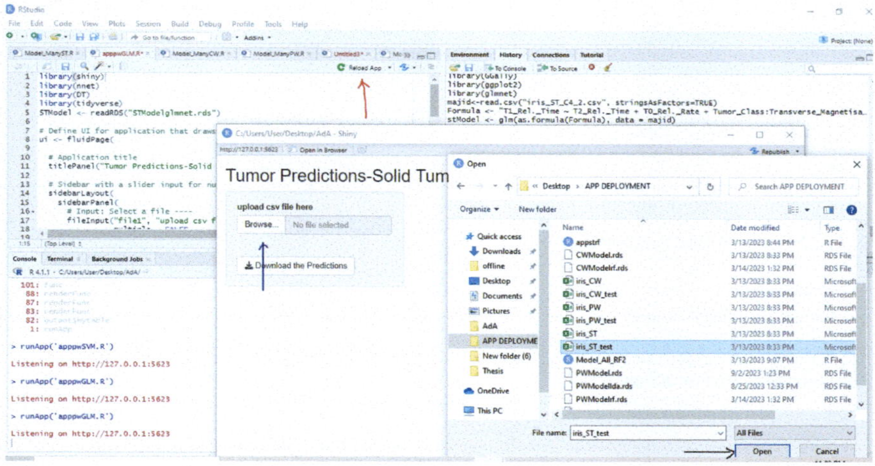

Fig. 11.51 Solid tumor dataset for app deployment

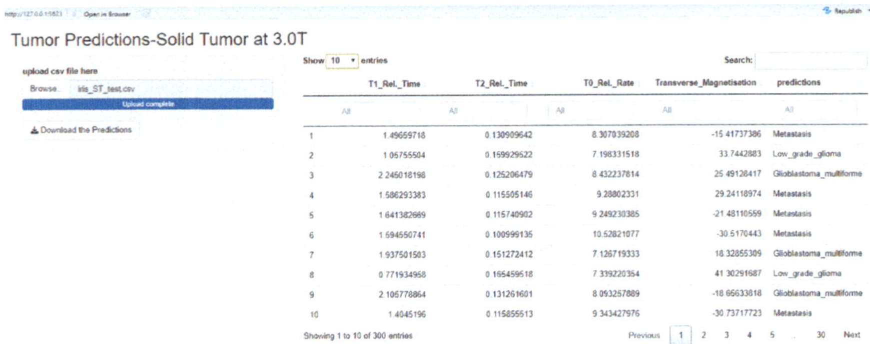

Fig. 11.52 Solid tumor white matter tumor predictions column (1–10)

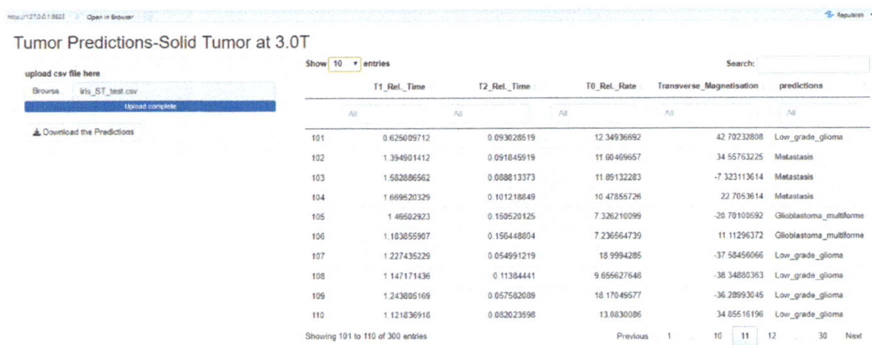

Fig. 11.53 Solid tumor predictions column (101–110)

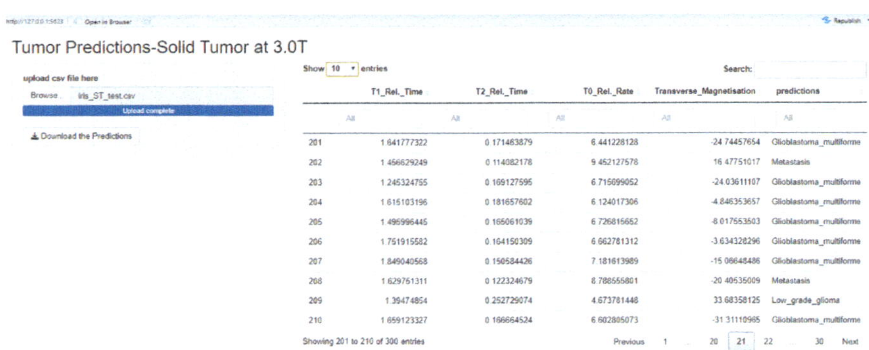

Fig. 11.54 Solid tumor predictions column (201–210)

11.8 Conclusion

This research has shown how MR fingerprinting can be used to quantitatively evaluate glioblastomas, metastases, and lower grade gliomas. MRF-based relaxometry can show quantitative differences between contralateral regions of glioblastomas and metastases and to peritumoral regions of glioblastomas and lower grade gliomas as well as between solid tumor regions of lower grade gliomas and metastases. Three models of artificial neural networks (Garson, SNNs and PNNs) were used. On the other hand, nine model which include KNN, CART, NB, SVM, RF, GLMNET, TREE, C50 and SVMR were trained and used for deployment. These classifiers were used to train 80% of the dataset and an accuracy of 79% and 68% kappa for nb, 70% accuracy and 55% kappa for svm, and 72% accuracy and 58% kappa for glmnet for CW, PW and ST respectively. Since the nb, svm, and glmnet for CW, PW and ST, respectively, had the highest accuracy and kappa. They were used to test and the remaining 20% of the data was also used to deploy the model. The shiny in R shows how the data were predicted according to their tumor classes.

In clinical settings, MRF has the potential to produce faster, cheaper and more efficient quantitative maps of brain tumors with the availability of relaxometry data, software and models. Larger sample studies are required to further explore the effectiveness of this technique.

Magnetic resonance fingerprinting has the capacity to provide high reproducible, accurate and rapid quantitative imaging of multiple MR parameters at once, as well as enable the development of dependable and standardized imaging biomarkers and their multi parametric combinations, which can be validated in prospectively collected studies due to their strong reproducibility across sites and even across different MR vendor.

Due to the impractically long scan times required, especially for biological samples and patients, the direct prediction of the oscillating, incoherent signal evolutions through the Bloch simulation offers us the possibility to collect new quantitative information.

MRF can result in the direct identification of a material, tissue, or pathology based only on its fingerprint because of its capacity to offer quantitative results across several parameters simultaneously. For instance, many cancer cells possess changes in a number of MR parameters (such as T1, T2, self-diffusion tensor, etc.), the combination of which (rather than any one parameter alone) could possibly characterize them as differentiated from all neighboring normal tissue types and therefore, as potentially distinguishable. In an ideal situation, every substance, tissue, and pathology would have their own unique signal evolution that would be independent of all other signal evolutions.

Acknowledgements The authors acknowledge the motivated research efforts of Abdumajid A. Yusuf as a promising young researcher.

References

Anitha, G.S., Kuldeep, S.: Neural network approach for processing substation alarms. Int. J. Power Electron. Control. Convert. **1**(1), 21–28 (2015)

Awojoyogbe, O.B.: A mathematical model of Bloch NMR equations for quantitative analysis of blood flow in blood vessels with changing cross-section I. Physica A **303**, 163–175 (2002)

Awojoyogbe, O.B.: A Mathematical model of Bloch NMR equations for quantitative analysis of blood flow in blood vessels with changing cross-section II. Physica A **323**, 534–550 (2003)

Awojoyogbe, O.B.: Analytical solution of the time dependent Bloch NMR equations: a translational mechanical approach. Physica A **339**, 437–460 (2004)

Awojoyogbe, O.B., Dada, O.M., Faromika, O.P., Dada, O.E.: Mathematical concept of the Bloch flow equations for general magnetic resonance imaging: a review. Concepts Magn. Reson. Part A **38**(3), 85–101 (2011)

Badve, C., Yu, A., Dastmalchian, S., Rogers, M., Ma, D., Jiang, Y., Gulani, V.: MR fingerprinting of adult brain tumors: initial experience. Am. J. Neuroradiol. **38**(3), 492–499 (2017)

Badve, C., Yu, A., Rogers, M., Ma, D., Liu, Y., Schluchter, M., Sunshine, J., Griswold, M., Gulani, V.: Simultaneous T1 and T2 brain relaxometry in asymptomatic volunteers using magnetic resonance fingerprinting. Tomography **1**(2), 136–144 (2015)

Boux, F., Forbes, F., Arbel, J., Lemasson, B., Barbier, E.L.: Bayesian inverse regression for vascular magnetic resonance fingerprinting. IEEE Trans. Med. Imaging **40**(7), 1827–1837 (2021)

Buonincontri, G., Sawiak, S.J.: MR fingerprinting with simultaneous B1 estimation. Magn. Reson. Med. **76**(4), 1127–1135 (2016)

Chen, Y., Chen, M.H., Baluyot, K.R., Potts, T.M., Jimenez, J., Lin, W.: UNC/UMN Baby Connectome Project Consortium: MR fingerprinting enables quantitative measures of brain tissue relaxation times and myelin water fraction in the first five years of life. *Neuroimage* **186**, 782–793 (2019)

Chen, Y., Jiang, Y., Pahwa, S., Ma, D., Lu, L., Twieg, M.D., Gulani, V.: MR fingerprinting for rapid quantitative abdominal imaging. Radiology **279**(1), 278 (2016)

Christen, T., Pannetier, N.A., Ni, W.W., Qiu, D., Moseley, M.E., Schuff, N., Zaharchuk, G.: MR vascular fingerprinting: a new approach to compute cerebral blood volume, mean vessel radius, and oxygenation maps in the human brain. Neuroimage **89**, 262–270 (2014)

Cloos, M.A., Assländer, J., Abbas, B., Fishbaugh, J., Babb, J.S., Gerig, G., Lattanzi, R.: Rapid radial T1 and T2 mapping of the hip articular cartilage with magnetic resonance fingerprinting. J. Magn. Reson. Imaging **50**(3), 810–815 (2019)

Cohen, O., Zhu, B., Rosen, M.S.: MR fingerprinting deep reconstruction network (DRONE). Magn. Reson. Med. **80**(3), 885–894 (2018)

Cruz, G., Jaubert, O., Schneider, T., Botnar, R.M., Prieto, C.: Rigid motion-corrected magnetic resonance fingerprinting. Magn. Reson. Med. **81**(2), 947–961 (2019)

Dada, M.O., Awojoyogbe, B.O.: Computational MRI based on Bloch's NMR flow equation, MRI fingerprinting and python deep learning for classifying adult brain tumors. In: Computational Molecular Magnetic Resonance Imaging for Neuro-oncology. Biological and Medical Physics, Biomedical Engineering. Springer, Cham (2021)

Dinh, L., Sohl-Dickstein, J., Bengio, S.: Density estimation using real NVP. In: 5th International. In Conference on Learning Representations, ICLR (2017)

Kreyszig, E.: Advanced Engineering Mathematics. John Wiley and Sons (2011)

Ropella-Panagis, K.M., Seiberlich, N., Gulani, V.: Magnetic resonance fingerprinting: implications and opportunities for PET/MR. IEEE Trans. Radiat. Plasma Med. Sci. **3**(4), 388–399 (2019)

Chapter 12
Fundamental Quantum Mechanical Dynamics in Computational Magnetic Resonance Imaging from Low Field to High Field

Abstract In recent years, it has been discovered that heterogeneous diseases such as cardiovascular diseases, diabetes, hepatitis and cancer may be attacked effectively by developing specific treatment plans that are generally referred to as personalized treatments. This new approach in medicine involves using novel methodology as a bridge between diagnostic and therapeutic applications to form a single agent, allowing for diagnosis, drug delivery and treatment response monitoring. This method is time- and cost-effective because it is a one in all procedure which combines diagnosis and therapy. This novel methodology involving atomic nuclei could be studied with NMR since their unique resonant frequencies lead to the production of distinct spectra for diagnostic imaging. Atomic nuclei are quantum objects and most of their properties can be accurately explained only by resorting to quantum mechanical formulation. In this study, we have solved the Bloch NMR flow equation quantum mechanically to describe the evolution of magnetic resonance imaging from low magnetic field to high magnetic field. We must note that, the expression $\gamma(B_o + g) \ll \frac{\hbar}{\beta m_o \alpha^2}$ gives the order of magnitude of the low static magnetic field B_o and $\gamma(B_o + g) \gg \frac{\hbar}{\beta m_o \alpha^2}$ for high static magnetic field B_o where m_o is the mass of the particle and β is a dimensionless constant. Spatial magnetic field gradients (**g**) are usually required to alter the polarising magnetic field such that the field seen by each nucleus within an object are location-dependent. In accordance with the uncertainty relation, a particle confined in a volume with radius of the order α would have the momentum of the order $\frac{\hbar}{\alpha}$. This expression equals the minimum depth of a spherical potential well of radius α at which a bound state of a particle with discrete energy level appears.

Keywords Quantum mechanical dynamics · Computational magnetic resonance imaging · Tissue disease condition · Heterogeneous diseases · Perturbation theory · Low and high static magnetic fields

B. O. Awojoyogbe and M. O. Dada, *Digital Molecular Magnetic Resonance Imaging*, Series in BioEngineering, https://doi.org/10.1007/978-981-97-6370-2_12

12.1 Introduction

The real world runs on quantum physics. Computational methods based on quantum concepts should in many situations be our best tools for understanding the real world around us. No computational concept has the working memory to handle all the possible permutations of molecular behavior by using any known methods. We do not have a good way to create these computational spaces with classical mechanics, which limits their usefulness without quantum computation.

It is very interesting to note from Dada and Awojoyogbe (2021a, 2021b) that the atomic nuclei within the molecules that are precessing have a memory of their unique nature and their environment. This is the information they give out as energy and is reconstructed as images. Once placed within a static magnetic field (B_0), depending on their orientation, atomic nuclei undergo excitations with energy difference (ΔE). The change in energy, ΔE is a factor on which the strength of the resonance signals depends. The relationship between this energy and the resonance frequency (f_0) is expressed by the following equation (Slichter 2013):

$$\Delta E = hf_0, \tag{12.1}$$

where the term, h represents the Planck's constant. The static magnetic field (B_0) and the constant gyromagnetic ratio (γ) determine the magnitude of ΔE and f_0. Gyromagnetic ratio is a constant, which is unique for different kinds of atomic nucleus. Equation (12.1) is known as the Larmor relationship and it relates this unique constant to the magnetic field and resonant frequency. This equation is the basis on which the principles of magnetic resonance are based (Cowan 1997):

$$f_0 = \left(\frac{\gamma}{2\pi}\right)B_0, \tag{12.2}$$

A combination of Eq. (12.1) and (12.2) is given as:

$$\Delta E = hf_0 = h\left(\frac{\gamma}{2\pi}\right)B_0. \tag{12.3}$$

The phenomenon of nuclear magnetic resonance was first observed and shown by Rabi (1937). Edward Purcell and Felix Bloch, though working independently, demonstrated how the detection of nuclear magnetic resonance signals could be achieved with the use of radio waves. The 1952 Nobel Prize for Physics was awarded to these scientists in recognition of this achievement (Castelli et al. 2011).

The discovery of NMR has led to development of nuclear magnetic spectroscopy or magnetic resonance spectroscopy (MRS), without which chemistry would not be modernized (Dawson and Lauterbur 2008). Within the static magnetic field, B_0 the atomic nuclei are able to respond to a time-dependent magnetic field B_1 according to their unique Larmor resonance frequency (f_0) with which their corresponding nuclear energy is changed. When the applied oscillatory field is removed (turned off), the atomic nuclei gives off radio waves as the nuclear spins return to their equilibrium states (Haacke et al. 1999). These emitted signals lead to the NMR spectra used for imaging. As mentioned above, various atomic nuclei could be studied with NMR since their unique resonant frequencies lead to the production of distinct spectra for diagnostic imaging.

The largest value of gyromagnetic ratio belongs to the hydrogen atom. This means that according to Eq. (12.3), the hydrogen nuclear spins produce the largest energy and hence, the largest resonant signal at any specific magnetic field strength (H_0). This is due to the abundance of hydrogen atom in nature. In addition to this, the human body is made up of about two-thirds water and the incredible convergence of signal strength and natural abundance is responsible for significantly high resolution being associated with the hydrogen nucleus. In fact, most clinical applications of MRI are reliant on proton resonance, f. Meanwhile, in the studies and imaging of tissue metabolism, other nuclei such as ^{31}P and ^{23}Na have proved to be very important and useful (Cowan 1997).

Spatial magnetic field gradients (**g**) are usually required to alter the polarising magnetic field such that the field seen by each nucleus within an object are location-dependent (the vector G represents gradient field that is dependent on its three components g_x, g_y and g_z). For instance, if g_x has linear behaviour along the x-axis, an extra magnetic field ($g_x x$) is produced at a given point x. In 3D space, gradient fields g_x, g_y and g_z at unique point (x, y, z) influences Eq. (12.3) as follows Cowan 1997; Haacke et al. 1999):

$$f_0 = \frac{\gamma}{2\pi}(B_0 + g), \tag{12.4}$$

$$E = h\frac{\gamma}{2\pi}(B_0 + g) = \hbar(B_0 + g), \tag{12.5}$$

where for 2-D imaging, **g** is usually g_x and g_y, with g_z added for three dimensional imaging. The use of spatial gradients results in atomic nuclei in various spatial locations experiencing distinct values of the term $(B_0 + g)$, the problem of reconstructing an image becomes a problem of interpreting the spectral frequencies (Dawson and Lauterbur 2008). New MRI sequences and methods developed from time to time with each possessing diverse strengths and weaknesses according to the clinical image required or desired. Although current practices of MRI rely on two dimensional image acquisitions, the methodology employed could as well be used for three or more dimensions. However, the only hindrance to achieving this is the number of resonance signals, which can be obtained. How well the imaging hardware and image sequence software are able to cope with the requirements (Haacke et al. 1999) can be guided by the fundamental Bloch NMR flow equation (Awojoyogbe 2022, 2023, 2024).

When the RF B_1 field spatially varies with x we derive the equation (Awojoyogbe 2007):

$$\frac{d^2\psi}{dx^2} + G^2(x)\psi = 0 \tag{12.6}$$

where

$$G^2(x) = \frac{\gamma^2 B_1^2(x)}{v^2} + \frac{T_g - T_R}{v^2} \tag{12.7}$$

The following assumptions will be made.

$$G^2(x) = \frac{2m_o\beta}{\hbar^2}[E - U(x)] + T_R - T_g \tag{12.8}$$

$$\frac{\gamma^2 B_1^2(x)}{v^2} = \frac{2m_o\beta}{\hbar^2}[E - U(x)]$$

where ħ is the reduced Planck's constant, m_o is particle mass, β is a dimensionless weighting parameter while E and U(x) are the total energy and potential energy of the particle respectively. Based on the uncertainty principle, a wave packet is assumed to initially describe the particle (electron) under study (Dada and Awojoyogbe 2021a, 2021b). Then, when the particle encounters a force (diseases such as cardiovascular diseases, diabetes, hepatitis and cancer, so its potential energy is no longer zero), the force modifies the wave packet. Finding such propagation techniques, and applying them appropriately can provide useful techniques to find solutions to biological, medical and physical problems which otherwise could not be easily solved.

For simplicity, let us suppose that the RF B_1 field is designed such that $G^2(x)$ vanishes at a single point $x = x_o$ only. Then

$$\begin{aligned} G^2(x) &> 0;\ x > x_0 \\ G^2(x) &< 0;\ x < x_0 \end{aligned} \tag{12.9}$$

The implication of this condition according to Eq. (12.7) is:

$$\begin{aligned} \gamma^2 B_1^2(x) &> T_R - T_g;\ x > x_0 \\ \gamma^2 B_1^2(x) &< T_R - T_g;\ x < x_0 \end{aligned} \tag{12.10}$$

Equation (12.1j) is related to the linearity condition in nuclear magnetic resonance.

12.1.1 Perturbation Theory for NMR Wave Equation

In this chapter, we consider the case when the potential energy of a particle in an external magnetic field B_o can be dealt with as a perturbation. Here the unperturbed Schröndinger equation is an equation of the freely moving particle:

$$\frac{-\hbar^2}{2\beta m_o}\nabla^2\psi^{(0)} = \mathrm{E}^0\psi^{(0)} \tag{12.11}$$

If we introduce the notation

$$\kappa^2 = \frac{2\beta m_o E^{(o)}}{\hbar^2} = \frac{p^2}{\hbar^2} \tag{12.12}$$

where p^2 is the momentum of the particle). Equation (12.11) can be given the form:

$$(\nabla^2 + k^2)\psi^{(0)} = 0 \tag{12.13}$$

A plane wave is the solution of this equation, namely,

$$\psi^{(0)} = e^{ikr} \tag{12.14}$$

$(\mathrm{k} = \frac{p}{\hbar})$.

Adding the perturbation operator $\tilde{\mathrm{V}}$, whose role is played by the potential energy U, to the operator $\hat{\mathrm{H}}_0 = (-\hbar, 2\beta m_0)\nabla^2$ in Eq. (12.11), we obtain the equation:

$$\left(-\frac{\hbar^2}{2\beta m_o}\nabla^2 + U\right)\psi = \mathrm{E}\psi \tag{12.15}$$

To find the psi-function in the first approximation, we introduce $\psi = \psi^{(0)} + \Delta\psi^{(1)}$ and $\mathrm{E} = E^{(0)}$ into Eq. (12.15):

$$\left(-\frac{\hbar^2}{2\beta m_o}\nabla^2 + U\right)\left(\psi^{(0)} + \Delta\psi^{(1)}\right) = E^{(0)}\left(\psi^{(0)} + \Delta\psi^{(1)}\right) \tag{12.16}$$

Based on Eq. (12.2a), we obtain:

$$-\frac{\hbar^2}{2\beta m_o}\nabla^2\psi^{(1)} + U\psi^{(0)} + U\Delta\psi^{(1)} = E^{(0)}\Delta\psi^{(1)}$$

Owing to the small values of the quantities U and $\Delta\psi^{(1)}$, we may disregard their product. The equation remaining after this has been done can be reduced to the following form with the aid of Eq. (12.12):

$$\nabla^2\Delta\psi^{(1)}(r) + k^2\Delta\psi^{(1)}(r) = \frac{2\beta m_o}{\hbar^2}U(\mathrm{r})\psi^{(0)}(r) \tag{12.17}$$

Let us turn to Eq. (3.408) for the scalar transverse magnetization $\mathrm{M_y(x,t)} = \varphi(r,t)$ produced by charges of density known from electrodynamics:

$$-4\pi\rho(r,t) = \frac{M_o\gamma B_1(x,t)}{T_1} \tag{12.18}$$

$$\nabla^2\varphi(r,t) - \frac{1}{c^2}\frac{\partial^2\varphi(r,t)}{\partial t^2} = 4\pi\rho(r,t) \tag{12.19}$$

The retarded potential,

$$\varphi(r,t) = \int \frac{\rho(r\prime, t - R/c)}{R} dV\prime \tag{12.20}$$

is the solution of this equation. We have introduced the notation

$$R = |-r\prime| \tag{12.21}$$

Consider the case when charge density at each point varies according to a harmonic law:

$$\rho(r,t) = \rho(r)e^{-i\omega t} \tag{12.22}$$

Equation (12.9) becomes:

$$\varphi(r,t) = \int \frac{\rho_0(r\prime)e^{-i\omega}(t - R/c)}{R} dV\prime = \varphi_o(r)e^{-i\omega t} \tag{12.23}$$

where

$$\varphi_o(r) = \int \frac{\rho_0(r\prime)e^{i\omega}R/c}{R} dV\prime \tag{12.24}$$

Introducing the functions in Eqs. (12.23) and (12.22) into Eq. (12.18) and cancelling the factor $e^{-i\omega t}$ in all the terms, we obtain a differential equation for $\varphi_o(r)$:

$$\nabla^2\varphi_o(r) + \frac{\omega^2}{c^2}\varphi_o(r) = 4\pi\rho_0(r) \tag{12.25}$$

Equation (12.24) is a solution of Eq. (12.19) which we are interested in is similar to Eq. (12.25). Consequently, its solution can be obtained by replacing in Eq. (12.24) ω/c with k and $\varphi_o(r\prime)$ with:

$$(\beta m_0/2\pi\hbar^2)U(\mathrm{r}\prime)\psi^{(0)}(r\prime) = -(\beta m_0/2\pi\hbar^2)U(r\prime)e^{ikr\prime cos\theta} \tag{12.26}$$

θ is the angle between the vectors k and $r\prime$. As a result, we obtain:

$$\Delta\psi^{(1)}(r) = -\frac{\beta m_o}{2\pi\hbar^2}\int \frac{1}{|r - r\prime|}U(r\prime)e^{ik(r\prime cos\theta + |r - r\prime|)}dV\prime \tag{12.27}$$

We have replaced R with $|r - r\prime|$. Recall that this function is a first order correction to the unperturbed functions $\psi^{(0)}(r)$ due to the perturbation $\tilde{V} = U(r)$. Let us find the conditions in which the potential U(r) can be considered as perturbation. We shall proceed from the fact that the condition for the applicability of the perturbation theory consists in the requirement $\left|\Delta\psi^{(1)}\right| \ll \left|\psi^{(0)}\right|$. Since the magnitude of the unperturbed function in Eq. (12.4) is unity, we arrive at the condition $\left|\Delta\psi^{(1)}\right| \ll 1$ 1, that is:

$$\frac{m_o}{2\pi\hbar^2}\left|\int \frac{1}{|r - r\prime|} U(r\prime) e^{ik(r\prime cos\theta + |r - r\prime|)} dV\prime\right| \ll 1 \quad (12.28)$$

12.2 Discussion

Let the function U(r) noticeably differ from zero only when r $\lesssim \alpha$ (the wavy line signifies that we are dealing not with exact values but the order of magnitude of the relevant quantities). Hence, the main contribution to the Eq. (12.28) is made by the part of the integral that relates to the volume of radius α. it thus follows that $\Delta\psi^{(1)}$ diminishes with r so that if the conditions in Eq. (12.28) is observed at r = 0, it will be observed at any r. the condition in Eq. (12.28) can thus be replaced with condition

$$\frac{\beta m_o}{2\pi\hbar^2}\left|\int \frac{1}{r\prime} U(r\prime) e^{ikr\prime(cos\theta + 1)} dV\prime\right| \ll 1 \quad (12.29)$$

We can assess the integral in two limiting cases:

1. At such small energies that k$\alpha \ll 1$. Accordingly,

$$E \ll \frac{\hbar^2}{\beta m_o \alpha^2} \quad (12.30)$$

Based on equation (12.5), we can write:

$$\gamma(B_o + g) \ll \frac{\hbar}{\beta m_o \alpha^2} \quad (12.31)$$

(we have dropped the unimportant factor two in the denominator.

2. At such great energies that k$\alpha \gg 1$. Accordingly,

$$E \gg \frac{\hbar^2}{\beta m_o \alpha^2} \quad (12.32)$$

Based on equation (12.1e), we can write:

$$\gamma(B_o + g) \gg \frac{\hbar}{\beta m_o \alpha^2} \tag{12.33}$$

It is important to note that, the expression $\frac{\hbar}{\beta m_o \alpha^2}$ gives the order of magnitude of the static magnetic field B_o that, in accordance with the uncertainty relation, a particle confined in a volume with a radius of the order α would have the momentum of the order $\frac{\hbar}{\alpha}$. It can also be shown that this expression equals the minimum depth of a spherical potential well of radius α at which a bound state of a particle with discrete energy level appears.

In the first case, the exponential factor in the integrand in the entire region of integration virtually equals unity so that Eq. (12.31) transforms into:

$$\frac{\beta m_o}{2\pi\hbar^2} \int\limits_{r\prime \leq \alpha} \left| \frac{U(r\prime)dV\prime}{r\prime} \right| \approx \frac{\beta m_o}{2\pi\hbar^2} \left| \langle U \rangle \right| \int\limits_{r' \leq \alpha} \frac{dV\prime}{r\prime} = \frac{\beta m_o}{2\pi\hbar^2} |\langle U \rangle| 2\pi\alpha^2 \ll 1$$

where $\langle U \rangle$ is the value of the potential energy averaged over the region of radius α. The inequality obtained leads to the condition

$$|\langle U \rangle| \ll \frac{\hbar^2}{\beta m_o \alpha^2} \tag{12.34}$$

The condition must be observed for energies satisfying Eq. (12.32).

In the second case, the exponential factor in Eq. (12.31) varies very rapidly and in the integration over the region $r' \leq \alpha$ it completes many cycles of change. Consequently, we replace the comparatively slowly varying factor $U(r')$ with its mean value $\langle U \rangle$ and put it outside the integration. As a result, the condition in Eq. (12.31) becomes:

$$\frac{\beta m_o}{\hbar^2} |\langle U \rangle| \left| \int\limits_{r\prime \leq \alpha} e^{ikr\prime(cos\theta+1)} sin\theta d\theta r\prime dr\prime \right| \ll 1 \tag{12.35}$$

(we have substituted $2\pi sin\theta d\theta r'^2 dr\prime$ for $dV\prime$ and cancelled 2π in the number and denominator; here θ is the angle between the vectors k and $r\prime$ measured from the fixed vector k). Evaluation of integral yields:

$$\int_0^a r\prime dr\prime \int_0^a e^{i\hbar r\prime(cos\theta+1)} sin\theta d\theta = \frac{1}{ik} \int_0^a \left(e^{2i\hbar r\prime} - 1\right) dr\prime = \frac{1}{ik} \left[\frac{e^{2i\hbar a} - 1}{2ik} - a \right] \approx \frac{a}{ik}$$

In this expression, we have disregarded the first term in the brackets because k is very large according to the condition stated in the equation above). Introducing the value of the integral into inequality in Eq. (12.35), we obtain: $\frac{\beta m_o}{\hbar^2} |\langle U \rangle| \frac{a}{k} \ll 1$.

Hence,

$$|\langle U\rangle| \ll \frac{k\hbar^2}{\beta m_o\alpha} \tag{12.36}$$

12.3 Conclusion

We have solved the Bloch NMR flow equation to describe the evolution of the wave like properties and find the wave functions which can be useful to understand quantum mechanical dynamics in computational magnetic resonance imaging. The rationale, opportunities, and challenges of low-field MRI are revealed in the language of quantum mechanics. The important quantum mechanical considerations for low-field imaging are explored, summarizing limitations and potential future developments that may enhance the clinical adoption of this technology.

For example, the perturbation function U must satisfy this condition at energy values determined by Eq. (12.33). We must note that the condition (29) is weaker than Eq. (12.34). Indeed the right hand side of Eq. (12.36) is $k\alpha$ times larger than the right hand side of Eq. (12.34), and in case 2 the product ka is very large. Hence, it can be considered as a perturbation at low energies of a particle, this is also at high energies. The condition in Eq. (12.36) can be represented in the form:

$$|\langle U\rangle| \ll \frac{\hbar v}{\alpha} \tag{12.37}$$

where $v = \hbar k/\beta m_0 = p/\beta m_0$ is the velocity of a particle.

For a Coulomb field, the potential $U = \alpha/r$ diminishes slowly such that we cannot introduce the concept of the effective radius α of region of intervention. Taking arbitrarily a certain quantity α, the value of the Coulomb potential average over a sphere of radius α is:

$$\langle U\rangle = \frac{1}{\frac{4}{3}\pi a^3}\int_0^a \frac{\alpha}{r}4\pi r^2 dr \approx \frac{\alpha}{a}$$

The condition in Eq. (12.37) for a Coulomb field can thus be written as:

$$\alpha \ll \hbar v \tag{12.38}$$

Particularly, if the potential is set up by a nucleus with the charge Ze, while the particle has the charge e, the condition in Eq. (12.38) becomes:

$$Ze^2 \ll \hbar v \tag{12.39}$$

Low-field MRI provides opportunities to develop a more compact, and portable system in comparison to current clinical MRI scanners (1.5 T field strengths and above). Equation (12.32) can be a fundamental principle for Physics-informed neural network for digital low-field portable magnetic resonance imaging (MRI) scanners which can be more accessible, cost-effective, and sustainable with lower carbon emissions. Clinical high-field MRI scanners based on Eq. (12.33) can be quantum mechanically designed. Although, the images produced by low field scanners have relatively poor image quality, lower signal-to-noise ratio, and limited spatial resolution, these can be improved by the application of artificial intelligence-based technology. The formulation of the parameter β is an interesting dimensionless mechanism that may provide a switch between low-field and high-field MRI. This parameter can be used to tailor laboratory RF pulse to quantum RF energy absorptions.

To appreciate the significance of this result, we consider a hydrogen molecule, in the rectangular solvated cavity model of 1 cm long. Putting the appropriate numbers in Eq. (12.32 and 12.33) show that in this case:

1. At low static magnetic field,

$$(\mathrm{B_o} + g) \ll 1 \tag{12.40}$$

2. At high static magnetic field,

$$(\mathrm{B_o} + g) \gg 1 \tag{12.41}$$

In Eqs. (12.40 and 12.41), the value of the gradient g, is in the order of 0.002 Tm^{-1} and the dimensionless parameter β is in the order of 0.576×10^{-11}. Based on Eq. (12.34), the potential can be estimated as:

$$|\langle U \rangle| \ll 1.6 \times 10 - 37\mathrm{J} \tag{12.42}$$

In Eq. (12.42), the value of the dimensionless parameter β is in the order of 8. It may be interesting to note that when the particle (a healthy cell) encounters a force (diseases such as cardiovascular diseases, diabetes, hepatitis and cancer, so its potential energy is no longer zero but in the order of Eq. (12.42)). The force modifies the wave packet in terms of low and high static magnetic fields as shown in Eqs. (12.41) and (12.42). Applying such propagation techniques appropriately can provide useful techniques to find solutions to biological, medical and physical problems in low and high income settings.

Acknowledgements The authors acknowledge the motivated research efforts of Raymond Confidence (Department of Physics, Federal University of Technology, Minna, Nigeria) as a promising young researcher.

References

Awojoyogbe, O.B.: A mathematical model of the Bloch NMR equations for quantitative analysis of blood flow in blood vessels with changing cross-section—I. Phys. A **303**(1–2), 163–175 (2002)

Awojoyogbe, O.B.: A mathematical model of Bloch NMR equations for quantitative analysis of blood flow in blood vessels of changing cross-section—PART II. Phys. A **323**, 534–550 (2003)

Awojoyogbe, O.B.: Analytical solution of the time-dependent Bloch NMR flow equations: a translational mechanical analysis. Phys. A **339**(3–4), 437–460 (2004)

Awojoyogbe, O.B.: A quantum mechanical model of the Bloch NMR flow equations for electron dynamics in fluids at the molecular level. Phys. Scr. **75**(6), 788 (2007)

Castelli, D.D., Gianolio, E., Aime, S.: MRI contrast agents: state of the art and new trends. In: Bioinorganic Medicinal Chemistry, pp. 223–251 (2011).

Cowan, B.P.: Nuclear Magnetic Resonance and Relaxation, 1st edn. Cambridge University Press, Cambridge (1997)

Dada, M.O., Awojoyogbe, B.O.: Fundamental physics of nuclear magnetic resonance. In: Computational Molecular Magnetic Resonance Imaging for Neuro-Oncology, pp. 11–114 (2021)

Dada, M.O., Awojoyogbe, B.O.: Analysis of the hydrogen-like atom for neuro-oncology based on Bloch's NMR flow equation. In: Computational Molecular Magnetic Resonance Imaging for Neuro-Oncology, pp. 219–246 (2021)

Dawson, J., Lauterbur, P.C.: Magnetic resonance imaging. Scholarpedia **3**(7), 3381 (2008) (revision #123947). https://doi.org/10.4249/scholarpedia.3381

Haacke, E.M., Brown, R.W., Thompson, M.R., Venkatesan, R.: Magnetic Resonance Imaging: Physical Principles and Sequence Design, pp. 39–40. Wiley, New York (1999)

Rai, A.J., Yee, J., Fleisher, M.: Biomarkers in the era of personalized medicine—a multiplexed SNP assay using capillary electrophoresis for assessing drug metabolism capacity. Scand. J. Clin. Lab. Invest. **70**(sup242), 15–18 (2010)

Slichter, C.P.: Principles of magnetic resonance, vol. 1. Springer Science & Business Media Rabi, I. I., Berlin, Germany (1937) (Space quantization in a gyrating magnetic field. Phys. Rev. 51(8), 652)

Chapter 13
Artificial Intelligence-Based Classification of Brain Tumors

Abstract The most difficult aspect of diagnosing a brain tumor is the tumor's location, size, and shape fluctuations. The diagnosis of a brain tumor depends on the nature and location of the tumor so that doctors can forecast the patient's survival chances and make treatment selections ranging from surgery to radiotherapy and chemotherapy In addition, pathologists, who are highly trained in the field of neurosurgery, visually examine patients to diagnose brain tumors in clinics. Nonetheless, this process is performed manually, which is not only time-consuming but also laborious and prone to human error. Therefore, the need for a model that could accurately classify brain tumors is crucial. The goal of this chapter is to develop and implement a transfer learning approach for the artificial intelligence-based classification of brain tumors using MRI images by: (i) pre-processing magnetic resonance scans of brain tumors according to their class (glioma, meningioma, or pituitary tumors) (ii) develop a transfer learning model capable of accurately classifying the various types of brain tumors (iii) develop an easy-to-use web application/GUI based on the trained model.

Keywords Artificial intelligence · Classification of brain tumors · Magnetic resonance scans imaging · Transfer learning · Web application/GUI

13.1 Introduction

Researchers in the past few years have proposed a variety of techniques with solutions for diagnosing brain tumors using MRI images. These techniques range from classic machine learning methods to deep learning models. In image processing techniques, artificial intelligence and deep learning are primarily used to segment, identify, and classify MRI images as well as to classify and diagnose brain tumors. Many methods for classifying brain tumors on MRI brain images have been proposed recently and are based on deep transfer learning models. Bulla et al. (2020) created a hybrid model based on CNN to categorise the different types of brain tumors. Based on the MRI brain image dataset, Khan et al. (2020) proposed an automated brain tumor detection mechanism using CNN with transfer learning models. The authors' analysis

B. O. Awojoyogbe and M. O. Dada, *Digital Molecular Magnetic Resonance Imaging*, Series in BioEngineering, https://doi.org/10.1007/978-981-97-6370-2_13

of the pre-processing of MRI image data improves the classification accuracy for diagnosing brain tumor disease (Hatami et al. 2019). The three types of tumors found in brain MRI images were classified by researchers (Hidayatullah et al. 2019) using a new CNN-based model.

To categorise and find anomalies in biological images, numerous machine learning and deep learning models (Harish and Baskar 2020; Ahmad et al. 2021) were applied. The process for the early diagnosis of brain tumors is described in the model in Khan et al. (2020), which calls for the extraction and concatenation of features from numerous levels. The three deep learning models used to validate this model before it was trained were DensNet201 and Inception-v3. These two models were used to investigate two different potential directions for categorising BT. A model for the automated identification of brain tumors was created (Khan et al. 2020) and presented using a dataset of 253 images and pre-trained versions of VGG-16 and Inception-V3, 155 images of malignant tumors and 98 images of healthy tissue are included in this dataset. Due to the dataset's insufficient size, it was unable to fine-tune the CNNs using it, and the test dataset was too small to evaluate the model's effectiveness.

A model for the automated identification of brain tumors was proposed (Amin et al. 2020, 2021) using VGG-16 and the BRaTs dataset the model's accuracy was improved to 84% by using transfer learning and fine-tuning over the course of 50 epochs. Overfitting in neural networks is a problem that Srivastava et al. addressed by presenting a dropout strategy (Srivastava et al. 2014). This method entails randomly removing the connections and the units.

Convolutional neural networks were chosen as the learning technique by Dvořák and Menze (2015) because they are efficient at handling feature correlations when used with the freely accessible BRATS2014 data set, which contains three different multimodal segmentation tasks. For the brain tumor segmentation data set, which contained 254 multimodal volumes and took just 13 s to process, they evaluated the method and were able to obtain cutting-edge results. Irsheidat and Duwairi (2020) used artificial convolutional neural networks in their research study. The outcomes showed that the CNN model produced good outcomes.

The time and data required to train a deep neural network can vary substantially based on several variables, including the size of the network, the complexity of the task it is being trained for, the quality of the data, and the computational resources available for training. In general, training a deep neural network can take a lot of time and computational resources, especially if you are training on a large dataset or if you are training for a complex task. Additionally, the amount of data necessary for training can vary considerably. Certain tasks, such as picture classification, may necessitate extremely large datasets (tens of thousands or even millions of samples) for optimal performance. Brain tumors are one of the most challenging and life-threatening diseases, affecting millions of people worldwide. Accurate detection of brain tumors is crucial for effective treatment planning, and early detection can significantly improve patient outcomes. However, detecting brain tumors from medical images is a complex and time-consuming task, requiring specialized skills and knowledge. Therefore, the development of an accurate and efficient brain tumor detection system is of paramount importance.

In recent years, deep learning techniques have shown great potential in medical image analysis, including brain tumor detection. However, the success of these techniques heavily relies on the availability of large amounts of annotated medical images, which are often limited in medical imaging datasets. Transfer learning is a promising approach that can leverage pre-trained deep learning models on large-scale datasets to enhance the performance of medical image analysis tasks with limited training data. Therefore, the research problem addressed in this final year project is to develop an accurate brain tumor detection system using a transfer learning approach. The proposed system aims to leverage pre-trained deep learning models on largescale datasets to enhance the performance of brain tumor detection on limited medical image datasets.

Magnetic resonance imaging (MRI) is a medical imaging technology that employs a magnetic field and radio waves generated by a computer to obtain detailed images of organs and tissues in the body. Researchers depend on brain magnetic resonance imaging (MRI) as one of the best imaging modalities for detecting brain cancers and modelling tumor growth during both the detection and therapy stages. High-resolution MRI pictures have a significant impact on the field of automatic medical image processing since they can provide a wealth of information on the brain's structure and tissue abnormalities. Because tumors vary in size, appearance, and location, tumor segmentation and categorization are a tough undertaking. Due to its complex anatomy, it is difficult to fully segment and diagnose a brain tumor with a single modality scan. MRI's capacity to record several pictures, known as multimodality imaging, can offer the detailed structure of the brain required for the efficient classification of brain tumors. MRI produces images of inside organs using magnetic fields and radio waves that are both powerful and efficient. MRI gives more accurate information regarding interior organs than CT or EEG scans, and is, therefore more effective. In recent years, artificial intelligence and deep learning have led to important advancements in medical science, such as medical image processing, which enables clinicians to reliably and swiftly identify disease. It was time intensive. Therefore, computer-assisted technology is required to overcome these constraints, as the Medical Field requires efficient and dependable ways to identify life threatening disorders, such as cancer, which is the top cause of death among patients worldwide.

The process of correctly identifying a brain tumor can take a significant amount of time, and the results can vary widely from one physician to the next. Automating this segmentation process will assist the doctors to come up with a treatment plan for the tumor more quickly, ultimately speeding up the diagnosing process. The automation of this procedure is accomplished by constructing a model with the use of a machine-learning approach known as transfer learning.

This model can appropriately categorize a case into the categories of Glioma Tumor, No Tumor, Meningioma Tumor, or Pituitary Tumor by thoroughly analysing the input brain picture with 99.9% accuracy. The model is taught using a dataset that has been labeled to indicate whether it contains examples of brain tumors, and the learned model will be able to properly predict the class to which the input brain MRI picture belongs.

13.2 Python Modules/Packages

The standard library in Python is very large and provides a wide range of facilities. The library includes built-in modules (written in C) that give access to system functionality like file I/O that Python programmers would not otherwise have access to, as well as modules are written in Python that offer standardised solutions for many issues that arise in routine programming. By abstracting away platform-specifics into platform-neutral APIs, some of these modules are specifically created to promote and improve the portability of Python programmes. In the fields of machine learning, data science, data visualization, etc., python libraries are extremely important. Some popular examples of python libraries are TensorFlow, Matplotlib, Pandas, Numpy, and Scikit-learn, etc.

13.3 Streamlit

Streamlit is an open-source Python library that makes it easy to create and share beautiful, custom web apps for machine learning and data science (https://docs.streamlit.io/).

13.4 Jupyter Notebook

Jupyter notebook was created to create open-source software, open standards, and services for interactive computing in dozens of programming languages. The jupyter notebook can be run on a local desktop without the need for internet connectivity, or it can be installed on a remote server (for example, Anaconda) and accessed via the internet. It offers an easy-to-use, efficient, document-focused interface.

13.5 MRI Brain Tumor Dataset

The MRI data were obtained from Kaggle, the dataset consists of 3,264 images grouped into training and testing. Kaggle is a crowd-sourced platform to attract, nurture, train and challenge data scientists from all around the world to solve data science, machine learning and predictive analytics problems. It has over 536,000 active members from 194 countries, and it receives close to 150,000 submissions per month. There are 3264 MRI brain images in total, including 926 gliomas, 937 meningiomas, 901 pituitary and 500 neutral (no tumor) tumor images. Figure 13.1 shows examples of the various classes of brain tumor images. The images is of size

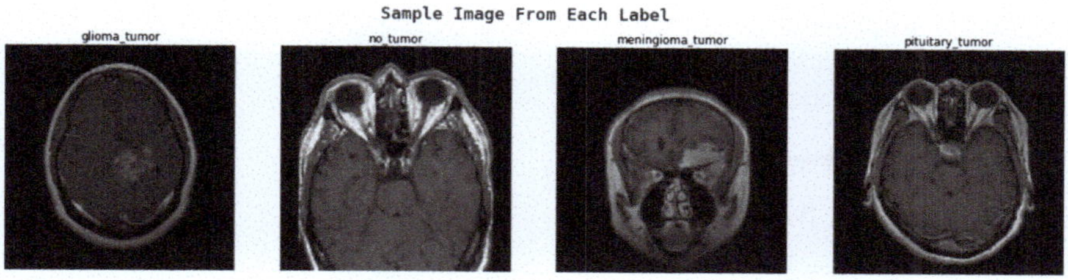

Fig. 13.1 Dataset sample

Table 13.1 A description of the brain data set

Tumor class	Images	Class label	Testing data	Training data
Glioma	926	0	100	826
No tumor	500	1	105	395
Meningioma	937	2	115	822
Pituitary	901	3	74	827

150×150 rs. In this study, 2870 MRI images (87.93%) were used for training, while 394 were used for testing (12.1%) (Table 13.1).

13.5.1 Data Processing

This dataset includes 3264 images of different types of BTs. Each image was transformed to a grayscale format. These data, which have been pre-processed, are one of the things that the neural network uses as input, along with the label of the image. Label 0 describes an image of a glioma; Label 1 describes an image with no tumor; Label 2 depicts the meningioma; and Label 3 depicts a pituitary gland. However, a dataset consisting of 3264 MRI images is needed, in order to train the EfficientNetB0 transfer learning model effectively, which has four million and fifty-four thousand six hundred and ninety-five (4,054,695) parameters.

13.5.2 Criteria for Exclusion from the Study

Individuals or subjects with implanted metal clips or connectors of the kind that may concentrate radiofrequency fields or damaged tissue from twisting in a magnetic field were excluded from the study. MR scanning is contraindicated in these people or subjects.

These instances were considered.

i Metallic Implant
ii Electronic devices (Pacemaker or defibrillator implanted in the heart)
iii Dental implants or dentures

13.6 Proposed Model

We used a transfer learning model of EfficientNet-B0 which is a convolutional neural network that is trained on more than a million images from the ImageNet database for the segmentation and classification of three types of brain tumors (pituitary, meningioma, and glioma) in order to further enhance the classification performance of BTs and shorten training time.

13.6.1 Deep Learning

Deep learning is a subset of machine learning that is essentially a three- or more-layered neural network. These neural networks attempt to simulate the behaviour of the human brain, albeit with limited success, allowing it to "learn" from large amounts of data. While a single-layer neural network can still make approximate predictions, additional hidden layers can help optimise and refine for accuracy (Fig. 13.2).

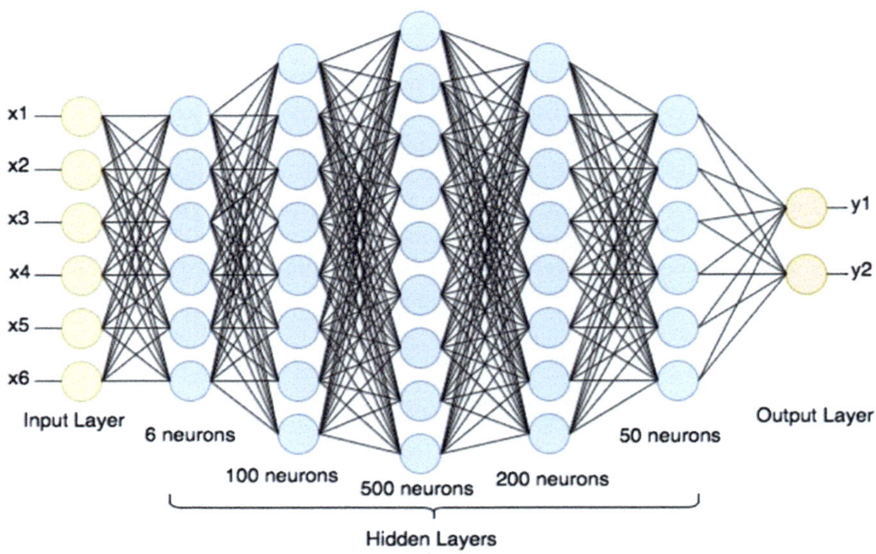

Fig. 13.2 Architecture of deep neural network

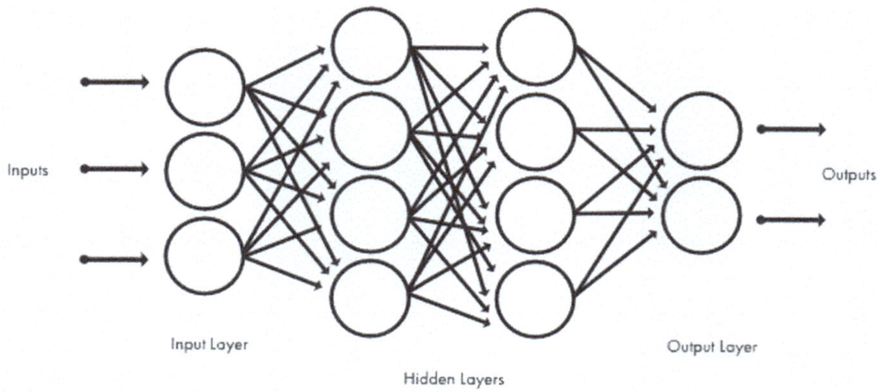

Fig. 13.3 Convolutional neural network (CNN)

13.6.2 Convolutional Neural Network (CNN)

The structure of the convolution neural network employed in the model developed in this study is shown in Figs. 13.3 and 13.4.

13.6.3 Global Average Pooling

Global average pooling is a deep learning method frequently used for computer vision tasks like image classification to obtain a global representation of the image, it involves averaging all the feature maps in a convolutional neural network (CNN) layer. A pooling operation called global average pooling applies across the entire feature map. It specifically entails averaging each feature map to produce a single value for each channel. By doing so, the spatial dimension of the feature maps is reduced to one, making it easier for later fully connected layers to work with them.

13.6.4 Dropout Layer

Convolutional neural networks (CNNs) can use the regularisation method known as dropout to avoid over fitting. When a model learns the noise or random fluctuations in the training data instead of the underlying patterns that generalise to new, untried data, overfitting takes place. Dropout operates by arbitrarily eliminating (setting to zero) a portion of the network's units (neurons) during training. This encourages the network to learn more reliable and generalizable representations of the input data and helps prevent it from relying too heavily on any one feature.

```
Model: "model"
__________________________________________________________________________________________________
 Layer (type)                   Output Shape         Param #     Connected to
==================================================================================================
 input_3 (InputLayer)           [(None, 150, 150, 3  0           []
                                )]

 rescaling_2 (Rescaling)        (None, 150, 150, 3)  0           ['input_3[0][0]']

 normalization_2 (Normalization  (None, 150, 150, 3)  7          ['rescaling_2[0][0]']
 )

 stem_conv_pad (ZeroPadding2D)  (None, 151, 151, 3)  0           ['normalization_2[0][0]']

 stem_conv (Conv2D)             (None, 75, 75, 32)   864         ['stem_conv_pad[0][0]']

 stem_bn (BatchNormalization)   (None, 75, 75, 32)   128         ['stem_conv[0][0]']

 stem_activation (Activation)   (None, 75, 75, 32)   0           ['stem_bn[0][0]']

block1a_dwconv (DepthwiseConv2  (None, 75, 75, 32)  288          ['stem_activation[0][0]']
D)

block1a_bn (BatchNormalization  (None, 75, 75, 32)  128          ['block1a_dwconv[0][0]']
)

block1a_activation (Activation  (None, 75, 75, 32)  0            ['block1a_bn[0][0]']
)

block1a_se_squeeze (GlobalAver  (None, 32)          0            ['block1a_activation[0][0]']
agePooling2D)

block1a_se_reshape (Reshape)   (None, 1, 1, 32)     0            ['block1a_se_squeeze[0][0]']

block1a_se_reduce (Conv2D)     (None, 1, 1, 8)      264          ['block1a_se_reshape[0][0]']

block1a_se_expand (Conv2D)     (None, 1, 1, 32)     288          ['block1a_se_reduce[0][0]']

 block1a_se_excite (Multiply)   (None, 75, 75, 32)  0           ['block1a_activation[0][0]',
                                                                  'block1a_se_expand[0][0]']

 block1a_project_conv (Conv2D)  (None, 75, 75, 16)  512         ['block1a_se_excite[0][0]']

 block1a_project_bn (BatchNorma  (None, 75, 75, 16)  64         ['block1a_project_conv[0][0]']
 lization)

 block2a_expand_conv (Conv2D)   (None, 75, 75, 96)  1536        ['block1a_project_bn[0][0]']

 block2a_expand_bn (BatchNormal  (None, 75, 75, 96)  384        ['block2a_expand_conv[0][0]']
 ization)

 block2a_expand_activation (Act  (None, 75, 75, 96)  0          ['block2a_expand_bn[0][0]']
 ivation)

 block2a_dwconv_pad (ZeroPaddin  (None, 77, 77, 96)  0          ['block2a_expand_activation[0][0]
 g2D)                                                           ']
```

Fig. 13.4 CNN model training

```
block2a_dwconv (DepthwiseConv2  (None, 38, 38, 96)  864         ['block2a_dwconv_pad[0][0]']
D)

block2a_bn (BatchNormalization  (None, 38, 38, 96)  384         ['block2a_dwconv[0][0]']
)

block2a_activation (Activation  (None, 38, 38, 96)  0           ['block2a_bn[0][0]']
)

block2a_se_squeeze (GlobalAver  (None, 96)          0           ['block2a_activation[0][0]']
agePooling2D)

block2a_se_reshape (Reshape)   (None, 1, 1, 96)     0           ['block2a_se_squeeze[0][0]']

block2a_se_reduce (Conv2D)     (None, 1, 1, 4)      388         ['block2a_se_reshape[0][0]']

block2a_se_expand (Conv2D)     (None, 1, 1, 96)     480         ['block2a_se_reduce[0][0]']

block2a_se_excite (Multiply)   (None, 38, 38, 96)  0           ['block2a_activation[0][0]',
                                                                 'block2a_se_expand[0][0]']

block2a_project_conv (Conv2D)  (None, 38, 38, 24)  2304        ['block2a_se_excite[0][0]']

block2a_project_bn (BatchNorma  (None, 38, 38, 24)  96          ['block2a_project_conv[0][0]']
lization)

block2b_expand_conv (Conv2D)   (None, 38, 38, 144)  3456        ['block2a_project_bn[0][0]']

block2b_expand_bn (BatchNormal  (None, 38, 38, 144)  576         ['block2b_expand_conv[0][0]']
ization)

block2b_expand_activation (Act  (None, 38, 38, 144)  0           ['block2b_expand_bn[0][0]']
ivation)

block2b_dwconv (DepthwiseConv2  (None, 38, 38, 144)  1296        ['block2b_expand_activation[0][0]
D)                                                              ']

block2b_bn (BatchNormalization  (None, 38, 38, 144)  576         ['block2b_dwconv[0][0]']
)

block2b_activation (Activation  (None, 38, 38, 144)  0           ['block2b_bn[0][0]']
)

block2b_se_squeeze (GlobalAver  (None, 144)         0           ['block2b_activation[0][0]']
agePooling2D)

block2b_se_reshape (Reshape)   (None, 1, 1, 144)    0           ['block2b_se_squeeze[0][0]']

block2b_se_reduce (Conv2D)     (None, 1, 1, 6)      870         ['block2b_se_reshape[0][0]']

block2b_se_expand (Conv2D)     (None, 1, 1, 144)    1008        ['block2b_se_reduce[0][0]']

block2b_se_excite (Multiply)   (None, 38, 38, 144)  0           ['block2b_activation[0][0]',
                                                                 'block2b_se_expand[0][0]']
```

Fig. 13.4 (continued)

```
block2b_project_conv (Conv2D)  (None, 38, 38, 24)   3456       ['block2b_se_excite[0][0]']

block2b_project_bn (BatchNorma  (None, 38, 38, 24)  96         ['block2b_project_conv[0][0]']
lization)

block2b_drop (Dropout)         (None, 38, 38, 24)   0          ['block2b_project_bn[0][0]']

block2b_add (Add)              (None, 38, 38, 24)   0          ['block2b_drop[0][0]',
                                                                'block2a_project_bn[0][0]']

block3a_expand_conv (Conv2D)   (None, 38, 38, 144)  3456       ['block2b_add[0][0]']

block3a_expand_bn (BatchNormal  (None, 38, 38, 144)  576       ['block3a_expand_conv[0][0]']
ization)

block3a_expand_activation (Act  (None, 38, 38, 144)  0         ['block3a_expand_bn[0][0]']
ivation)

block3a_dwconv_pad (ZeroPaddin  (None, 41, 41, 144)  0        ['block3a_expand_activation[0][0]
g2D)                                                          ']

block3a_dwconv (DepthwiseConv2  (None, 19, 19, 144)  3600     ['block3a_dwconv_pad[0][0]']
D)

block3a_bn (BatchNormalization  (None, 19, 19, 144)  576      ['block3a_dwconv[0][0]']
)

block3a_activation (Activation  (None, 19, 19, 144)  0        ['block3a_bn[0][0]']
)

block3a_se_squeeze (GlobalAver  (None, 144)          0        ['block3a_activation[0][0]']
agePooling2D)

block3a_se_reshape (Reshape)   (None, 1, 1, 144)    0         ['block3a_se_squeeze[0][0]']

block3a_se_reduce (Conv2D)     (None, 1, 1, 6)      870       ['block3a_se_reshape[0][0]']

block2b_se_expand (Conv2D)     (None, 1, 1, 144)    1008      ['block2b_se_reduce[0][0]']

block2b_se_excite (Multiply)   (None, 38, 38, 144)  0         ['block2b_activation[0][0]',
                                                               'block2b_se_expand[0][0]']

block2b_project_conv (Conv2D)  (None, 38, 38, 24)   3456      ['block2b_se_excite[0][0]']

block2b_project_bn (BatchNorma  (None, 38, 38, 24)  96        ['block2b_project_conv[0][0]']
lization)

block2b_drop (Dropout)         (None, 38, 38, 24)   0         ['block2b_project_bn[0][0]']

block2b_add (Add)              (None, 38, 38, 24)   0         ['block2b_drop[0][0]',
                                                               'block2a_project_bn[0][0]']

block3a_expand_conv (Conv2D)   (None, 38, 38, 144)  3456      ['block2b_add[0][0]']

block3a_expand_bn (BatchNormal  (None, 38, 38, 144)  576      ['block3a_expand_conv[0][0]']
ization)
```

Fig. 13.4 (continued)

```
 block6d_drop (Dropout)          (None, 5, 5, 192)    0           ['block6d_project_bn[0][0]']

 block6d_add (Add)               (None, 5, 5, 192)    0           ['block6d_drop[0][0]',
                                                                   'block6c_add[0][0]']

 block7a_expand_conv (Conv2D)    (None, 5, 5, 1152)   221184      ['block6d_add[0][0]']

 block7a_expand_bn (BatchNormal  (None, 5, 5, 1152)   4608        ['block7a_expand_conv[0][0]']

 ization)

 block7a_expand_activation (Act  (None, 5, 5, 1152)   0           ['block7a_expand_bn[0][0]']
 ivation)

 block7a_dwconv (DepthwiseConv2  (None, 5, 5, 1152)   10368       ['block7a_expand_activation[0][0]
 D)                                                               ']

 block7a_bn (BatchNormalization  (None, 5, 5, 1152)   4608        ['block7a_dwconv[0][0]']
 )


 block7a_se_squeeze (GlobalAver  (None, 1152)          0           ['block7a_activation[0][0]']
 agePooling2D)

 block7a_se_reshape (Reshape)   (None, 1, 1, 1152)    0           ['block7a_se_squeeze[0][0]']

 block7a_se_reduce (Conv2D)     (None, 1, 1, 48)      55344       ['block7a_se_reshape[0][0]']

 block7a_se_expand (Conv2D)     (None, 1, 1, 1152)    56448       ['block7a_se_reduce[0][0]']

 block7a_se_excite (Multiply)   (None, 5, 5, 1152)    0           ['block7a_activation[0][0]',
                                                                   'block7a_se_expand[0][0]']

 block7a_project_conv (Conv2D)  (None, 5, 5, 320)     368640      ['block7a_se_excite[0][0]']

 block7a_project_bn (BatchNorma  (None, 5, 5, 320)    1280        ['block7a_project_conv[0][0]']
 lization)

 top_conv (Conv2D)              (None, 5, 5, 1280)    409600      ['block7a_project_bn[0][0]']

 top_bn (BatchNormalization)    (None, 5, 5, 1280)    5120        ['top_conv[0][0]']

 top_activation (Activation)    (None, 5, 5, 1280)    0           ['top_bn[0][0]']

 global_average_pooling2d (Glob  (None, 1280)         0           ['top_activation[0][0]']
 alAveragePooling2D)

 dropout (Dropout)              (None, 1280)          0           ['global_average_pooling2d[0][0]'
                                                                  ]

 dense (Dense)                  (None, 4)             5124        ['dropout[0][0]']

==================================================================================================
Total params: 4,054,695
Trainable params: 4,012,672
Non-trainable params: 42,023
```

Fig. 13.4 (continued)

13.6.5 Dense Layer

A dense layer, which is a fully connected layer, is used to categorise the features that the convolutional layers have extracted. This is the output layer which classifies the image into 1 of the 4 possible classes.

13.6.6 Softmax

For multi-class classification tasks, a Convolutional Neural Network (CNNoutput) layer frequently employs the activation function known as Softmax. The softmax function transforms a CNN's output from the top layer into a probability distribution over all conceivable classes.

13.7 EfficientNetB0

A convolutional neural network named EfficientNet-b0 was trained using more than a million pictures from the ImageNet database. The network can categorise images into 1000 different object categories, including many animals, a keyboard, a mouse, and a pencil. The network has therefore acquired rich feature representations for a variety of images. The network accepts images with a resolution of 224 by 224 (Figs. 13.5, 13.6 and Table 13.2).

The Max Pooling, fully connected, softmax, input, convolutional, activation, normalisation, and classification layers are among the layers of the proposed learning model (EfficientNetB0). The images are fed into the input layer. The input images for the EfficientNetB0 learning model are 224 × 224 × 3. The width, height, and grayscale channel size of the image are indicated by those three digits (1 for grayscale images, and 3 for colour images). Prior to any further processing, the images were first sent to an input layer. The input image matrix and a filter are needed as inputs for the mathematical operation that makes up the convolutional layer. A feature map was produced as an output after the filter multiplied the input image. The mathematical expression of the convolution layer is as follows in Equation (13.1).

$$\mathrm{Y} = \sum (X_i * W_i) + \mathrm{b} \tag{13.1}$$

```
effnet = EfficientNetB0(weights='imagenet',include_top=False,input_shape=(image_size,image_size,3))

Downloading data from https://storage.googleapis.com/keras-applications/efficientnetb0_notop.h5
16711680/16705208 [==============================] - 16s 1us/step
16719872/16705208 [==============================] - 16s 1us/step
```

Fig. 13.5 Imagenet weights

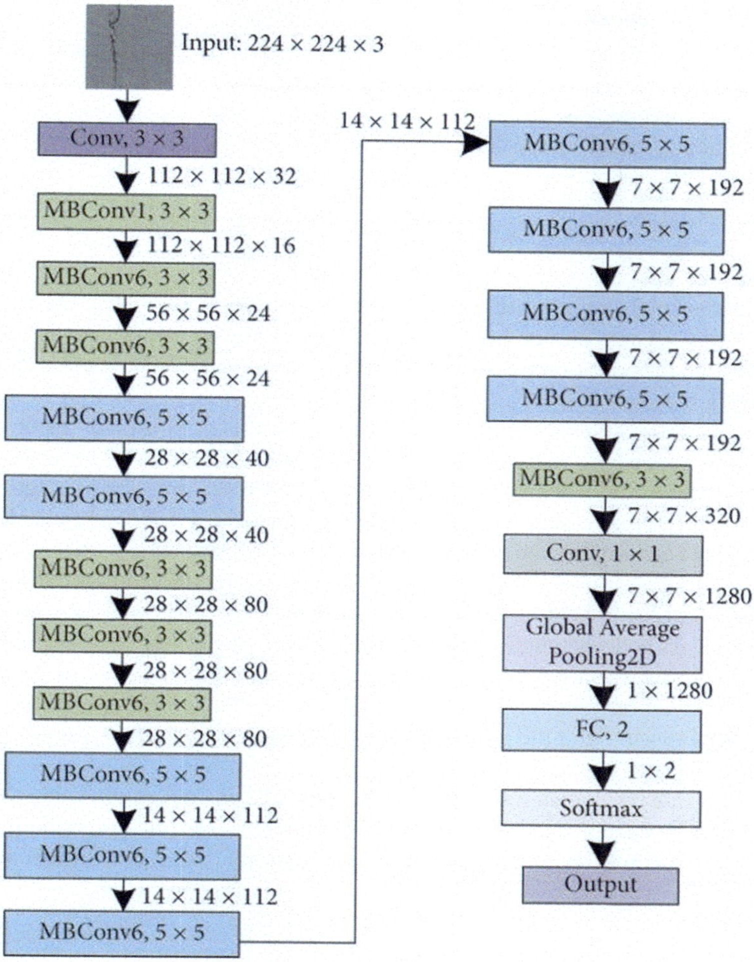

Fig. 13.6 EfficientNetB0 Architecture

Table 13.2 The specifications of the layers in the proposed learning model

Layer	Filter size	Number of filter
Convolution layer	3 × 3	
Batch_norm_layer	–	–
Soft-max layer	–	–
MBConv6 layer	3 × 3	
Glob_Avg_Pooling	7 × 7	
FC layer		
SoftMax		
Classification layer		

where W_i represent the weights, X_i represent the input, b shows bias, and the feature map in both layers represented by Y. The activation layer includes an activation function that gives nonlinearity to the neural network. Rectifier linear units (ReLUs) are used because they increase the training speed. Equation (13.2) shows the mathematical equation for the ReLU activation.

$$R(x) = f(x) = \{x, x > 0$$

$$0, x < 0 \tag{13.2}$$

A batch normalisation layer is applied to the outputs to normalise the parameters the proposed convolution layers produced. Normalization shortens the training period for the proposed model, which increases learning efficiency and speeds up the learning process.

The convolutional layer's primary drawback is that it only picks up features that are location dependent. As a result, the classification is incorrect if the feature's position in the image slightly changes. By making the representation more compact and invariant to minute changes and insignificant details, Max-Pooling enables the network to get around this constraint. To connect the features, maximum pooling and average pooling were used. The convolutional layers' learned features are sent to the fully connected layer. When a layer is described as "fully connected," it means that every one of its nodes is connected to the nodes in the layer above it. The primary goal of this layer is to assign input images to the appropriate classes. In this layer, a softmax activation function is employed. During training, the Loss function (H) must be minimised. After the image has passed through each of the prior layers, the output is calculated.

The error rate is calculated after it is compared to the desired output using the loss function. Iterations of this process are carried out until the loss function is minimised. The loss function used was the categorical cross-entropy (CCE) or Softmax loss.

$$output$$

$$\text{Loss} = -\sum i = size_1 \quad yi \, log\hat{y}i \tag{13.3}$$

where y_i represents the true class distribution and $\hat{y}_i$ represents predicted class distribution among I number of samples. The output of the fully connected layer is further normalised by the activation function. The network performs the probabilistic computation, and softmax produces the output with positive values for each category. The last layer of the model that will be displayed is the classification layer. This layer is in charge of combining all of the inputs to produce the output. The application of the softmax activation function resulted in the acquisition of a probability distribution.

13.7.1 ImageNet

ImageNet is a database of images that have been arranged according to the WordNet hierarchy (currently only the nouns), with thousands of images representing each node of the hierarchy. The project has significantly advanced the fields of computer vision and deep learning. For non-commercial use, the data is freely available to researchers.

13.8 Transfer Learning Setting (Method)

A model created for one task is used as the basis for another using the machine learning technique known as transfer learning. Pre-trained models are frequently used as the foundation for deep learning tasks in computer vision and natural language processing because they save both time and money compared to developing neural network models from scratch and because they perform vastly better on related tasks (Fig. 13.7).

13.8.1 Precision/Positive Predictive Value

The precision measures the model's accuracy in classifying a sample as positive.

$$PPV = \frac{TP}{TP + FP} \tag{13.4}$$

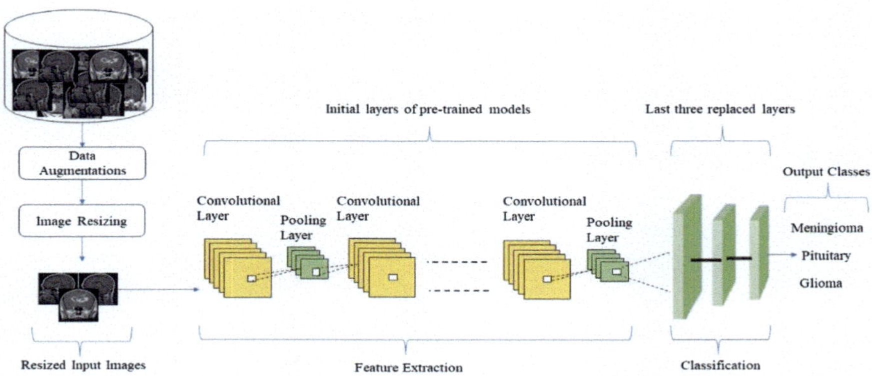

Fig. 13.7 Transfer learning

13.8.2 Recall/Negative Predictive Value

The recall measures the model's ability to detect Positive samples. The higher the recall, the more positive samples detected.

$$NPV = \frac{TP}{TP + FN} \tag{13.5}$$

13.9 CNN Model (Training and Testing Result)

The process and results from CNN model training are shown as follows (Fig. 13.8).

13.9.1 Training Metrics Visualizations

Model Accuracy

From Fig. 13.9 represents the performance evaluation of the convolutional neural network with training and validation loss and accuracy. At 10 epochs, the training accuracy is 99.8%, and validation accuracy is 97.6%. Similarly, the training loss is 0.009, and 0.08 is the validation loss.

13.9.2 Model Loss

From Fig. 13.10, at 10 epochs, the training loss is 0.009, and 0.08 is the validation loss.

13.9.3 Model Evaluation

See Fig. 13.11.

```
Epoch 1/12
83/83 [==============================] - ETA: 0s - loss: 0.4997 - accuracy: 0.8218
Epoch 1: val_accuracy improved from -inf to 0.89456, saving model to effnet.h5
83/83 [==============================] - 383s 4s/step - loss: 0.4997 - accuracy: 0.8218 - val_loss: 0.4103 - val_accuracy: 0.8946 - lr: 0.0010
Epoch 2/12
83/83 [==============================] - ETA: 0s - loss: 0.1904 - accuracy: 0.9361
Epoch 2: val_accuracy improved from 0.89456 to 0.91497, saving model to effnet.h5
83/83 [==============================] - 308s 4s/step - loss: 0.1904 - accuracy: 0.9361 - val_loss: 0.3225 - val_accuracy: 0.9150 - lr: 0.0010
Epoch 3/12
83/83 [==============================] - ETA: 0s - loss: 0.1605 - accuracy: 0.9478
Epoch 3: val_accuracy did not improve from 0.91497
83/83 [==============================] - 361s 4s/step - loss: 0.1605 - accuracy: 0.9478 - val_loss: 0.3986 - val_accuracy: 0.8844 - lr: 0.0010
Epoch 4/12
83/83 [==============================] - ETA: 0s - loss: 0.1097 - accuracy: 0.9682
Epoch 4: val_accuracy did not improve from 0.91497


Epoch 4: ReduceLROnPlateau reducing learning rate to 0.0003000000142492354.
83/83 [==============================] - 418s 5s/step - loss: 0.1097 - accuracy: 0.9682 - val_loss: 0.2302 - val_accuracy: 0.9116 - lr: 0.0010
Epoch 5/12
83/83 [==============================] - ETA: 0s - loss: 0.0432 - accuracy: 0.9860
Epoch 5: val_accuracy improved from 0.91497 to 0.95578, saving model to effnet.h5
83/83 [==============================] - 453s 5s/step - loss: 0.0432 - accuracy: 0.9860 - val_loss: 0.1389 - val_accuracy: 0.9558 - lr: 3.0000e-04
Epoch 6/12
83/83 [==============================] - ETA: 0s - loss: 0.0324 - accuracy: 0.9902
Epoch 6: val_accuracy improved from 0.95578 to 0.96939, saving model to effnet.h5
83/83 [==============================] - 457s 6s/step - loss: 0.0324 - accuracy: 0.9902 - val_loss: 0.1158 - val_accuracy: 0.9694 - lr: 3.0000e-04
Epoch 7/12
83/83 [==============================] - ETA: 0s - loss: 0.0151 - accuracy: 0.9970
Epoch 7: val_accuracy improved from 0.96939 to 0.97619, saving model to effnet.h5
83/83 [==============================] - 379s 5s/step - loss: 0.0151 - accuracy: 0.9970 - val_loss: 0.1043 - val_accuracy: 0.9762 - lr: 3.0000e-04


Epoch 8/12
83/83 [==============================] - ETA: 0s - loss: 0.0114 - accuracy: 0.9970
Epoch 8: val_accuracy did not improve from 0.97619
83/83 [==============================] - 317s 4s/step - loss: 0.0114 - accuracy: 0.9970 - val_loss: 0.1021 - val_accuracy: 0.9762 - lr: 3.0000e-04
Epoch 9/12
83/83 [==============================] - ETA: 0s - loss: 0.0156 - accuracy: 0.9947
Epoch 9: val_accuracy did not improve from 0.97619

Epoch 9: ReduceLROnPlateau reducing learning rate to 9.000000427477062e-05.
83/83 [==============================] - 310s 4s/step - loss: 0.0156 - accuracy: 0.9947 - val_loss: 0.0890 - val_accuracy: 0.9762 - lr: 3.0000e-04
Epoch 10/12
83/83 [==============================] - ETA: 0s - loss: 0.0092 - accuracy: 0.9977
Epoch 10: val_accuracy did not improve from 0.97619
83/83 [==============================] - 312s 4s/step - loss: 0.0092 - accuracy: 0.9977 - val_loss: 0.0830 - val_accuracy: 0.9762 - lr: 9.0000e-05

Epoch 11/12
83/83 [==============================] - ETA: 0s - loss: 0.0082 - accuracy: 0.9981
Epoch 11: val_accuracy did not improve from 0.97619

Epoch 11: ReduceLROnPlateau reducing learning rate to 2.700000040931627e-05.
83/83 [==============================] - 342s 4s/step - loss: 0.0082 - accuracy: 0.9981 - val_loss: 0.0849 - val_accuracy: 0.9728 - lr: 9.0000e-05
Epoch 12/12
83/83 [==============================] - ETA: 0s - loss: 0.0066 - accuracy: 0.9989
Epoch 12: val_accuracy did not improve from 0.97619
83/83 [==============================] - 335s 4s/step - loss: 0.0066 - accuracy: 0.9989 - val_loss: 0.0890 - val_accuracy: 0.9728 - lr: 2.7000e-05
```

Fig. 13.8 CNN model training process

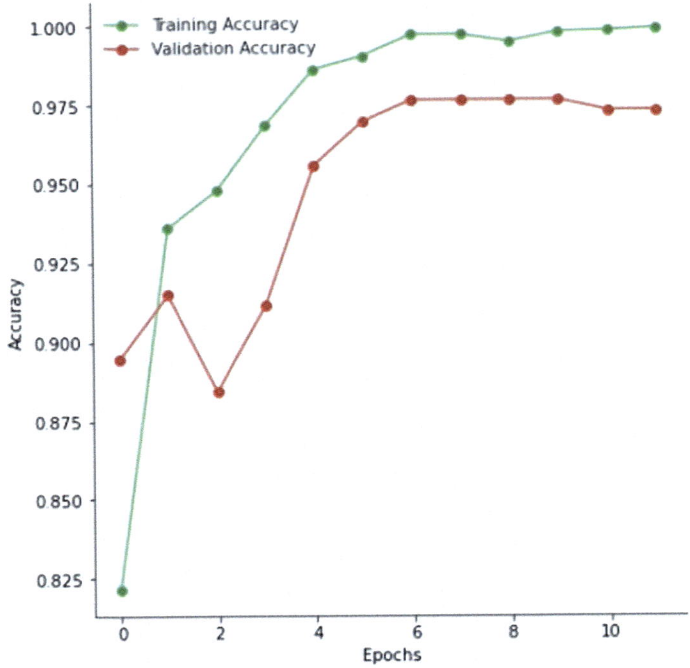

Fig. 13.9 Model accuracy

13.10 Classification Report on Tested Data

The report of tissue classification as obtained from model implementation with the test dataset is given in Table 13.3. Furthermore, the trained model returned accuracy 98%, precision 98%, recall 98% and F1-measure 98%.

13.10.1 Evaluation Metrics

a. **True Positive (TP):** A test result that correctly indicates the presence of a condition or characteristic
b. **True Negative (TN):** A test result that correctly indicates the absence of a condition or characteristic
c. **False Positive (FP):** A test result which wrongly indicates that a particular condition or attribute is present
d. **False Negative (FN):** A test result which wrongly indicates that a particular condition or attribute is absent

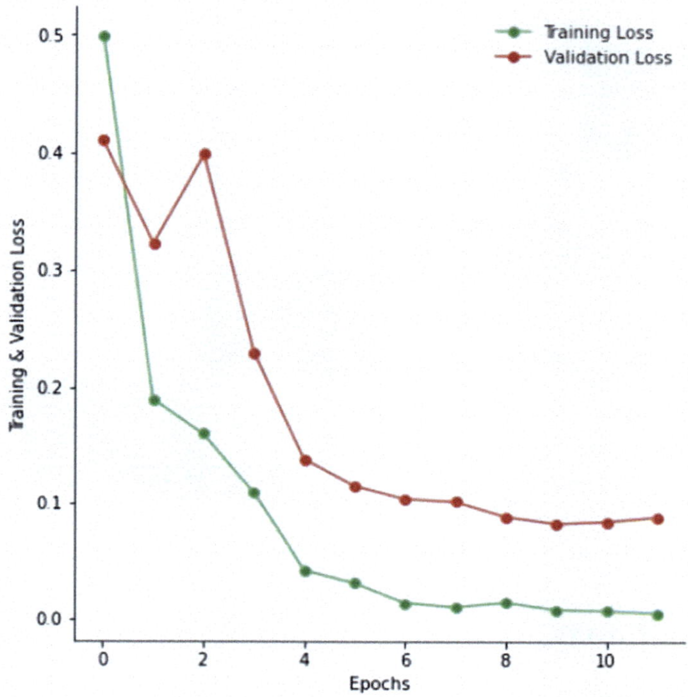

Fig. 13.10 Model loss

```
Epoch 12/12
83/83 [==============================] - ETA: 0s - loss: 0.0066 - accuracy: 0.9989
Epoch 12: val_accuracy did not improve from 0.97619
83/83 [==============================] - 335s 4s/step - loss: 0.0066 - accuracy: 0.9989 - val_loss: 0.0890 - val_accuracy:
0.9728 - lr: 2.7000e-05
```

Fig. 13.11 Model result on the validation dataset

Table 13.3 Classification report

	Precision	Recall	F1-score	Support
Glioma tumor	0.99	0.95	0.97	93
No tumor	0.96	1.00	0.98	51
Meningioma tumor	0.97	0.98	0.97	96
Pituitary tumor	0.99	1.00	0.99	87
Accuracy			0.98	327
Macro avg			0.98	327
Weighted avg	0.98	0.98	0.98	327

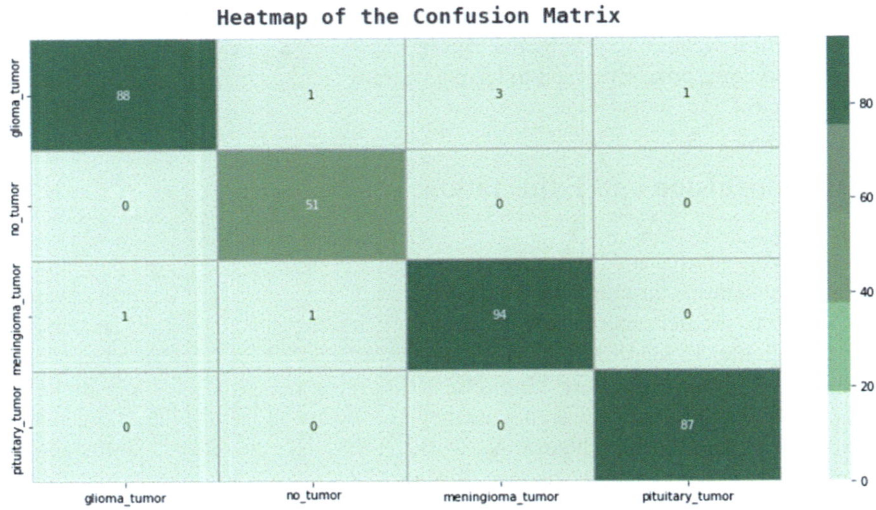

Fig. 13.12 Confusion matrix of the model according to tumor class

13.10.2 F1-Score

F1 score is a combination of precision and recall. A perfect model achieves an F1 score of 1.0.

$$F1score = \frac{2 \times precision \times recall}{precision \times recall} \tag{13.6}$$

13.10.3 Support

From Fig. 13.12 the number of samples each metric was calculated on is as follows:

i Glioma Tumor metrics was calculated with a **support** of 93
ii No Tumor metrics was calculated with a **support** of 51
iii Meningioma Tumor metrics was calculated with a **support** of 96
iv Pituitary Tumor metrics was calculated with a **support** of 87

13.10.4 Accuracy

Accuracy is a metric that generally describes how the model performs across all classes. It is useful when all classes are of equal importance.

$$Accuracy = \frac{TP + TN}{TP + TN + FP + FN} \tag{13.7}$$

(i) Macro average: the average precision, recall and F1 score between classes.
(ii) Weighted average: the weighted average precision, recall and F1 score between classes. Weighted means each metric is calculated with respect to how many samples there are in each set.

13.11 Confusion Matrix on the Test Data

13.12 Model Deployment Process with Streamlit

i. On the anaconda prompt run "pip install Streamlit", this is an installation of Streamlit library. Streamlit is an open-source Python library that makes easy to create and share beautiful, custom web apps for machine learning and data science. In just few minutes you can build and deploy powerful data apps, we then wrote a python code where the model saved was called and widgets to make our GUI web application interactive were included.
ii. Write a python program with the model of best accuracy score and then save it as a.hdf5 file using TensorFlow. HDF5 is a general-purpose library and file format for storing scientific data. HDF5 can store two primary types of objects: datasets and groups. A dataset is essentially a multidimensional array of data elements, and a group is a structure for organizing objects in an HDF5 file.
iii. Create a Streamlit app. This can be done writing a python code using Spyder IDE that creates the user interface and allows that machine learning model to take data and give feedbacks from the model. Save the APP as a.py file.
iv. On anaconda prompt change the environment directory to where your files saved. Then run the line, Streamlit run then paste the path of your.py file. ***Streamlit run "C:\Users\lenovo\Documents\Project\New Project\APP Deployment\brian_tumor_APP.py"***
v. Register github then create a repository and upload the files which includes a profile, requirement.txt, app.py, setup.sh, runtime_txt, classifier.pkl and README.md (Figs. 13.13, 13.14, 13.15, 13.16 and 13.17).

13.13 Model Deployment Process with DigitalOcean

i. Create a Streamlit app. This can be done writing a python code using Spyder IDE that creates the user interface and allows that machine learning model to take data and give feedbacks from the model. Save the APP as a.py file.

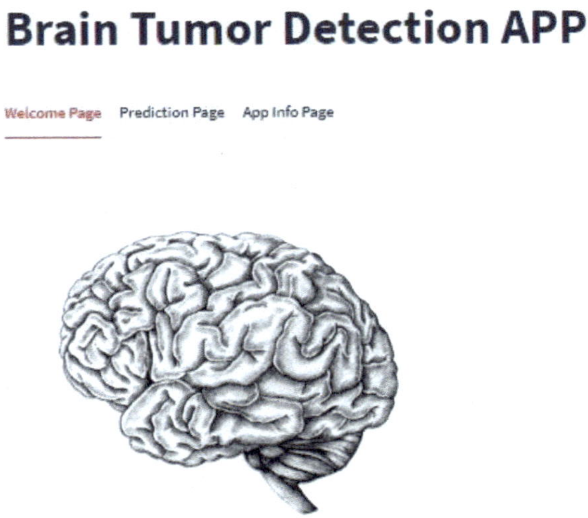

Fig. 13.13 Brain tumor app welcome page

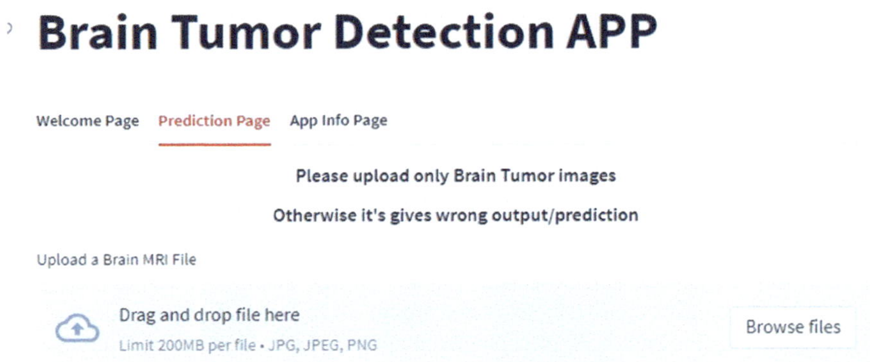

Fig. 13.14 Brain tumor app prediction page

ii. Register github then create a repository and upload the files which includes a profile, requirement.txt, app.py, setup.sh, runtime_txt, classifier.pkl and README.md.
iii. Sign up on DigitalOcean cloud platform, login and then navigate to a new project.
iv. Click on create and select App, on the resource page select Github and the app repository, fill in the environment variables, and click create.
v. Wait for some minutes as Digital Ocean configures the console needed for the successful deployment of the app.
vi. Copy the link created and share with the public (Figs. 13.18 and 13.19).

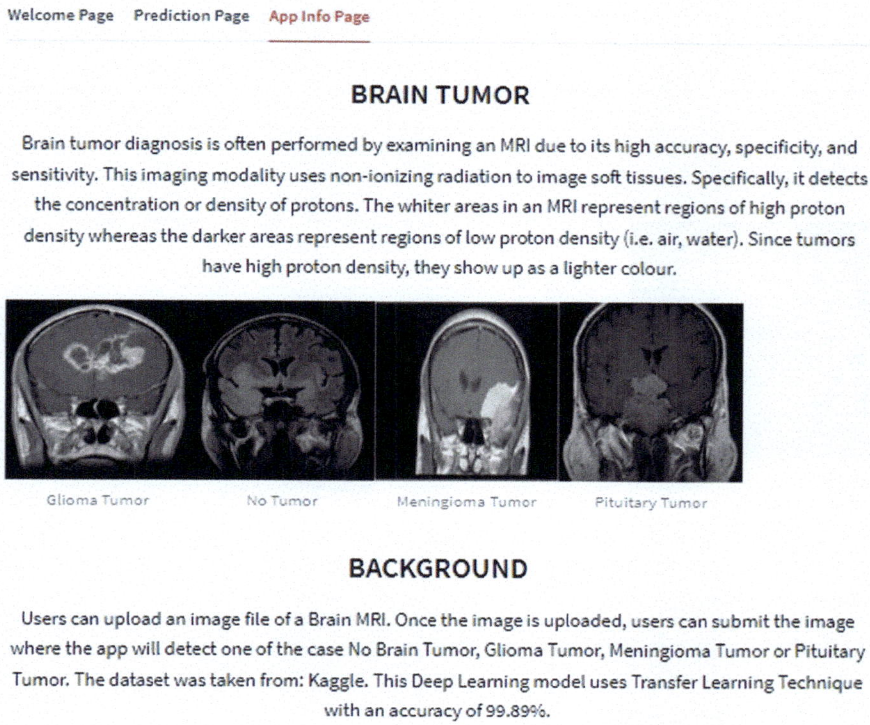

Fig. 13.15 Brain tumor app info page

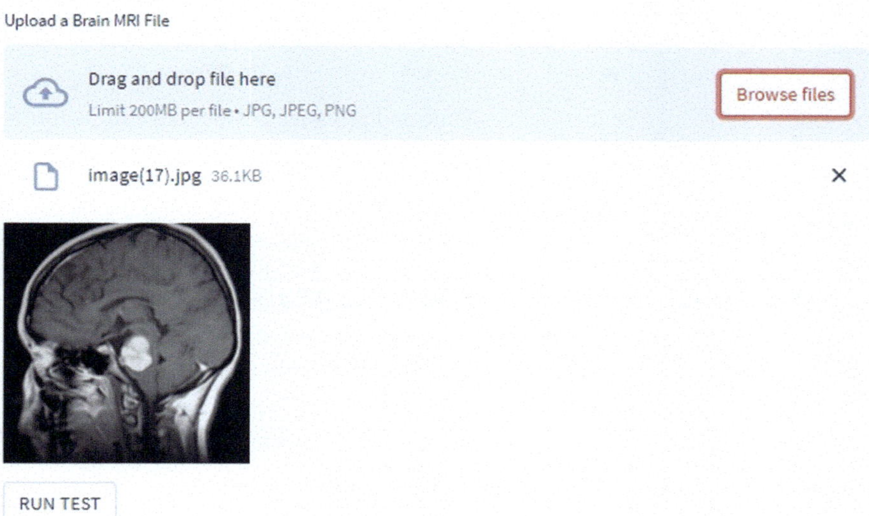

Fig. 13.16 MR scan testing

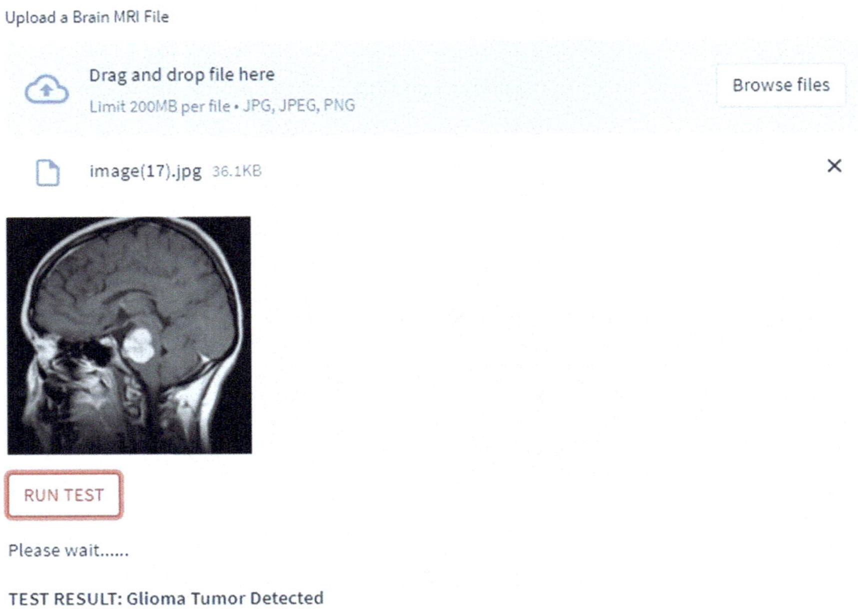

Fig. 13.17 Glioma tumor detection

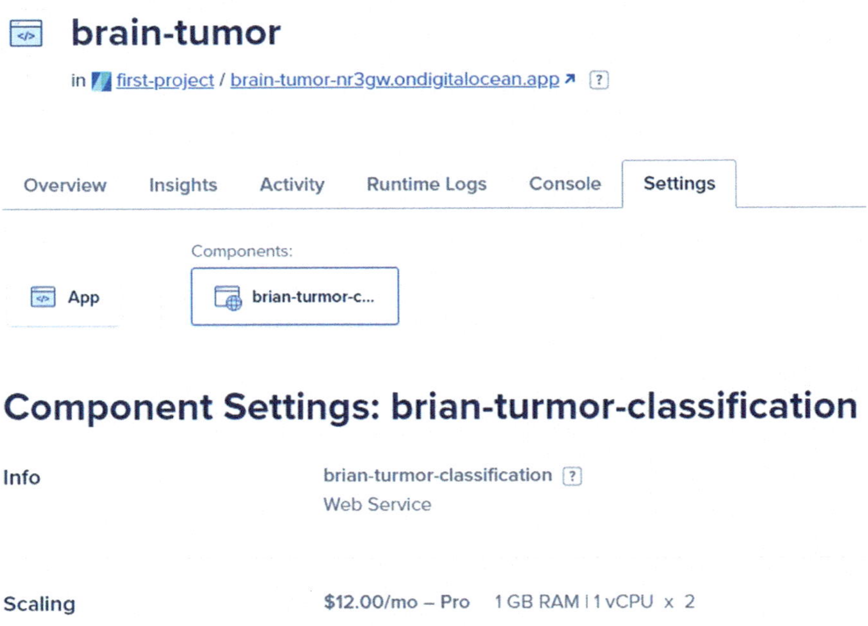

Fig. 13.18 DigitalOcean deployment page

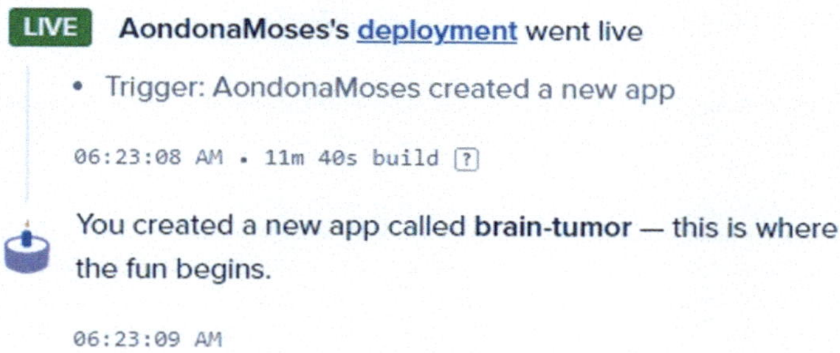

Fig. 13.19 Deployment successful message

13.14 Conclusion

The complicated structure of the brain makes finding a brain tumor challenging. Every organ in the body has a function that the brain manages. Deep learning and machine learning methods are essential for automatically classifying early-stage brain tumors. The likelihood of a patient surviving is increased thanks to these systems' quick diagnosis capabilities. Additionally, these techniques aid specialists and radiologists in their choice of diagnoses and courses of action. This study proposed the use of EfficientNetB0 transfer learning architecture to classify the brain tumors tri-classification (pituitary, meningioma, and glioma), using the MR brain tumor images dataset Accuracy, precision, recall, and F1-measure are the four metrics that are used to assess the performance of the model. The suggested model yields the best outcome, achieving 98% accuracy, 98.0% precision, 98.0% recall, and 98.0% F1 measure. The outcomes demonstrated that the suggested model is the most effective at identifying MRI brain images. Comparing our system to related works, it performs better in terms of classification accuracy, precision, recall, and F1 score. It exhibits strong performance with few training samples and epochs, which helps shorten processing time. It is advantageous to implement transfer learning techniques in medical application as it can be used safely with high accuracy, little data, and less time spent on training. Model performance callback functions could be implemented during training to monitor and save the best-performing models. Brain tumor specialists could make available well-labeled data with all features to further advance research. It would be interesting to consider a system to categorise medical images from various modalities, like X-rays, PET scans, and computed tomography (CT) scans. This will be the focus of our next studies where for learning and experimenting, we would offer a platform, TensorFlow Quantum (TFQ), for prototyping hybrid quantum-classical AI models.

Acknowledgements The authors acknowledge the motivated research efforts of Iorumbur Aondona Moses (Department of Physics, Federal University of Technology, Minna, Nigeria) as a promising young researcher.

References

Ahmad, I., Ullah, I., Khan, W.U., Ur Rehman, A., Adrees, M.S., Saleem, M.Q., Cheikhrouhou, O., Hamam, H., Shafiq, M.: Efficient algorithms for E-healthcare to solve multiobject fuse detection problem. J. Healthcare Eng. **2021**, 1–16 (2021)

Amin, J., Sharif, M., Raza, M., Saba, T., Sial, R., Shad, S.A.: Brain tumor detection: a long short-term memory (LSTM)-based learning model. Neural Comput. Appl. **32**, 15965–15973 (2020)

Amin, J., Sharif, M., Haldorai, A., Yasmin, M., Nayak, R.S.: Brain tumor detection and classification using machine learning: a comprehensive survey. Compl. Intell. Syst. 1–23 (2021)

Bulla, P., Anantha, L., Peram, S.: Deep neural networks with transfer learning model for brain tumors classification. Traitement Du Sig. **37**(4), 593–601 (2020)

Dvořák, P., & Menze, B.: Structured prediction with convolutional neural networks for multimodal brain tumour segmentation. In: Medical Computer Vision: Algorithms for Big Data: International Workshop, MCV 2015, Held in Conjunction with MICCAI 2015, Munich, Germany, October 5–9, 2015, Revised Selected Papers 18, pp. 59–71. Springer International Publishing (2015)

Harish, P., Baskar, S.: MRI based detection and classification of brain tumor using enhanced faster R-CNN and Alex Net model. Mater. Today Proc. **11**, 495 (2020)

Hatami, T., Hamghalam, M., Reyhani-Galangashi, O., Mirzakuchaki, S.: A machine learning approach to brain tumors segmentation using adaptive random forest algorithm. In: 2019 5th Conference on Knowledge Based Engineering and Innovation (KBEI), pp. 076–082. IEEE (2019)

Hidayatullah, P., Mengko, T. E. R., Munir, R., Barlian, A.: A semiautomatic sperm cell data annotator for convolutional neural network. In: 2019 5th International Conference on Science in Information Technology (ICSITech) (pp. 211–216). IEEE (2019)

Irsheidat, S., Duwairi, R.: Brain tumor detection using artificial convolutional neural networks. In: 2020 11th International Conference on Information and Communication Systems (ICICS), pp. 197–203. IEEE (2020)

Khan, H.A., Jue, W., Mushtaq, M., Mushtaq, M.U.: Brain tumor classification in MRI image using convolutional neural network. Math. Biosci. Eng. **17**(5), 6203–6216 (2020)

Srivastava, N., Hinton, G., Krizhevsky, A., Sutskever, I., Salakhutdinov, R.: Dropout: a simple way to prevent neural networks from overfitting. J. Mach. Learn. Res. **15**(1), 1929–1958 (2014)

Chapter 14
General Conclusion

Abstract The field of 3D rendering is always changing as new techniques are developed to address the shortcomings of earlier strategies. Neural Radiance Fields (NeRFs) are one such invention that makes use of a neural network to understand the distribution of light and density within a scene, allowing for the production of photorealistic renderings from any angle. NeRFs may be able to overcome obstacles like occlusion and oversimplification, opening the door for a new era in medical visualization, according to early indications. This innovative method has the power to completely transform the 3D rendering industry and provide more precise and thorough visualizations for applications such as medical imaging. NeRFs have the potential to greatly improve our understanding of complex 3D environments and open up new possibilities for a wide range of applications as they continue to develop and get better. NeRF techniques have great potential to advance the visualization and quantification of complex pathologies affecting important organs such as the brain and chest by allowing MRI's exceptional soft tissue perspectives to be unlocked. Coordinate-based radiance fields tailored to MRI constraints may be able to unlock next-generation 3D perspectives for precision diagnostics, clarify spatial disease spread patterns, and break through long-standing obstacles to seamless perspective rendering.

Keywords 3D rendering techniques · Volume rendering · NeRFs in medical imaging

14.1 Introduction

Magnetic resonance imaging (MRI) has become an essential modality in modern medical diagnostics due to accurate capturing of anatomical cross-sections without ionizing radiation with excellent soft tissue contrast (McRobbie et al. 2017). Still, it has long been a challenge to visualize complex structures in 3D from traditional 2D slice acquisitions (Bilic et al. 2023). Reconstructing geometric complexity, visibility, and lighting realistically for clinical use is a challenge for surface and volume

B. O. Awojoyogbe and M. O. Dada, *Digital Molecular Magnetic Resonance Imaging*, Series in BioEngineering, https://doi.org/10.1007/978-981-97-6370-2_14

rendering techniques (Tewari et al. 2022). This complicates the spatial understanding of multi-tissue complex diseases. The recent development of Neural Radiance Fields (NeRFs) in computer vision and graphics shows great promise for removing obstacles in 3D modeling for medical applications (Xie et al. 2022). NeRFs are a type of neural network that optimizes a multilayer perceptron network to map a continuous 5D coordinate scene representation. It was first demonstrated for photorealistic novel view synthesis from RGB images (Mildenhall et al. 2021). The three coordinates are 3D location, 2D view direction, and density-encoded radiance. Differentiable volumetric raymarching can render scenes at test time from any perspective thanks to the learned implicit scene encoding. Coordinate-based neural rendering approaches, like those used in Med-NeRF (Zhang et al. 2023) has shown promise in the direct generation of 3D anatomical reconstructions from sparse 2D scans. Better generalization from fewer data is demonstrated by custom constraints. Volumetric visualization also supports image-guided interventions through augmented reality interfaces. Nevertheless, there is little talk of MRI data synergies and the majority of demonstrations are still limited to CT scans.

14.2 Introduction to 3D Rendering Techniques

In the physical world, light sources emit photons that typically travel in straight lines until they come into contact with a surface or a volume. When a photon encounters a surface, it can either be absorbed, reflected, or transmitted. Some of these photons may reach the retina of an observer, where they are converted into a signal that is then interpreted by the brain, resulting in the formation of an image. Similarly, photons may be captured by the sensor of a camera, leading to the creation of a visual representation of the environment (Oborn et al. 2017). In both cases, the resulting image is a 2D depiction of the 3D environment. Computer graphics can mimic and replicate the process of photons interacting with a 3D environment to create an image. Here, a 3D geometric model takes the place of the real world, and different algorithms are used to simulate how light interacts with the model. Rendering is the process of creating an image through the simulation of light behavior. The process of rendering entails determining each pixel's color and brightness in the finished picture by analyzing how light interacts with the three-dimensional model. This entails taking into consideration the surfaces' inherent material qualities, the behavior of light sources, and the results of shadows and reflections. Creating a realistic and aesthetically pleasing representation of the virtual scene is the aim. To do this, a variety of rendering methods can be used, each with advantages and disadvantages of their own. For instance, ray tracing is a technique that replicates how light rays move through a 3D scene, giving realistic lighting and reflections. However, rasterization is a quicker method that uses textures and polygons to turn the 3D scene into 2D pictures (Taibo et al. 2014). Rendering includes not only how light interacts with surfaces but also how light behaves when it passes through participating media, like

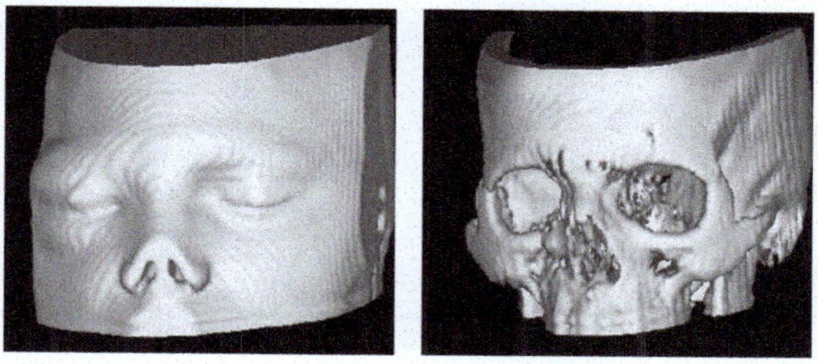

Fig. 14.1 Surface rendering of human head CT data, corresponding to (left) skin and (right) skull

water, smoke, or fog. Although it increases rendering complexity, this is necessary to produce realistic and captivating visuals.

The rendering field continues to advance with the development of new algorithms and technologies. This includes real-time rendering, which seeks to produce images at interactive frame rates, and physically-based rendering, which attempts to accurately reproduce the physical properties of light and materials. All things considered, rendering is crucial to computer graphics and visual effects because it enables the creation of engrossing and realistic images for a range of applications, such as medical physics, virtual reality, video games, and movies.

14.2.1 Surface Rendering

An established visualization method for producing three-dimensional (3D) images from sectional image data is surface rendering (Schreyer and Warfield 2002). It has been demonstrated that the three-dimensional visualization of sectional two-dimensional radiological data is useful for surgical planning, interdisciplinary communication, and diagnosis. It is crucial to remember that CT or MR acquisitions, which are frequently mistaken for 2D images, really show a slice of a particular volume with a particular thickness (Segaran et al. 2021). Because these images are three-dimensional, the pixels within them are called voxels. Since these images are in three dimensions, it is not too difficult to achieve three-dimensional visualization (Fig. 14.1).

14.2.2 Volume Rendering

A potent method for producing a 3D representation of MRI data in medical imaging is volume rendering. Volume rendering, as opposed to conventional 2D imaging, makes use of the full volume of data, providing a more thorough understanding of internal structures and spatial relationships. This is accomplished by giving voxels—the 3D equivalent of pixels—different opacity values according to their intensity. Structures that might not be clearly discernible in conventional 2D images can now be

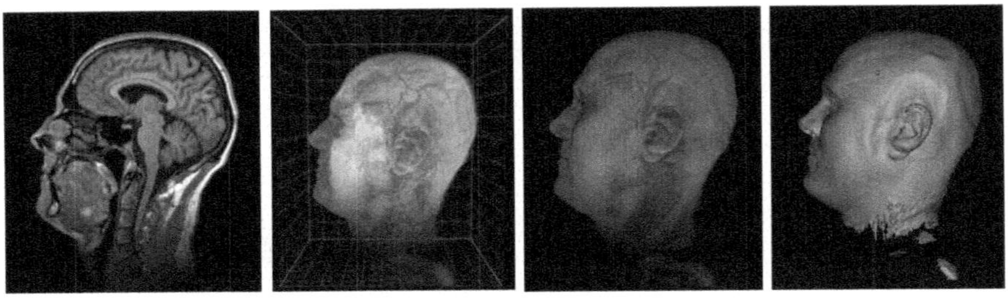

Fig. 14.2 Example of volume rendering

visualized. While there are many different algorithms used in volume rendering, two popular techniques are ray casting and splatting. While splatting entails projecting the volume data onto 2D planes and blending the contributions of voxels within each plane, raycasting entails tracing rays through the volume data to determine the contribution of each voxel to the final image. All things considered, volume rendering is a useful technique in medical imaging that helps with the diagnosis and treatment of a number of illnesses and provides a deeper comprehension of intricate anatomical structures. It offers a more thorough and in-depth view of the internal structures, which is beneficial for surgical planning and medical decision-making (Fig. 14.2).

14.3 Overview of Neural Radiance Fields (NeRFs)

With the use of sophisticated machine learning, a method known as neural radiance fields (NeRFs) creates 3D representations of objects or scenes from 2D images. By encoding the entire object or scene into an artificial neural network, the method creates new 3D views from various angles by predicting the light intensity or radiance at any point in the 2D image. The NeRF model can render photorealistic views of the scene from new perspectives because it can learn the geometry of the scene, camera poses, and object reflectance properties. Initially released in 2020 (Mildenhall et al. 2021).

NeRFs are superior to conventional 3D reconstruction methods in a number of ways. NeRFs learn directly from the input images, in contrast to conventional approaches that rely on heuristics and assumptions about scene geometry and lighting NeRFs are able to capture complex lighting effects and scene geometry with high fidelity because they learn directly from the input images. NeRFs can also produce new perspectives on a scene that match the input images, allowing for realistic rendering of scenes from any angle. Modeling 3D scenes can be done with greater flexibility and scalability thanks to the use of artificial neural networks in NeRFs. NeRFs can capture a wide range of scene complexities and lighting conditions by training the neural network on a diverse set of images. This versatility makes NeRFs

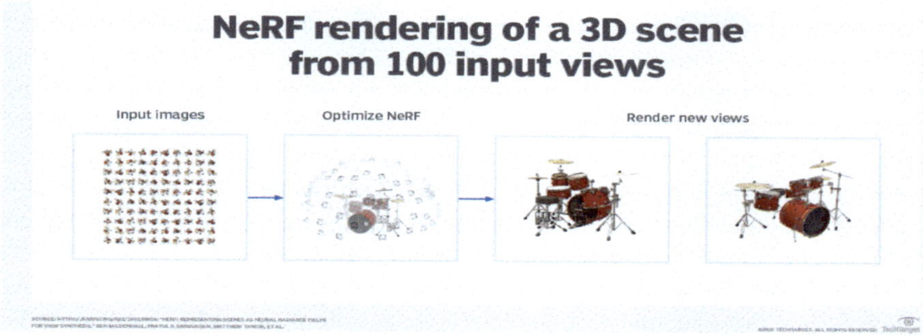

Fig. 14.3 NeRF rendering of a 3D scene from 100 input view

suitable for a multitude of applications, including digital content creation, robotics, and virtual and augmented reality.

The creation of 3D models for virtual worlds, video games, and visual effects in motion pictures can be streamlined with the use of NeRFs. Furthermore, NeRFs' photorealistic rendering powers can improve virtual and augmented reality apps' immersive experiences by giving them a more realistic and visually appealing appearance. To sum up, Neural Radiance Fields, or NeRFs, are a strong and adaptable method for producing excellent 3D reconstructions from 2D pictures. NeRFs have the potential to completely change the way we create and interact with 3D content in a variety of fields, including computer graphics, computer vision, virtual and augmented reality, and more, thanks to their capacity to capture intricate lighting effects and scene geometry (Fig. 14.3).

14.4 Applications of NeRFs in Medical Imaging

Although NeRFs have shown great promise in computer graphics, there is still much to learn about applying them to medical imaging, especially with data from Magnetic Resonance Imaging (MRI). This section examines pertinent research that addresses NeRFs' possible uses in medical imaging settings. Guidance from studies investigating the integration of NeRFs with other imaging modalities could be useful in customizing this method to the complexities of MRI data.

- **Medical Imaging**: NeRFs have the potential to completely change how internal organs and tissues are seen and analyzed in the medical field. NeRFs are able to produce precise 3D models from medical images such as MRIs and CT scans, giving medical professionals realistic and in-depth depictions of human anatomy. This can ultimately result in better patient outcomes and the advancement of medical research and practice through improved surgical preparation, personalized treatment planning, and improved diagnostic accuracy.

- **Surgery Planning**: The capacity of NeRFs to create three-dimensional models of abnormalities and tumors from medical images has great promise for improving medical interventions. NeRFs can help with preoperative planning by giving surgeons precise 3D representations, which improves their ability to visualize and plan complex procedures. Additionally, these models can play a key role in the development and refinement of novel surgical techniques, which will ultimately result in better surgical outcomes, lower surgical risks, and advancements in the field of medical interventions for a variety of ailments.
- **Telemedicine**: The ability to remotely create 3D models of patients' bodies using NeRFs has the potential to completely change how accessible healthcare is. Through the utilization of this technology, healthcare providers can virtually treat patients in underserved or rural areas, circumventing geographical obstacles. It also makes it possible to collaborate and consult with experts without difficulty, which eliminates the need for travel. This improves patient outcomes and closes healthcare disparities by increasing access to medical expertise and boosting the efficacy and efficiency of healthcare delivery.
- **Education and Training**: Transformative educational opportunities can be provided by using NeRFs to generate interactive 3D models of medical anatomy and surgical procedures. These models offer dynamic and immersive visualizations that improve comprehension of intricate anatomical structures and surgical techniques, making them effective training tools for surgeons and medical students. NeRF-generated models have the potential to improve medical education by providing a more engaging learning environment. This can lead to better knowledge retention and skill development, which in turn can improve surgical outcomes and patient care.

14.5 Challenges in Using NeRF in Medicine

Neural networks have a lot of potential for use in medicine, but there are still some issues that need to be resolved. These issues include the neural network's need for a lot of data to be trained, the time and computational resources needed for this process, and the interpretability of the models that are produced. It will be imperative to resolve these issues if NeRFs are to be widely used in medical applications.

- **Training NeRFs can be computationally costly**: Large image datasets are necessary for NeRF training, which can be resource- and time-intensive. The procedure has significant processing and storage requirements, which raises the cost of operation. Furthermore, scalability and efficiency issues may arise due to the intricacy of NeRF training algorithms. Reducing these computational obstacles will be essential to increasing the usability and accessibility of NeRFs for medical applications.

- **NeRFs have the potential to be sensitive to noise**: This poses a serious problem because inaccurate training data can undermine the validity of the 3D models that are produced. NeRF-generated models may not be as useful in medical applications due to distorted representations of anatomical structures or diseases caused by incomplete or noisy data. Robust data preprocessing, noise reduction methods, and advances in neural network training are needed to address this sensitivity and guarantee the fidelity and accuracy of the resulting 3D models.
- **NeRFs aren't readily accessible yet**: NeRFs are a relatively new technology, which contributes to their limited availability in the medical field. For medical professionals, this means that there is a lack of widespread integration and accessibility. To overcome this obstacle and establish NeRFs as a dependable and easily accessible tool in healthcare, more investigation, development, and validation will be needed.

As research in this area continues, we can expect to see even more innovative applications for NeRFs in the years to come.

14.6 Conclusion

This book is important because it has the potential to completely change medical imaging, particularly as it relates to computational molecular magnetic resonance imaging (MRI). Developing and applying Neural Radiance Fields (NeRFs) to three-dimensional MRI data rendering is an attempt to overcome important drawbacks of conventional 2D visualization techniques.

- **Enhancing the capabilities of MRI visualization:** A level of anatomical visualization that has never before been possible for researchers and clinicians could be made possible by the incorporation of NeRFs into the MRI rendering process. This development enables a more thorough understanding of the intricate structures found within the human body, surpassing the constraints of traditional 2D methods.
- **Improvement of diagnostic accuracy:** Improved diagnostic accuracy may result from NeRF-enabled enhanced three-dimensional rendering. Medical professionals can improve their diagnostic skills and lower the risk of error by gaining a deeper understanding of the spatial relationships and subtle abnormalities by presenting a more realistic representation of anatomical features.
- **Encouraging Surgical Education and Planning:** The results of this book may have a big influence on medical education and surgical planning. NeRF-produced three-dimensional renderings can be extremely useful for preoperative planning, helping surgeons to more precisely visualize and plan intricate procedures. Moreover, medical practitioners and students' comprehension can be improved by using these visualizations in educational settings.

- **Advancing Imaging Technologies:** The effective use of NeRFs in the rendering of MRIs advances the continuing development of imaging technologies. This broadens the toolkit that medical professionals can use and sets the stage for upcoming advancements in medical imaging that could impact a wider range of diagnostic modalities.
- **Improving Patient Experience and Care**: In the end, this book's importance goes beyond how it might affect patient care. More accurate and immersive visualization and interpretation of MRI data can result in earlier and more accurate diagnoses, which can speed up timely interventions and ultimately improve patient outcomes. Furthermore, a patient population that is better informed and involved may benefit from the improved visualization experience.

In conclusion, by pushing the limits of conventional MRI visualization methods, this book has significant implications for everything from clinical diagnostics to medical education. The expected results could completely change the medical imaging landscape and lead to innovations that will help patients and healthcare providers alike.

Acknowledgements The authors acknowledge the motivated research efforts of Ismail O. Muhammedawal as a promising young researcher.

References

Bilic, P., Christ, P., Li, H.B., Vorontsov, E., Ben-Cohen, A., Kaissis, G., Szeskin, A., Jacobs, C., Mamani, G.E.H., Chartrand, G., Menze, B.: The liver tumor segmentation benchmark (lits). Med. Image Anal. **84**, 102680 (2023)

McRobbie, D.W., Moore, E.A., Graves, M.J., Prince, M.R.: MRI from Picture to Proton. Cambridge University Press (2017)

Mildenhall, B., Srinivasan, P.P., Tancik, M., Barron, J.T., Ramamoorthi, R., Ng, R.: Nerf: representing scenes as neural radiance fields for view synthesis. Commun. ACM **65**(1), 99–106 (2021)

Oborn, B.M., Dowdell, S., Metcalfe, P.E., Crozier, S., Mohan, R., Keall, P.J.: Future of medical physics: real-time MRI-guided proton therapy. Med. Phys. **44**(8), e77–e90 (2017)

Schreyer, A.G., Warfield, S.K.: Surface rendering. In: 3D Image Processing: Techniques and Clinical Applications, pp. 31–34. Springer Berlin Heidelberg, Berlin, Heidelberg (2002)

Segaran, N., Saini, G., Mayer, J.L., Naidu, S., Patel, I., Alzubaidi, S., Oklu, R.: Application of 3D printing in preoperative planning. J. Clin. Med. **10**(5), 917 (2021)

Taibo, J., Jaspe, A., Seoane, A., Agus, M., Hernández, L.: Practical line rasterization for multi-resolution textures. In: STAG, pp. 9–18 (2014)

Tewari, A., Thies, J., Mildenhall, B., Srinivasan, P., Tretschk, E., Yifan, W., Lassner, C., Sitzmann, V., Martin-Brualla, R., Lombardi, S., Golyanik, V.: Advances in neural rendering. In: Computer Graphics Forum, vol. 41, no. 2, pp. 703–735 (2022)

Xie, Y., Takikawa, T., Saito, S., Litany, O., Yan, S., Khan, N., Tombari, F., Tompkin, J., Sitzmann, V., Sridhar, S.: Neural fields in visual computing and beyond. In: Computer Graphics Forum, vol. 41, no. 2, pp. 641–676 (2022)

Zhang, H., Zhu, L., Xiang, Y., Zheng, J., Song, A.: Haptic rendering of neural radiance fields. In: Proceedings of the 36th Annual ACM Symposium on User Interface Software and Technology, pp. 1–10 (2023)

Chapter 15
Democratizing Access to Magnetic Resonance Imaging: Capacity Building in Quantum Computing

Abstract Independent of artificial intelligence, quantum computing is going to be a big deal. However, coupled with AI, it will be a game changer. Quantum computing, an entirely new way of processing information, when combined with AI, has the potential, without exaggeration, to bring about a new computing revolution. The field of AI, particularly machine learning, is poised to experience a significant leap forward due to quantum computing. This advancement is largely attributed to quantum-specific algorithms tailored to enhance computational processes. A wide range of sectors in healthcare, are expected to benefit from integrating AI with quantum computing. This synergy promises more efficient problem-solving and innovation within medical fields. The fusion of AI with quantum computing is not just about incremental changes but about paving the way for transformative advancements. This combination will revolutionize our approach to magnetic resonance imaging and its applications in various fields of medicine. This chapter focuses international attention to democratizing access to magnetic resonance imaging powered by quantum computing.

Keywords Magnetic resonance imaging · Quantum computing · Artificial intelligence

15.1 Introduction

Modeling the behavior of individual atoms in a molecule is a complex problem, because of all the different electrons interacting with one another. There are some complex problems that we do not know how to solve with classical computing at any scale. Nevertheless, the real world runs on quantum physics. Computational methods that make calculations by using the quantum concepts should in many situations be our best tools for understanding the real world around us. No computational concept has the working memory to handle all the possible permutations of molecular behavior by using any known methods. We do not have a good way to create these

B. O. Awojoyogbe and M. O. Dada, *Digital Molecular Magnetic Resonance Imaging*, Series in BioEngineering, https://doi.org/10.1007/978-981-97-6370-2_15

computational spaces with classical mechanics, which limits their usefulness without quantum computation.

15.2 Building Capacity for Digital Molecular Magnetic Resonance Imaging

In recent years, it has been discovered that heterogeneous diseases such as cardiovascular diseases, diabetes, hepatitis and cancer may be attacked effectively by developing specific treatment plans that are generally referred to as personalized treatments. This new approach in medicine involves using novel methodology as a bridge between diagnostic and therapeutic applications to form a single agent, allowing for diagnosis, drug delivery and treatment response monitoring at the molecular level. This method is time and cost effective because it is a one in all procedure which combines diagnosis and therapy. This novel methodology involving atomic nuclei could be studied with NMR/MRI since their unique resonant frequencies lead to the production of distinct spectra for diagnostic imaging. Atomic nuclei are quantum objects and most of their properties can be accurately explained only by resorting to quantum mechanical formulation as developed by Awojoyogbe (2007). Based on our proposal for capacity building in Digital Molecular Magnetic Resonance Imaging-powered by quantum computing, we have quantum mechanically described the evolution of magnetic resonance imaging. According to the uncertainty principle, a wave packet is assumed to initially describe the particle (electron or normal cell) under study. Then, when the particle encounters a force (diseases such as cardiovascular diseases, diabetes, hepatitis and cancer), so its potential energy is no longer zero, the force modifies the wave packet. Finding such propagation techniques, and applying them appropriately can provide useful computational tools to find solutions to biological, medical and physical problems which otherwise could not be easily solved. Laboratory for capacity building in quantum computing based on mathematical algorithms, quantum physics concepts, and novel computational methods earlier formulated by Dada and Awojoyogbe (2021a, 2021b, 2021c) may take a new approach to solve these complex problems—creating multidimensional computational spaces. This may turn out to be a much more efficient way of solving complex NMR/MRI problems. Towards this goal, we have introduced the possibility of advanced digital signal process technique and modular designed structure to generate arbitrary pulse sequence in order to meet the requirement of digital magnetic resonance imaging methods for medical use. We described the principle of artificial intelligence (AI) in medicine, neural network in scientific computing, physics-informed neural networks and the use of partial differential equations (PDEs), and evolution of physics-informed neural networks powered by quantum computing. This is an entirely new way of processing information. When quantum computing is combined with AI, it has the potential, without exaggeration, to bring about a new computing revolution. Transformation may be an understatement.

Although, quantum computing is not a term that is particularly familiar to those outside the technology and physics worlds, but it soon will be (Anazodo et al. 2023). This form of computing, still in its relative infancy, originates out of quantum mechanics, the behavior of nature at the scale of atoms and subatomic particles.

15.3 Prospect of Quantum Artificial Intelligence (QAI) in Digital Molecular Magnetic Resonance Imaging

Quantum Artificial Intelligence (QAI)-and in particular, its subset research area Quantum Machine Learning (QML)-will use new quantum-designed algorithms with superpowers that result in much more powerful AI models (Behera 2023). As QAI gradually emerges, we can expect significant improvements in the speed, efficiency, and accuracy of AI. We should also anticipate the emergence of completely new and surprising medical capabilities.

But quantum computing isn't confined anymore to research labs, although technological hurdles must still be overcome, usage is still fairly specialized and often experimental in nature. Practical uses today can be found in pharmaceutical development. Rather than playing catch-up when QAI arrives, organizations should have the technology on their radar and follow developments closely. Those lucky enough to have an innovation lab or equivalent, should consider some experimental and research work, if only for the purposes of acquiring skills and knowledge. The high level of interest from industry and the pace of scientific research clearly suggest that QAI is coming, and the question will be whether you are ready. It's probably closer than you think and it's likely going to be a very big deal.

At first, the benefits will be reaped by certain industries and needs, particularly those with optimization requirements. But later, as QAI becomes more mainstream, almost every enterprise will need to have a strategy.

15.4 Conclusion and Expectation

In predictive analytics and deep learning, quantum computing offers the potential to process vast datasets far more efficiently. This efficiency could significantly advance AI's ability to predict outcomes from large and complex data sets, such as personalized medicine. In summary, the proposed laboratory for research and capacity building in **Digital Molecular Magnetic Resonance Imaging** will be a digitally generated magnetic resonance signal using the fundamental Bloch NMR flow model equation, Physics-informed Neural Networks (PNN) and Artificial Intelligence to design and deploy magnetic resonance systems with the capacity to work in a digital manner with features of low cost, high performance and accuracy. Based on the preliminary research activities (Awojoyogbe 2007; Dada and Awojoyogbe 2021a,

2021b, 2021c) and this book, the unprecedented computational power of quantum computing will break current paradigms. That is why working to ensure an open future for quantum computing is crucial and requires immediate action. Quantum computing has emerged as a transformative force, promising to redefine the boundaries of computation as we know it. Quantum computing opens up new problem-solving capabilities such as quantum medicine, quantum drugs, quantum medical imaging, and MRI quantum computing (digital MRI).

As we peer into this quantum horizon in MRI, a new initiative involving International Society for Magnetic Resonance in Medicine (ISMRM), World Molecular Imaging Society (WMIS), top related research institutes and companies should be motivated to establish Quantum MRI computing laboratories in strategic regions to democratize access to quantum computing that may be able to tackle some of the most pressing global challenges in modern medicine bridging physics and digital health.

Acknowledgements The authors acknowledge the unique, commendable and aggressive efforts of Dr. Uduna Unazodo (Department of Neurology and Neurosurgery, McGill University, Montreal, Canada) in democratizing access to Magnetic Resonance Imaging in Africa.

References

Anazodo, U.C., Ng, J.J., Ehiogu, B., Obungoloch, J., Fatade, A., Mutsaerts, H.J., Secca M.F., Diop, M., Opadele, A., Alexander, D.C., Dada, M.O., Ogbole, G., Nunes, R., Figueiredo, P., Figini, M., Aribisala, B., Awojoyogbe, B. O., Aduluwa, H., Sprenger, C., Wagner, R., Olakunle, A., Romeo, D., Sun, Y., Fezeu, F., Orunmuyi, A.T., Geethanath, S., Gulani, V., Nganga, E., Adeleke, S., Ntobeuko, N., Minja, F.J., Webb, A.G., Asllani, I., Dako, F., Consortium for Advancement of MRI Education and Research in Africa (CAMERA): A framework for advancing sustainable magnetic resonance imaging access in Africa. NMR Biomed. **36**(3), 1 (2023)

Awojoyogbe, O.B.: A quantum mechanical model of the Bloch NMR flow equations for electron dynamics in fluids at the molecular level. Phys. Scr. **75**(6), 788 (2007)

Behera, A.: The Role of Quantum Computing in Advancing Artificial Intelligence (2023). https://www.analyticsvidhya.com/blog/2023/12/the-role-of-quantum-computing-in-advancing-artificial-intelligence/

Dada, M.O., Awojoyogbe, B.O.: Fundamental physics of nuclear magnetic resonance. In: Computational Molecular Magnetic Resonance Imaging for Neuro-Oncology, pp. 11–114 (2021)

Dada, M.O., Awojoyogbe, B.O.: Analysis of the hydrogen-like atom for neuro-oncology based on Bloch's NMR flow equation. In: Computational Molecular Magnetic Resonance Imaging for Neuro-Oncology, pp. 219–246 (2021)

Dada, M.O., Awojoyogbe, B.O.: Quantum mechanical model of the Bloch NMR flow equations for transport analysis of quantum-drugs in microscopic blood vessels applicable in nanomedicine. In: Computational Molecular Magnetic Resonance Imaging for Neuro-Oncology, pp. 247–267 (2021)

The manufacturer's authorised representative in the EU is Springer Nature Customer Service Centre GmbH, Europaplatz 3, 69115 Heidelberg, Germany. If you have any concerns regarding our products, please contact ProductSafety@springernature.com

Printed and bound by CPI Group (UK) Ltd, Croydon, CR0 4YY
30/09/2026
02225304-0013